WELDING METALLURGY AND WELDABILITY OF STAINLESS STEELS

WELDING METALLURGY AND WELDABILITY OF STAINLESS STEELS

JOHN C. LIPPOLD
DAMIAN J. KOTECKI

A JOHN WILEY & SONS, INC., PUBLICATION

Published by John Wiley & Sons, Inc., Hoboken, New Jersey.
Published simultaneously in Canada.

Library of Congress Cataloging-in-Publication Data:

Lippold, John C.
Welding metallurgy and weldability of stainless steels / John C. Lippold, Damian J. Kotecki.
p. cm.
"A Wiley-Interscience publication."
Includes bibliographical references and index.
ISBN 0-471-47379-0 (cloth : alk. paper)
1. Steel, Stainless—Welding. 2. Steel, Stainless—Metallurgy. I. Kotecki, Damian J. II. Title.

TS227.2.L57 2005
620.1'7—dc22

2004058043

Printed in the United States of America
10 9 8 7 6 5 4 3 2 1

To my parents, Carl and Hilda (Schmieder) Lippold,
who have been a source of strength, encouragement,
and inspiration throughout my life. *John C. Lippold*

To Kate (Zander), my bride of more than 40 years,
without whose support, companionship, and love,
my education, my career, and my personal life
would have been much less. *Damian J. Kotecki*

CONTENTS

PREFACE xv

1 INTRODUCTION 1

1.1 Definition of a Stainless Steel / 2
1.2 History of Stainless Steel / 2
1.3 Types of Stainless Steel and Their Application / 4
1.4 Corrosion Resistance / 5
1.5 Production of Stainless Steel / 6
References / 7

2 PHASE DIAGRAMS 8

2.1 Iron–Chromium System / 9
2.2 Iron–Chromium–Carbon System / 10
2.3 Iron–Chromium–Nickel System / 12
2.4 Phase Diagrams for Specific Alloy Systems / 15
References / 18

3 ALLOYING ELEMENTS AND CONSTITUTION DIAGRAMS 19

3.1 Alloying Elements in Stainless Steels / 19
3.1.1 Chromium / 20
3.1.2 Nickel / 20
3.1.3 Manganese / 21

3.1.4 Silicon / 21
3.1.5 Molybdenum / 22
3.1.6 Carbide-Forming Elements / 22
3.1.7 Precipitation-Hardening Elements / 23
3.1.8 Interstitial Elements: Carbon and Nitrogen / 23
3.1.9 Other Elements / 24
3.2 Ferrite-Promoting Versus Austenite-Promoting Elements / 24
3.3 Constitution Diagrams / 25
3.3.1 Austenitic–Ferritic Alloy Systems: Early Diagrams and Equivalency Relationships / 25
3.3.2 Schaeffler Diagram / 29
3.3.3 DeLong Diagram / 33
3.3.4 Other Diagrams / 34
3.3.5 WRC-1988 and WRC-1992 Diagrams / 40
3.4 Austenitic–Martensitic Alloy Systems / 43
3.5 Ferritic–Martensitic Alloy Systems / 46
3.6 Neural Network Ferrite Prediction / 50
References / 52

4 MARTENSITIC STAINLESS STEELS **56**

4.1 Standard Alloys and Consumables / 57
4.2 Physical and Mechanical Metallurgy / 59
4.3 Welding Metallurgy / 63
4.3.1 Fusion Zone / 63
4.3.2 Heat-Affected Zone / 67
4.3.3 Phase Transformations / 70
4.3.4 Postweld Heat Treatment / 71
4.3.5 Preheat, Interpass, and Postweld Heat Treatment Guidelines / 74
4.4 Mechanical Properties of Weldments / 77
4.5 Weldability / 77
4.5.1 Solidification and Liquation Cracking / 78
4.5.2 Reheat Cracking / 78
4.5.3 Hydrogen-Induced Cracking / 79
4.6 Supermartensitic Stainless Steels / 80
4.7 Case Study: Calculation of M_S Temperatures of Martensitic Stainless Steels / 84
References / 86

5 FERRITIC STAINLESS STEELS 87

5.1 Standard Alloys and Consumables / 88
5.2 Physical and Mechanical Metallurgy / 92
5.2.1 Effect of Alloying Additions on Microstructure / 95
5.2.2 Effect of Martensite / 95
5.2.3 Embrittlement Phenomena / 96
5.2.3.1 475°C Embrittlement / 97
5.2.3.2 Sigma and Chi Phase Embrittlement / 97
5.2.3.3 High-Temperature Embrittlement / 98
5.2.3.4 Notch Sensitivity / 103
5.2.4 Mechanical Properties / 104
5.3 Welding Metallurgy / 104
5.3.1 Fusion Zone / 104
5.3.1.1 Solidification and Transformation Sequence / 104
5.3.1.2 Precipitation Behavior / 109
5.3.1.3 Microstructure Prediction / 111
5.3.2 Heat-Affected Zone / 112
5.3.3 Solid-State Welds / 113
5.4 Mechanical Properties of Weldments / 114
5.4.1 Low-Chromium Alloys / 114
5.4.2 Medium-Chromium Alloys / 116
5.4.3 High-Chromium Alloys / 119
5.5 Weldability / 123
5.5.1 Weld Solidification Cracking / 123
5.5.2 High-Temperature Embrittlement / 124
5.5.3 Hydrogen-Induced Cracking / 126
5.6 Corrosion Resistance / 126
5.7 Postweld Heat Treatment / 130
5.8 Filler Metal Selection / 132
5.9 Case Study: HAZ Cracking in Type 436 During Cold Deformation / 132
5.10 Case Study: Intergranular Stress Corrosion Cracking in the HAZ of Type 430 / 135
References / 137

6 AUSTENITIC STAINLESS STEELS 141

6.1 Standard Alloys and Consumables / 143
6.2 Physical and Mechanical Metallurgy / 147
6.2.1 Mechanical Properties / 149

6.3 Welding Metallurgy / 151
6.3.1 Fusion Zone Microstructure Evolution / 153
6.3.1.1 Type A: Fully Austenitic Solidification / 154
6.3.1.2 Type AF Solidification / 155
6.3.1.3 Type FA Solidification / 155
6.3.1.4 Type F Solidification / 158
6.3.2 Interfaces in Single-Phase Austenitic Weld Metal / 162
6.3.2.1 Solidification Subgrain Boundaries / 162
6.3.2.2 Solidification Grain Boundaries / 163
6.3.2.3 Migrated Grain Boundaries / 163
6.3.3 Heat-Affected Zone / 164
6.3.3.1 Grain Growth / 165
6.3.3.2 Ferrite Formation / 165
6.3.3.3 Precipitation / 165
6.3.3.4 Grain Boundary Liquation / 166
6.3.4 Preheat and Interpass Temperature and Postweld Heat Treatment / 166
6.3.4.1 Intermediate-Temperature Embrittlement / 167
6.4 Mechanical Properties of Weldments / 168
6.5 Weldability / 173
6.5.1 Weld Solidification Cracking / 173
6.5.1.1 Beneficial Effects of Primary Ferrite Solidification / 175
6.5.1.2 Use of Predictive Diagrams / 177
6.5.1.3 Effect of Impurity Elements / 179
6.5.1.4 Ferrite Measurement / 181
6.5.1.5 Effect of Rapid Solidification / 182
6.5.1.6 Solidification Cracking Fracture Morphology / 186
6.5.1.7 Preventing Weld Solidification Cracking / 189
6.5.2 HAZ Liquation Cracking / 189
6.5.3 Weld Metal Liquation Cracking / 190
6.5.4 Ductility-Dip Cracking / 194
6.5.5 Reheat Cracking / 196
6.5.6 Copper Contamination Cracking / 199
6.5.7 Zinc Contamination Cracking / 200
6.5.8 Helium-Induced Cracking / 200
6.6 Corrosion Resistance / 200
6.6.1 Intergranular Corrosion / 201
6.6.1.1 Preventing Sensitization / 204
6.6.1.2 Knifeline Attack / 205
6.6.1.3 Low-Temperature Sensitization / 205

6.6.2 Stress Corrosion Cracking / 206
6.6.3 Pitting and Crevice Corrosion / 208
6.6.4 Microbiologically Induced Corrosion / 208
6.6.5 Selective Ferrite Attack / 209
6.7 Specialty Alloys / 211
6.7.1 Heat-Resistant Alloys / 211
6.7.2 High-Nitrogen Alloys / 214
6.8 Case Study: Selecting the Right Filler Metal / 220
6.9 Case Study: What's Wrong with My Swimming Pool? / 223
6.10 Case Study: Cracking in the Heat-Affected Zone / 224
References / 225

7 DUPLEX STAINLESS STEELS 230

7.1 Standard Alloys and Consumables / 231
7.2 Physical Metallurgy / 234
7.2.1 Austenite–Ferrite Phase Balance / 234
7.2.2 Precipitation Reactions / 237
7.3 Mechanical Properties / 237
7.4 Welding Metallurgy / 238
7.4.1 Solidification Behavior / 238
7.4.2 Role of Nitrogen / 240
7.4.3 Secondary Austenite / 244
7.4.4 Heat-Affected Zone / 246
7.5 Controlling the Ferrite–Austenite Balance / 250
7.5.1 Heat Input / 251
7.5.2 Cooling Rate Effects / 251
7.5.3 Ferrite Prediction and Measurement / 253
7.6 Weldability / 254
7.6.1 Weld Solidification Cracking / 254
7.6.2 Hydrogen-Induced Cracking / 254
7.6.3 Intermediate-Temperature Enbrittlement / 255
7.6.3.1 Alpha-Prime Embrittlement / 256
7.6.3.2 Sigma Phase Embrittlement / 256
7.7 Weld Mechanical Properties / 259
7.8 Corrosion Resistance / 261
7.8.1 Stress Corrosion Cracking / 261
7.8.2 Pitting Corrosion / 261
References / 262

8 PRECIPITATION-HARDENING STAINLESS STEELS 264

8.1 Standard Alloys and Consumables / 265
8.2 Physical and Mechanical Metallurgy / 267
8.2.1 Martensitic Precipitation-Hardening Stainless Steels / 269
8.2.2 Semi-Austenitic Precipitation-Hardening Stainless Steels / 274
8.2.3 Austenitic Precipitation-Hardening Stainless Steels / 276
8.3 Welding Metallurgy / 277
8.3.1 Microstructure Evolution / 278
8.3.2 Postweld Heat Treatment / 278
8.4 Mechanical Properties of Weldments / 279
8.5 Weldability / 280
8.6 Corrosion Resistance / 285
References / 285

9 DISSIMILAR WELDING OF STAINLESS STEELS 287

9.1 Applications of Dissimilar Welds / 287
9.2 Carbon or Low-Alloy Steel to Austenitic Stainless Steel / 288
9.2.1 Determining Weld Metal Constitution / 288
9.2.2 Fusion Boundary Transition Region / 291
9.2.3 Nature of Type II Boundaries / 294
9.3 Weldability / 296
9.3.1 Solidification Cracking / 296
9.3.2 Clad Disbonding / 298
9.3.3 Creep Failure in the HAZ of Carbon or Low-Alloy Steel / 299
9.4 Other Dissimilar Combinations / 301
9.4.1 Nominally Austenitic Alloys Whose Melted Zone Is Expected to Include Some Ferrite or to Solidify as Primary Ferrite / 301
9.4.2 Nominally Austenitic Alloys Whose Melted Zone Is Expected to Contain Some Ferrite, Welded to Fully Austenitic Stainless Steel / 301
9.4.3 Austenitic Stainless Steel Joined to Duplex Stainless Steel / 302
9.4.4 Austenitic Stainless Steel Joined to Ferritic Stainless Steel / 302
9.4.5 Austenitic Stainless Steel Joined to Martensitic Stainless Steel / 302
9.4.6 Martensitic Stainless Steel Joined to Ferritic Stainless Steel / 302
9.4.7 Stainless Steel Filler Metal for Difficult-to-Weld Steels / 303

9.4.8 Copper-Base Alloys Joined to Stainless Steels / 305

9.4.9 Nickel-Base Alloys Joined to Stainless Steels / 306

References / 307

10 WELDABILITY TESTING **309**

10.1 Introduction / 309

10.1.1 Weldability Test Approaches / 310

10.1.2 Weldability Test Techniques / 310

10.2 Varestraint Test / 311

10.2.1 Technique for Quantifying Weld Solidification Cracking / 312

10.2.2 Technique for Quantifying HAZ Liquation Cracking / 316

10.3 Hot Ductility Test / 319

10.4 Fissure Bend Test / 323

10.5 Strain-to-Fracture Test / 328

10.6 Other Weldability Tests / 329

References / 329

APPENDIX 1 NOMINAL COMPOSITIONS OF STAINLESS STEELS **331**

APPENDIX 2 ETCHING TECHNIQUES FOR STAINLESS STEEL WELDS **343**

AUTHOR INDEX **347**

SUBJECT INDEX **353**

PREFACE

The motivation for this book arose from our desire to develop a reference for engineers, scientists, and students that would represent an up-to-date perspective on the welding metallurgy aspects and weldability issues associated with stainless steels. Many excellent texts and reference books on these topics have preceded this book, including handbooks published by the American Welding Society and ASM International, as well as *Welding Metallurgy of Stainless and Heat-Resisting Steels* by R. J. Castro and J. J. de Cadenet (Cambridge: Cambridge University Press, 1968) and *Welding Metallurgy of Stainless Steels* by E. Folkhard (Berlin: Springer-Verlag, 1984; English translation, 1988). All were exceptional reference sources at the time of their publication and remain so today.

However, the technology of stainless steels has changed considerably even since the 1984 book by Folkhard. In particular, duplex stainless steels and the "super" grades of austenitic, martensitic, and duplex alloys have become prominent and widely used in engineering applications that require welding. Much additional research has been conducted on the standard austenitic grades. Accordingly, the authors believed that an update was due. Independently, each of the authors has taught courses and seminars on the subject of welding metallurgy and weldability of stainless steels, and the combination of these materials has led to this book.

The book begins with chapters introducing the history of stainless steels, relevant phase diagrams, and constitution diagrams. Then detailed chapters address the welding metallurgy and weldability of each of the five microstructural families of stainless steels (martensitic, ferritic, austenitic, duplex, and precipitation-hardening). Each of these chapters also contains a brief section describing the general physical and mechanical metallurgy of each of these systems and a discussion of weldability. Some of these chapters contain a few case studies that will allow the reader to see how the concepts described can be applied in real-world situations. Finally, there is a chapter specific to dissimilar metal joints.

Acknowledgments

Our special thanks go to Richard Avery, Dr. J. M. C. Farrar, Dr. John Grubb, Leo van Nassau, Dr. John Vitek, and Dr. Thomas Siewert, who carefully reviewed individual chapters and made many helpful suggestions that have been incorporated into the manuscript. We would also like to acknowledge Mikal Balmforth, whose comprehensive literature review of constitution diagrams contributed greatly to Chapter 3. We also wish to recognize Shu Shi and John McLane, whose painstaking metallographic work appears here along with that of the referenced authors. Finally, the authors are grateful to The Ohio State University and The Lincoln Electric Company for making available the time and providing the environment to prepare the lectures and seminars and the many hours of writing and editing that allowed this book to become a reality.

CHAPTER 1

INTRODUCTION

Stainless steels are an important class of engineering materials that have been used widely in a variety of industries and environments. Welding is an important fabrication technique for stainless steels, and numerous specifications, papers, handbooks, and other guidelines have been published over the past 75 years that provide insight into the techniques and precautions needed to weld these materials successfully. In general, the stainless steels are considered weldable materials, but there are many rules that must be followed to ensure that they can readily be fabricated to be free of defects and will perform as expected in their intended service. When the rules are not followed, problems are not uncommon during fabrication or in service. Often, the problems are associated with improper control of the weld microstructure and associated properties, or the use of welding procedures that are inappropriate for the material and its microstructure.

This book is designed to provide basic information regarding the welding metallurgy and weldability of the broad range of stainless steels that are currently available as engineering materials. This will include the various grades of stainless steels as classified by their microstructure: martensitic, ferritic, austenitic, and duplex (austenite plus ferrite); and the many specialty grades of stainless steel that have evolved since the early 1970s. In addition to providing a basic understanding of the metallurgical behavior of these steels, numerous examples in the form of case studies, failure investigations, and simple "how-to" explanations are included to assist the reader in applying these concepts for alloy selection and procedure specification for welded construction.

Welding Metallurgy and Weldability of Stainless Steels, by John C. Lippold and Damian J. Kotecki
ISBN 0-471-47379-0

1.1 DEFINITION OF A STAINLESS STEEL

Stainless steels constitute a group of high-alloy steels based on the Fe–Cr, Fe–Cr–C, and Fe–Cr–Ni systems. To be *stainless*, these steels must contain a minimum of 10.5 wt% chromium. This level of chromium allows formation of a passive surface oxide that prevents oxidation and corrosion of the underlying metal under ambient, noncorrosive conditions. Some steels that contain less than 11 wt% Cr, such as the 9 wt% Cr alloys used in power generation applications, are sometimes grouped with the stainless steels but are not considered here. It should also be pointed out that many 12 wt% Cr steels, or even those with much more Cr, will, in fact, exhibit "rust" when exposed to ambient conditions. This is because some of the Cr is tied up as carbides or other compounds, reducing the matrix Cr composition below the level that will support a continuous protective oxide. A classic example of this situation is hardfacing alloys of nominal composition 25 wt% Cr–4 wt% C, which rust due to the heavy intentional presence of chromium carbides. This book does not consider such alloys.

Corrosive media that attack and remove the passive oxide cause corrosion of stainless steels. Corrosion can take many forms, including pitting, crevice corrosion, and intergranular attack. These forms of corrosion are influenced by the corrosive environment, the metallurgical condition of the material, and the local stresses that are present. Engineers and designers must be very aware of the service environments and impact of fabrication practice on metallurgical behavior when selecting stainless steels for use in corrosive conditions.

Stainless steels also have good resistance to oxidation, even at high temperatures, and they are often referred to as *heat-resisting alloys*. Resistance to elevated-temperature oxidation is primarily a function of chromium content, and some high chromium alloys (25 to 30 wt%) can be used at temperatures as high as 1000°C (1830°F). Another form of heat resistance is resistance to carburization, for which stainless steel alloys of modest chromium content (about 16 wt%) but high nickel content (about 35 wt%) have been developed.

Stainless steels are used in a wide variety of applications, such as power generation, chemical and paper processing, and in many commercial products, such as kitchen equipment and automobiles. Stainless steels also find extensive use for purity and sanitary applications in areas such as pharmaceutical, dairy, and food processing.

Most stainless steels are weldable, but many require special procedures. In almost all cases, welding results in a significant alteration of the weld metal and heat-affected zone microstructure relative to the base metal. This can constitute a change in the desired phase balance, formation of intermetallic constituents, grain growth, segregation of alloy and impurity elements, and other reactions. In general, these lead to some level of degradation in properties and performance and must be factored into the design and manufacture.

1.2 HISTORY OF STAINLESS STEEL

The addition of chromium to steels and its apparent beneficial effect on corrosion resistance is generally attributed to the Frenchman Berthier, who in 1821 developed

Brief History of Stainless Steels

- *1821:* Frenchman Berthier experiments with Cr additions to steel.
- *1897:* German Goldschmidt develops technique to produce low-carbon, Cr-bearing alloys.
- *1904–1909:* 13 wt% and 17 wt% Cr alloys produced in France and Germany.
- *1913:* Englishman Brearly casts first commercial ingot at Thomas Firth and Sons, Cast No. 1008, August 20. Analysis (wt%): 0.24% C, 0.2% Si, 0.44% Mn, 12.86% Cr.
- *1916:* U.S. patent 1,197,256 for 9 to 16 wt% Cr steels with less than 0.7 wt% carbon, September 5.

a 1.5 wt% Cr alloy that he recommended for cutlery applications. Early experiments with these steels revealed, however, that with increased Cr, the formability of the steel deteriorated dramatically (due to the high carbon content of these early alloys), and interest in them waned until the early twentieth century. Interest in corrosion-resistant steels reemerged between 1900 and 1915 and a number of metallurgists are credited with developing corrosion-resistant alloys [1]. The apparent catalyst for this renewed activity was the development in 1897 by Goldschmidt in Germany of a technique for producing low-carbon Cr-bearing alloys [2]. Shortly thereafter, Guillet [3] (1904), Portevin [4] (1909), and Giesen [5] (1909) published papers describing the microstructure and properties of 13 wt% Cr martensitic and 17 wt% Cr ferritic stainless steels. In 1909, Guillet also published a study of chromium–nickel steels that were precursors of the austenitic grades of stainless steel. Another enabler for widespread production of stainless steels was the development of the direct-arc electric melting furnace by the Frenchman Heroult in 1899.

These laboratory studies sparked considerable interest in corrosion-resistant steels for industrial applications, and from 1910 to 1915 there was considerable effort to commercialize these alloys. The first reported commercial "stainless steel" alloys are attributed [6] to Harry Brearly, who was a metallurgist at Thomas Firth and Sons in Sheffield, England. Brearly came from a poor working-class family and started at Firth at the age of 12 as a bottle washer in the chemical laboratory. In 1907, at the age of 36, he was named head of the research laboratories. In May 1912, Brearly visited the Royal Small Arms factory in Enfield to investigate the failure of rifle barrels made of 5 wt% chromium steel due to internal corrosion. He concluded that higher chromium contents could possibly be a solution to the corrosion problem. He initially cast two steels with 10 and 15 wt% Cr and nominally 0.30 wt% C. Both of these were unsuccessful because of the excessively high carbon contents. In August 1913, however, an acceptable ingot was cast of composition (wt%) 12.86% Cr, 0.24% C, 0.20% Si, and 0.44% Mn. This material was used to make 12 experimental gun barrels, but the new barrels did not show the expected improvement. Some of this material was also made into cutlery blades and the *age of stainless steel* had begun.

The first stainless steel ingot was cast in the United States by Firth Sterling Ltd. in Pittsburgh on March 3, 1915. This eventually led to U.S. patent 1,197,256, assigned to Brearly for cutlery grade steel. It covered the composition range from 9 to 16 wt% chromium and less than 0.7 wt% carbon. Steels made under this patent soon came to be called *Firth Stainless*.

Although Brearly is widely recognized as the "inventor of stainless steels" based on his 1915 patent, it is clear that his invention would not have been possible without the ground-breaking research that occurred in France and Germany in the preceding decade. The work of contemporaries should also be recognized, including that of the Americans Dansitzen and Becket for ferritic alloys, and the Germans Maurer and Strauss [7] for austenitic alloys.

1.3 TYPES OF STAINLESS STEEL AND THEIR APPLICATION

Next to plain carbon and C–Mn steels, stainless steels are the most widely used steels. Because so many varieties of stainless steel are available, a wide range of desirable properties is achievable and they can be used in many different applications. Not surprisingly, considerable research has been conducted to define their microstructure and properties.

Unlike other material systems, where classification is usually by composition, stainless steels are categorized based on the metallurgical phase (or phases), which is predominant. The three phases possible in stainless steels are martensite, ferrite, and austenite. Duplex stainless steels contain approximately 50% austenite and 50% ferrite, taking advantage of the desirable properties of each phase. *Precipitation-hardenable* (PH) *grades* are termed such because they form strengthening precipitates and are hardenable by an aging heat treatment. PH stainless steels are further grouped by the phase or matrix in which the precipitates are formed: martensitic, semi-austenitic, or austenitic types.

The American Iron and Steel Institute (AISI) uses a system with three numbers, sometimes followed by a letter, to designate stainless steels: for example, 304, 304L, 410, and 430. Magnetic properties can be used to identify some stainless steels. The austenitic types are essentially nonmagnetic. A small amount of residual ferrite or cold working may introduce a slight ferromagnetic condition, but it is notably weaker than a magnetic material. The ferritic and martensitic types are ferromagnetic.

Types of Stainless Steel

- Martensitic (4XX)
- Ferritic (4XX)
- Austenitic (2XX,3XX)
- Duplex (austenite and ferrite)
- Precipitation hardened (PH)

Duplex stainless steels are relatively strongly magnetic, due to their high ferrite content.

Physical properties, such as thermal conductivity and thermal expansion, and mechanical properties vary widely for the different types and influence their welding characteristics. For example, austenitic stainless steels possess low thermal conductivity and high thermal expansion, resulting in higher distortion due to welding than other grades that are primarily ferritic or martensitic.

1.4 CORROSION RESISTANCE

In most cases, stainless steels are selected for their corrosion- or heat-resisting properties. By nature of the passive, Cr-rich oxide that forms, these steels are virtually immune to the general corrosion that plagues C–Mn and low-alloy structural steels. The stainless steels are, however, susceptible to other forms of corrosion, and their selection and application must be considered carefully based on the service environment. A detailed discussion of the various stainless steel corrosion mechanisms is beyond the scope of this book and the reader is referred to other texts to provide more detailed information [8–11]. A brief summary of the applicable corrosion mechanisms associated with welded stainless steels is included here. More detailed descriptions are provided in the appropriate chapters.

Two forms of localized corrosion that may occur in stainless steels are pitting corrosion and crevice corrosion. Mechanistically, they are similar and result in highly localized attack. As the term implies, *pitting corrosion* results from the local breakdown of the passive film and is normally associated with some metallurgical feature, such as a grain boundary or intermetallic constituent. Once this breakdown occurs, corrosive attack of the underlying material occurs and a small *pit* forms on the surface. With time, the solution chemistry within the pit changes and becomes progressively more aggressive (i.e., acidic). This results in rapid subsurface attack and a linking of adjacent pits that ultimately leads to failure. Pitting can be very insidious, since only a small pinhole may exist on the surface.

Crevice corrosion is similar mechanistically, but it does not require the presence of a metallurgical feature for initiation. Rather, as the term implies, a crevice consisting of a confined space must exist where a similar change in solution chemistry can occur. Crevice corrosion is common in bolted structures, where the space between the bolt head and the bolting surface can provide such a crevice. Both pitting and crevice corrosion occur readily in solutions containing chloride ions (such as seawater). Welding may result in the formation of microstructures that accelerate pitting attack or create crevices (lack of penetration, slag intrusions, etc.) that promote localized corrosion. Failure to remove oxides that form due to welding may also reduce corrosion resistance in certain media.

The most serious of all corrosion mechanisms in welded stainless steels, and the subject of many papers and reviews [11,12] is *intergranular attack* (IGA) and the associated phenomenon known as *intergranular stress corrosion cracking* (IGSCC). This form of attack is most common in the *heat-affected zone* (HAZ) of austenitic stainless steels and results from a metallurgical condition called *sensitization*. Sensitization occurs when grain boundary precipitation of Cr-rich carbides results in depletion of Cr in the region just adjacent to the grain boundary, making the microstructure sensitive to corrosive attack if the Cr content drops below 12 wt%. A similar phenomenon occurs in the HAZ of ferritic stainless steels. This mechanism is described in much more detail in the appropriate chapters addressing these alloy systems.

Transgranular stress corrosion cracking (TGSCC) is also a serious problem, especially with common austenitic stainless steels such as 304L and 316L. As the term implies, transgranular SCC has little or nothing to do with grain boundaries. It progresses along certain planes of atoms in each grain, often changing direction from one grain to another, branching as it progresses. The presence of chloride ions along with residual or applied stress promotes this form of cracking.

1.5 PRODUCTION OF STAINLESS STEEL

Stainless steel is produced in a wide variety of shapes and sizes, and the available forms of these alloys for commercial use are almost limitless. Some alloys, such as the ferritic and duplex grades, have some limits on formability, but in general almost any conceivable shape can be produced by a variety of casting, hot-working, and cold-working techniques.

Melting of stainless steels was revolutionized by the introduction of *argon–oxygen decarburization* (AOD) and *vacuum–oxygen decarburization* (VOD) techniques in the early 1970s [13,14]. Primary melting is usually conducted in an electric arc furnace, and then the molten *charge* is transferred to the refining vessel. The charge can contain from 1.5 to 2 wt% carbon. Using the AOD process, a mixture of argon and oxygen is injected into the molten steel. The oxygen combines with the carbon to form carbon monoxide, which is liberated from the molten metal. The mixture of argon and oxygen is controlled to achieve the desired carbon content. The process using VOD is similar, except that the argon carrier gas is not necessary and oxygen is injected directly into the melt. Using these processes, production of low-carbon stainless steels (less than 0.04 wt%) is now routine. As an added benefit, these processes also drastically reduce the residual sulfur content to as low as 0.001 wt%.

Following the refining process, the steel can be poured into ingot molds or transferred directly to a continuous casting machine. In the latter case, the steel can be cast directly to a slab that is nominally 13 to 25 cm (5 to 10 in.) thick and up to 2 m (6 ft) wide. Since the 1980s, continuous casting has expanded rapidly and is used extensively for the production of plate and sheet stock.

Once cast, the steel is subjected to various hot-working processes to reduce it to nearly final dimensions. These can include hot rolling, swaging, or extrusion. Many stainless steels can be used in the hot-rolled condition following appropriate heat treatment and pickling operations. Many grades are reduced further using cold-working techniques. For production of very thin sheet (less than 0.01 in.), a Sendzimir mill [15] is often used since very large reductions (up to 80%) are possible, reducing or eliminating the need for intermediate annealing treatments. Welding wire is produced using standard wire-drawing processes and commercial lubricants. Care must be taken to clean the wire properly to avoid contamination during welding. In extreme cases, the wire can actually be rolled to the desired diameter to avoid the surface "lapping" that can trap contamination. More details of these and other processes for the production of stainless steels can be found in the references [13,14].

REFERENCES

1. Castro, R. 1993. Historical background to stainless steels, in *Stainless Steels*, P. LaCombe, B. Baroux, and G. Beranger, eds., Les Éditions de Physique, Les Ulis, France, pp. 3–9.
2. Goldschmidt, H. 1897. *Elektrochemische Zeitschrift*, 4:143.
3. Guillet, L. 1904. *Revue de Metallurgie*, 1:155; 2(1905):350; 3(1906):372.
4. Portevin, A. 1909. Iron and Steel Institute, *Carnegie Scholarship Memoirs*, 1:230.
5. Giesen, W. 1909. Iron and Steel Institute, *Carnegie Scholarship Memoirs*, 1:1.
6. Stainless steel: the inventor, Harry Brearly, and his invention, *Materials Performance*, March 1990, pp. 64–68; reprinted from *1913–1988: 75 Years of Stainless Steel*, British Steel,
7. Maurer, E., and Strauss, B. 1920. *Kruppshe Monatsch*, pp. 120–146.
8. Fontana, M. G., and Green, N. D. 1978. *Corrosion Engineering*, 2nd ed., McGraw-Hill, New York.
9. Jones, D. A. 1995. *Principles and Prevention of Corrosion*, 2nd ed., Prentice Hall, Upper Saddle River, NJ.
10. LaCombe, P., Baroux, B., and Beranger, G., eds. 1993. *Stainless Steels*, Les Éditions de Physique, Les Ulis, France.
11. Sedriks, A. J. 1996. *Corrosion of Stainless Steels*, Wiley-Interscience, New York.
12. Scully, J. R. 2003. Corrosion and oxide films, in *Encyclopedia of Electrochemistry*, Vol. 3, M. Stratmann and G. S. Frankel, eds., Wiley-VCH, Weinheim, Germany, p. 344.
13. U.S. Steel. 1971. *The Making, Shaping, and Treating of Steel*, 9th ed., U.S. Steel Corporation, Pittsburgh, PA.
14. ASM. 1987. *ASM Metals Handbook*, 10th ed., Vol. 13, ASM International, Materials Park, OH.
15. Sendzimir, M. G. 1980. Cold mills for stainless steels, *Iron and Steel Engineer*, November, pp. 29–42.

CHAPTER 2

PHASE DIAGRAMS

Equilibrium phase diagrams can be used to describe phase transformations and phase stability in stainless steels. In this chapter the Fe–Cr binary system and the Fe–Cr–C and Fe–Cr–Ni ternary systems are described. The intention here is to provide the reader with a working knowledge of applicable equilibrium phase diagrams that can be used to predict microstructure evolution as they apply to the various grades of stainless steels. These diagrams can only approximate the actual microstructure that develops in welds since (1) stainless steel base and filler metals contain up to 10 alloying elements that cannot be accommodated easily with standard phase diagrams, and (2) phase diagrams are based on equilibrium conditions, while the rapid heating and cooling conditions associated with welding result in distinctly nonequilibrium conditions. For a more rigorous treatment of phase diagrams and phase stability associated with stainless steels, other sources are suggested [1–3].

Some of the limitations of classical phase diagrams have now been overcome by powerful computer programs that use thermodynamic information to construct phase diagrams for common alloy systems. These programs and the diagrams they create are only as good as the input data, but have proved to be reasonably accurate for iron-base systems. One of these software programs, ThermoCalc™, is described in this chapter, and some examples are provided that demonstrate how phase diagrams can be constructed for multicomponent systems.

Welding Metallurgy and Weldability of Stainless Steels, by John C. Lippold and Damian J. Kotecki
ISBN 0-471-47379-0

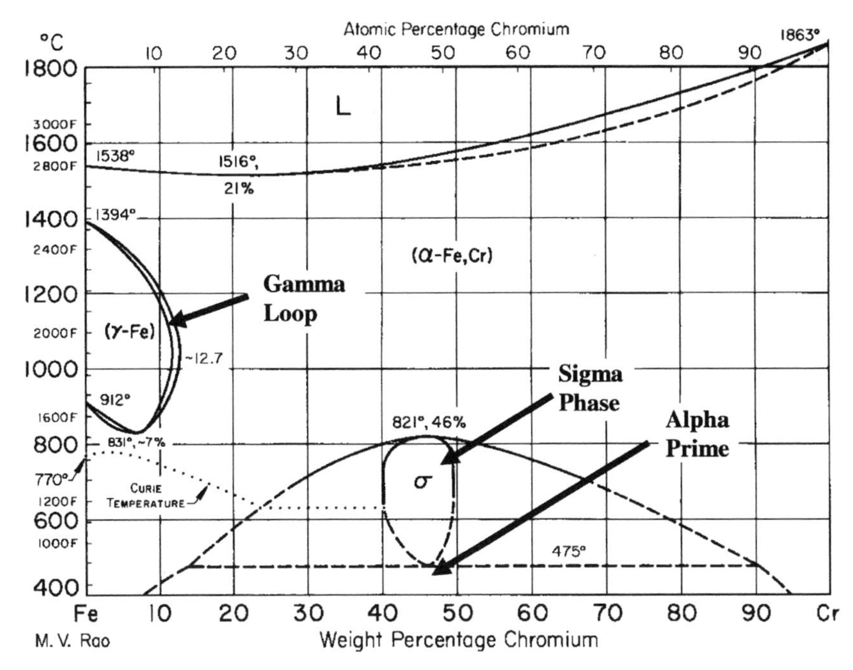

Figure 2.1 Fe–Cr equilibrium phase diagram. (From *Metals Handbook* [4]. Originally from Hansen [5]. Courtesy of McGraw-Hill.)

2.1 IRON–CHROMIUM SYSTEM

The iron–chromium phase diagram shown in Figure 2.1 is the starting point for describing stainless steel phase stability, since Cr is the primary alloying element. Note that there is complete solubility of Cr in iron at elevated temperatures, and solidification of all Fe–Cr alloys occurs as ferrite.* The solidification range for Fe–Cr alloys is very narrow. At low chromium concentrations a "loop" of austenite exists in the temperature range 912 to 1394°C (1670 to 2540°F). This is commonly referred to as the *gamma loop*. Alloys with greater than 12.7 wt% Cr will be fully ferritic at elevated temperature, while those with less than this amount of Cr will form at least some austenite at temperatures within the gamma loop. Alloys with less than about 12 wt% Cr will be completely austenitic at temperatures within the gamma loop. Upon rapid cooling, this austenite can transform to martensite.

A low-temperature equilibrium phase, called *sigma phase*, is present in the Fe–Cr system. This phase has a (Fe,Cr) stoichiometry and a tetragonal crystal structure. Sigma

*Ferrite is indicated on phase diagrams by the symbols α and δ. Based on the Fe–C system, δ-ferrite is considered high-temperature ferrite, and α-ferrite is low-temperature ferrite that forms from austenite. In Fe–Cr, Fe–Cr–C, and Fe–Cr–Ni systems, α and δ are often used interchangeably, since the ferrite that forms at elevated temperature is not fully transformed, and some (or all) can be retained at room temperature. In this book, α and δ are used interchangeably to designate the body-centered cubic (BCC) crystal structure characteristic of Fe-based alloys.

phase forms most readily in alloys exceeding 20 wt% Cr. Because sigma forms at low temperatures, the kinetics of formation are quite sluggish, and precipitation requires extended time in the temperature range 600 to 800°C (1110 to 1470°F). Because sigma is a hard, brittle phase, its presence in stainless steels is usually undesirable.

The diagram also contains a dotted horizontal line within the $\sigma + \alpha$ phase field at 475°C (885°F). A phenomenon known as 475°C *embrittlement* (or 885°F *embrittlement*) results from the formation of coherent Cr-rich precipitates within the alpha matrix. These precipitates are called *alpha-prime* (α'). They actually form within the temperature range 400 to 540°C (750 to 1000°F) and have been shown to have a severe embrittling effect in alloys with greater than 14 wt% Cr [2]. Formation of α' is also quite sluggish in Fe–Cr alloys, but its rate of formation can be accelerated by alloying additions.

2.2 IRON–CHROMIUM–CARBON SYSTEM

The addition of carbon to the Fe–Cr system significantly alters and complicates phase equilibrium. Since carbon is an austenite promoter it will expand the gamma loop, allowing austenite to be stable at elevated temperatures at much higher Cr contents. The effect of carbon on the expansion of the austenite phase field is shown in Figure 2.2. Note that even small amounts of carbon result in a dramatic expansion of the gamma loop. This is important for development of the martensitic stainless steels, since for martensite to form during cooling, these steels must be austenitic at elevated temperatures. For the ferritic grades, the size of the gamma loop must be controlled such that little or no austenite forms at elevated temperatures.

To allow the Fe–Cr–C ternary system to be viewed as a function of temperature, it is necessary to set one of the elements at a constant value. In this way, a *pseudobinary*

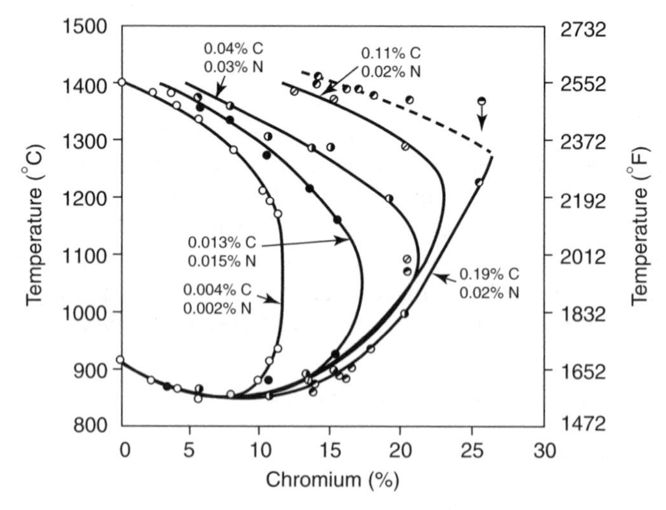

Figure 2.2 Effect of carbon on the expansion of the austenite phase field. (From Baerlacken et al. [6].)

phase diagram (or *isopleth*) can be constructed. Such a diagram is called *pseudobinary* because it represents a two-dimensional projection of a three-dimensional system. Because of this, the diagram cannot be used in the same manner as a binary diagram. For example, tie-lines cannot be used to predict the phase balance on a pseudobinary diagram because the diagram has depth (i.e., the tie-lines do not necessarily reside in the plane of the diagram). They are very useful for understanding phase equilibrium and phase transformations in three-component systems. Two pseudobinary diagrams based on the 13 wt% Cr and 17 wt% Cr systems with variable carbon are shown in Figure 2.3.

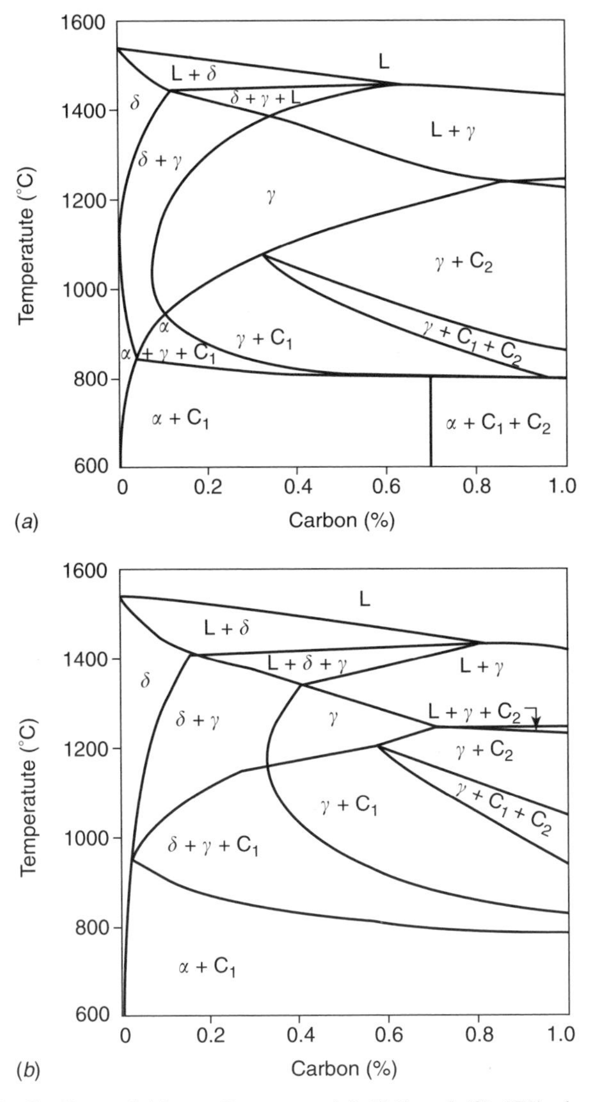

Figure 2.3 Fe–Cr–C pseudobinary diagram at (*a*) 13% and (*b*) 17% chromium. C_1 is a $(Cr,Fe)_{23}C_6$ carbide; C_2 is a $(Cr,Fe)_7C_3$ carbide. (From Castro and Tricot [7].)

The diagram is now much more complicated than the simple Fe–Cr diagram, primarily because of the introduction of carbon and the presence of additional two- and three-phase fields. Because of the addition of carbon, two different carbides, $(Cr,Fe)_{23}C_6$ and $(Cr,Fe)_7C_3$, now appear on the diagram.

For low-chromium ferritic and martensitic stainless steels, the 13 wt% Cr pseudobinary diagram can be used to explain phase stability and microstructure. At very low carbon contents (less than 0.1%), the ternary alloys are fully ferritic at elevated temperature. If cooled rapidly enough, the alloy will remain primarily ferritic. For the 13% Cr diagram, this is the basis for low-chromium ferritic stainless steels, such as Type 409.

At carbon contents above 0.1%, austenite will form at elevated temperatures, and mixtures of austenite and ferrite will be present at temperatures just below the solidification temperature range. Upon cooling, the structure will become fully austenitic at temperatures below 1200°C (2190°F). If the cooling rate is sufficiently rapid, this austenite will transform to martensite. This is the basis for low-chromium martensitic stainless steels such as Type 410. At lower carbon contents (e.g., 0.05 wt%), a mixture of austenite and ferrite will be present at elevated temperature, resulting in a microstructure consisting of ferrite and martensite upon rapid cooling. These microstructures are usually undesirable because of a loss in mechanical properties.

At higher constant chromium levels in the Fe–Cr–C system, the ferrite phase field expands and the austenite phase field shrinks, as represented in the 17 wt% constant chromium section in Figure 2.3*b*. This results from the ferrite-promoting effect of chromium. As a result, the ferrite formed at elevated temperature will be much more stable, and higher carbon contents are required to form elevated-temperature austenite. This diagram is the basis for medium-Cr ferritic stainless steels, such as Type 430, and medium-Cr, high-C martensitic alloys, such as Type 440.

2.3 IRON–CHROMIUM–NICKEL SYSTEM

Addition of nickel to the Fe–Cr system also expands the austenite phase field and allows austenite to be a stable phase at room temperature. This ternary system is the basis for the austenitic and duplex stainless steels. Both liquidus and solidus projections of the Fe–Cr–Ni system (Figure 2.4) are available [8] and can be used to describe the solidification behavior of alloys based on this system considered by viewing the liquidus and solidus surfaces, which define the start and completion of solidification, respectively. Note that the liquidus surface exhibits a single, dark line that runs from near the Fe-rich apex of the triangle to the Cr–Ni side. This line separates compositions that solidify as primary ferrite (above and to the left) from compositions that solidify as primary austenite. At approximately 48Cr–44Ni–8Fe, a ternary eutectic point exists.

The solidus surface exhibits two dark lines that run from near the Fe-rich apex to the Cr–Ni-rich side of the diagram. Between these two lines the austenite and ferrite phases coexist with liquid just above the solidus, but only with one another just below the solidus. This region separates the ferrite and austenite single-phase fields below the solidus. Note that these lines terminate at the ternary eutectic point. Arrows on these lines represent the direction of decreasing temperature.

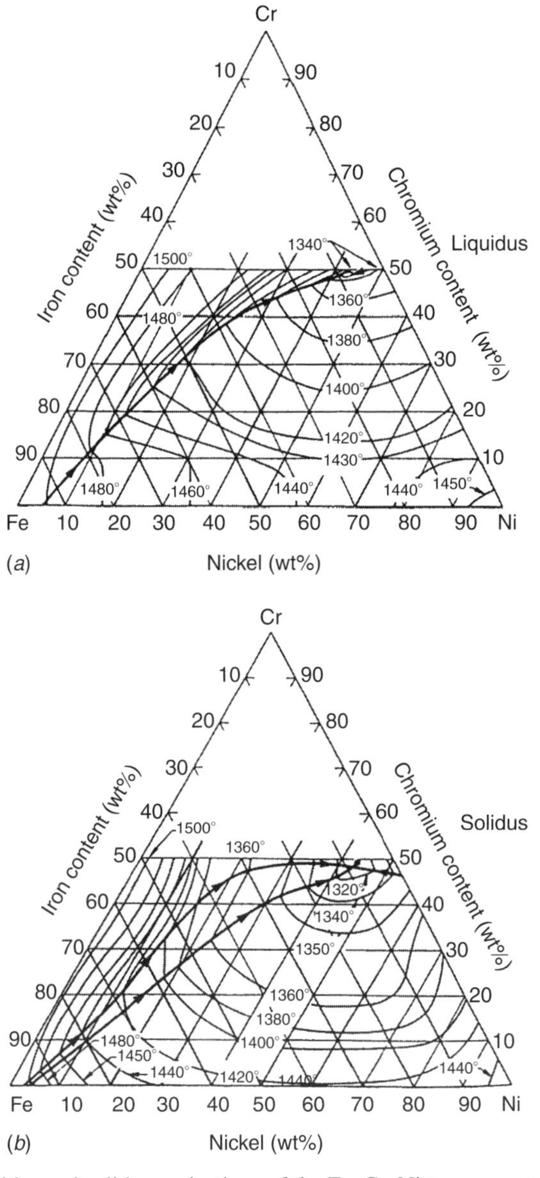

Figure 2.4 Liquidus and solidus projections of the Fe–Cr–Ni ternary system. (From *Metals Handbook* [8]. Courtesy of ASM International.)

By taking a constant-Fe section through the ternary phase diagram from the liquidus to room temperature, a pseudobinary Fe–Cr–Ni phase diagram can be generated. Two such diagrams have been constructed at 70 wt% Fe and at 60 wt% Fe based on isothermal ternary sections [9] and are shown in Figure 2.5. Because this is a ternary system, the phase fields exist in three dimensions, resulting in three-phase fields that cannot occur on a standard binary phase diagram.

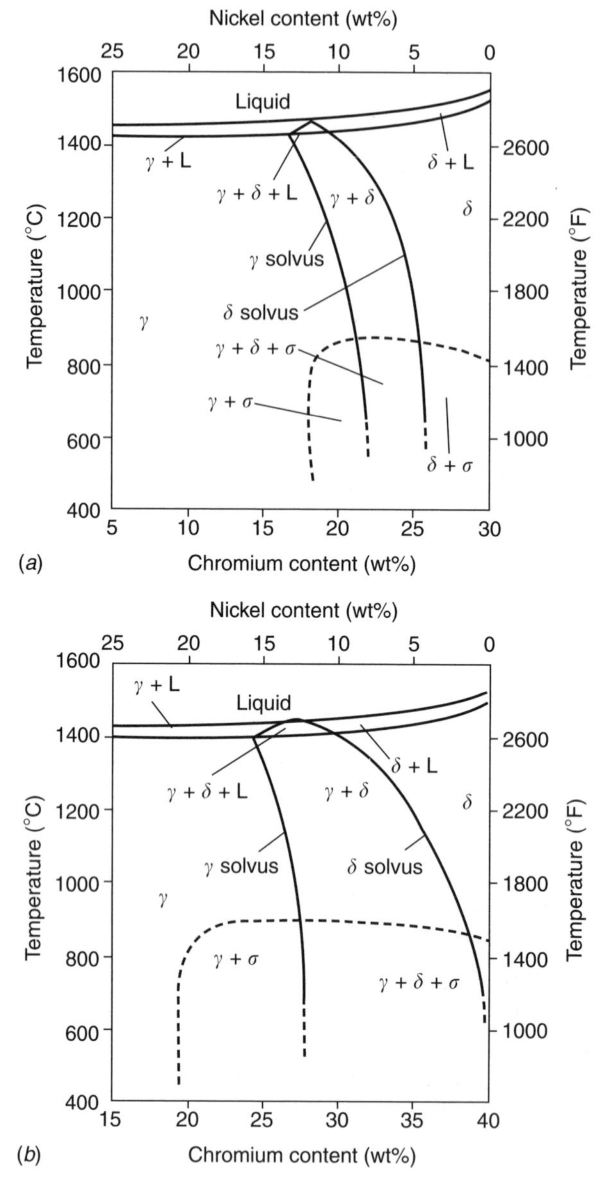

Figure 2.5 Pseudobinary sections of the Fe–Cr–Ni ternary system at (*a*) 70% Fe and (*b*) 60% Fe. (From Lippold and Savage [9]. Courtesy of the American Welding Society.)

Note the small triangular region between the solidus and liquidus lines. This is the three-phase austenite + ferrite + liquid region that separates alloys that solidify as austenite (to the left) from those that solidify as ferrite. In the solid state, the ferrite is stable at elevated temperature at chromium contents greater than 20 wt%.

As the temperature decreases, this ferrite will transform partially to austenite in the range 20 to 25 wt%. Alloys that solidify as austenite (to the left of the three-phase triangle) remain as austenite upon cooling to room temperature. Alloys that solidify as ferrite at compositions just to the right of the three-phase triangle, must cool through the two-phase austenite + ferrite region. This results in the transformation of some of the ferrite to austenite. At compositions farther to the right of the triangle (higher Cr/Ni ratios) ferrite will become increasingly stable, until ultimately a fully ferritic structure will exist toward the far right side of each diagram. These diagrams are used in Chapters 6 and 7 to explain the phase transformations and microstructure evolution in the austenitic and duplex stainless steels, respectively.

2.4 PHASE DIAGRAMS FOR SPECIFIC ALLOY SYSTEMS

A number of software packages are now available that allow phase diagrams to be constructed for specific alloy systems based on thermodynamic data. These packages take into account the interactions among multiple elements and then construct the phase equilibria over a range of temperature from the melting point to room temperature. One of the most widely used of these packages is ThermoCalc™. Software programs such as ThermoCalc™ are very useful for *predicting* microstructure evolution in stainless steel weld metal and heat-affected zones, but it should be recognized that the diagrams generated represent the equilibrium condition, which is usually not achieved in welds due to rapid heating and cooling rates.

An example of a phase diagram developed using ThermoCalc™ is shown in Figure 2.6. This represents a base alloy composition (wt%) of 12Cr–0.5Mn–0.5Si–0.1C that is typical of a Type 410 martensitic stainless steel. On the composition axis of this diagram, nickel varies from 0 to 5 wt%. Such a diagram is useful for determining how nickel affects the various phase fields, in particular how additions of nickel may affect the amount of ferrite that forms in weld metal or the heat-affected zone of this alloy. For example, if one compares Alloy A with 0.3 wt% Ni and Alloy B with 2.0 wt% Ni in Figure 2.6, it can be seen that the single phase ferrite (α) region near 1400°C (2550°F) disappears for Alloy B, and the temperature range over which the two-phase field ($\alpha + \gamma$) exists becomes progressively narrower with increasing Ni. These data can also be used to develop a plot of molar fraction of each phase as a function of temperature, as shown in Figure 2.7. Using approaches such as this, base and filler metal compositions can easily be evaluated with respect to phase stability and predictions with respect to microstructure evolution during welding and heat treatment can be made.

Another example is provided in Figure 2.8. This is the calculated phase diagram for the duplex stainless steel Alloy 2205 with nitrogen as the composition variable. A typical nitrogen composition for this alloy is 0.15 wt% N, as indicated by the dashed vertical line. With this diagram, it can be seen that the alloy is never fully ferritic at elevated temperature. Rather, there is always some austenite in the

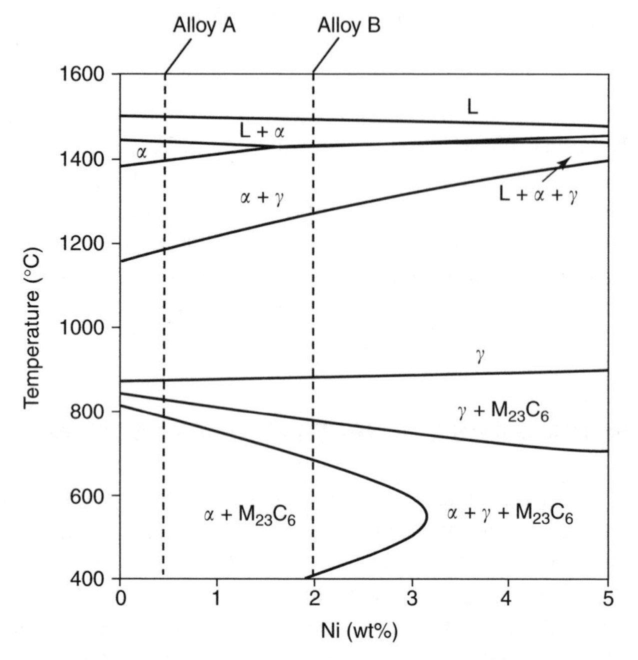

Figure 2.6 ThermoCalc™ phase diagram for a 12% Cr martensitic stainless steel, showing the effect of nickel additions. (Courtesy of Antonio Ramirez, Ohio State University, 2002.)

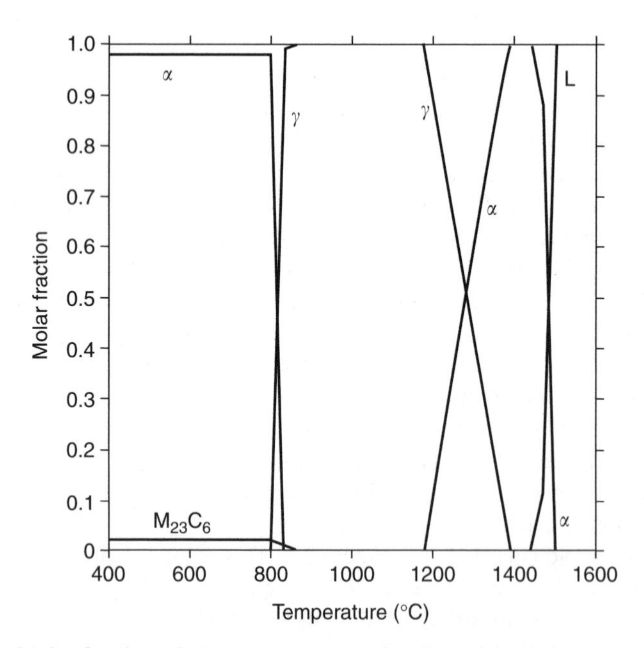

Figure 2.7 Molar fraction of phases present as a function of temperature for the 12% Cr alloy in Figure 2.6 at 0.3% Ni. (Courtesy of Antonio Ramirez, Ohio State University, 2002.)

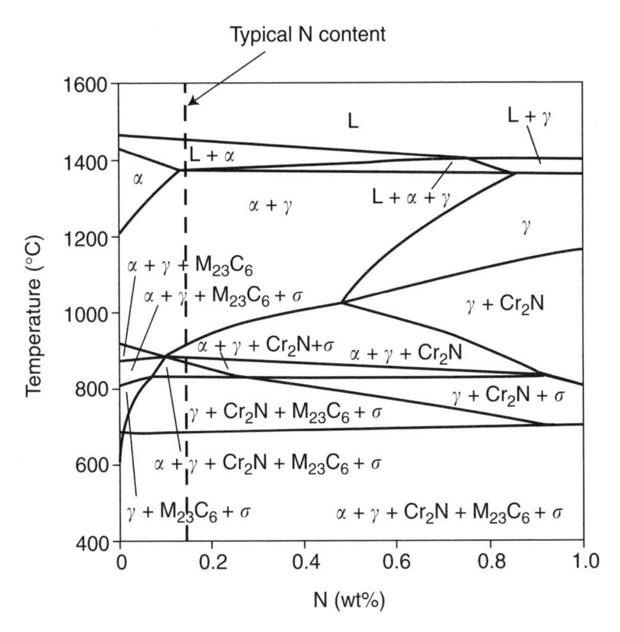

Figure 2.8 ThermoCalc™ phase diagram for duplex stainless steel Alloy 2205, showing the effect of nitrogen additions. (From Ramirez et al. [10].)

microstructure that can be effective in minimizing ferrite grain growth in the HAZ. It also shows the phase fields where carbide ($M_{23}C_6$), nitride (Cr_2N), and sigma phase are thermodynamically stable.

Similar to Figure 2.7, the fraction of phases as a function of temperature can also be estimated at a given nitrogen content (0.15 wt% N), as shown in Figure 2.9. Unlike Figure 2.7, the phase fraction shown is a volume fraction rather than a molar fraction. The carbide and nitride phases have been removed for clarity, since their volume fractions are very low. This type of diagram is very useful for determining the amount of each phase present at a given temperature. For example, at approximately 1375°C (2510°F) the microstructure consists of approximately 95% ferrite and 5% austenite.

Note that Figure 2.9 indicates that sigma phase is an equilibrium phase at temperatures below about 900°C (1650°F). Whereas sigma is very sluggish to form in iron–chromium alloys, it forms much more rapidly at about 700°C (1290°F) in iron–chromium–nickel stainless steels containing ferrite, especially those high in Cr and Mo. This feature of duplex stainless steels, together with their high ferrite content, makes them very susceptible to embrittlement from sigma formation. Avoidance of sigma formation from ferrite requires rapid cooling through the approximate temperature range 900°C (1650°F) to 500°C (930°F). This feature of duplex stainless steels also imposes severe restrictions on elevated temperature processing, including postweld heat treatment, as discussed in Chapter 7.

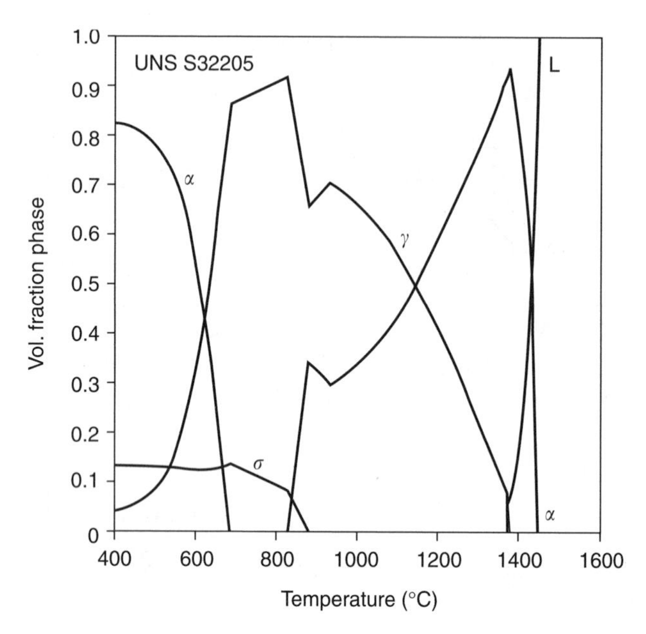

Figure 2.9 Volume fraction of phases present as a function of temperature for duplex stainless steel alloy 2205. (From Ramirez et al. [10].)

REFERENCES

1. Folkhard, E. 1984. *Welding Metallurgy of Stainless Steels*, Springer-Verlag, Berlin (in German; 1988, English translation).
2. Peckner, D., and Bernstein, I. M. 1977. *Handbook of Stainless Steels*, McGraw-Hill, New York.
3. Castro, R. J., and de Cadenet, J. J. 1974. *Welding Metallurgy of Stainless and Heat-Resisting Steels*, Cambridge University Press.
4. *ASM Metals Handbook*, 8th ed., Vol. 8, ASM International, Materials Park, OH, p. 291.
5. Hansen, M. 1958. *Constitution of Binary Alloys*, 2nd ed., McGraw-Hill, New York, 1958.
6. Baerlacken, E., et al. 1958. Investigations concerning the transformation behavior, notched impact toughness, and susceptibility to intercrystalline corrosion of iron–chromium alloys with chromium contents to 30%, *Stahl und Eisen*, 81(12), March.
7. Castro, R., and Tricot, R. 1962. Études des transformations isothermes dans les aciers inoxydables semi-ferritiques á 17% de chrome, *Memoires Scientifiques de la Revue de Metallurgie*, Part 1, 59:571–586; Part 2, 59:587–596.
8. ASM Metals Handbook, 8th Edition, Volume 8, p.424.
9. Lippold, J. C., and Savage, W. F. 1979. Solidification of austenitic stainless steel weldments, part 1: a proposed mechanism, *Welding Journal*, 58(12):362s–374s.
10. Ramirez, A. J., Brandi, S., and Lippold, J. C. 2003. The relationship between chromium nitride and secondary austenite precipitation in duplex stainless steels, *Metallurgical Transactions A*, 34A(8):1575–1597.

CHAPTER 3

ALLOYING ELEMENTS AND CONSTITUTION DIAGRAMS

Stainless steels contain a number of alloying additions (other than chromium, carbon, and nickel) that are intended to improve mechanical properties and corrosion resistance or to control the microstructure. This chapter provides an overview of the various alloying additions in stainless steels and summarizes the development of constitution diagrams for the prediction of microstructure based on composition.

3.1 ALLOYING ELEMENTS IN STAINLESS STEELS

Stainless steels are iron-base alloys with iron contents ranging from 50 to 88 wt% of the composition. The principal alloying additions to stainless steel are chromium and carbon for the ferritic and martensitic grades, with the addition of nickel for the austenitic and duplex grades. Essentially all of the stainless steels contain manganese and silicon as intentional additions. Other alloying additions include molybdenum, niobium, titanium, aluminum, copper, tungsten, nitrogen, and others to improve fabricability, develop special properties, enhance corrosion resistance, or influence microstructure. Impurity elements commonly found in stainless steels include nitrogen, oxygen, sulfur, and phosphorus. All of these alloy and impurity elements have some effect on weldability and performance. In most cases, the level of these elements in the base or filler metal is controlled by the material specification to assure that the steel performs as anticipated. This section provides some insight

Welding Metallurgy and Weldability of Stainless Steels, by John C. Lippold and Damian J. Kotecki
ISBN 0-471-47379-0

into the influence of these alloying elements on the properties and performance of stainless steels.

3.1.1 Chromium

Chromium is added primarily to provide corrosion protection to the steel. It is especially effective in oxidizing environments such as nitric acid. With the addition of chromium, an oxide of stoichiometry $(Fe,Cr)_2O_3$ forms on the steel surface. The presence of chromium increases the stability of the oxide since it has a much higher affinity for oxygen than does the iron. When the chromium level exceeds approximately 10.5 wt%, the steel is considered "stainless" under ambient conditions. Higher levels of chromium may be required for oxide stability in more aggressive environments.

Chromium is also a ferrite promoter. As shown in Chapter 2, iron–chromium alloys that contain more than 12 wt% chromium will be fully ferritic. In Fe–Cr–C and Fe–Cr–Ni–C alloys, increasing Cr will promote ferrite formation and retention in martensitic, austenitic, and duplex grades. In the ferritic alloys, chromium is the primary alloying element stabilizing the ferritic microstructure.

Chromium is also a strong carbide former. The most common Cr-rich carbide is $M_{23}C_6$, where the "M" is predominantly Cr but may also have some fraction of Fe and Mo present. In most stainless steel systems, this is normally the case, and the term $M_{23}C_6$ is used with the understanding that chromium is the predominant metallic element. This carbide is found in virtually all the stainless steels. It is also possible to form a Cr_7C_3 carbide, although this carbide type is not common. Other complex carbides and carbonitrides $[M_{23}(C,N)_6]$ are also possible [1]. Chromium also combines with nitrogen to form a nitride. The most common is Cr_2N, which has been observed in both ferritic and duplex grades.

Chromium is also a key ingredient in the formation of intermetallic compounds, many of which tend to embrittle the stainless steels. The most common is sigma phase (σ), which in the Fe–Cr system is a (Fe,Cr) compound that forms below 815°C (1500°F). Sigma phase can form in virtually any stainless steel but tends to be most common in the high-Cr austenitic, ferritic, and duplex alloys. Chromium is also present in chi (χ) and Laves intermetallic phases. These are discussed in more detail later in the appropriate chapters associated with specific alloy systems.

From a mechanical properties standpoint, chromium will provide some degree of solid solution strengthening since it is a substitutional atom in both the body-centered cubic (BCC) and face-centered cubic (FCC) crystal lattices. High chromium levels in ferritic alloys can result in very poor toughness and ductility, particularly when carbon and nitrogen are present. High-chromium ferritic grades must be treated very carefully, or carbon and nitrogen reduced to very low levels, to have acceptable mechanical properties in welded fabrication. This is discussed in detail in Chapter 5.

3.1.2 Nickel

The primary function of nickel is to promote the austenite phase such that predominantly austenitic or austenitic–ferritic alloys can be produced. By adding sufficient nickel, the austenite phase field can be greatly expanded such that austenite is stable to room temperature and below. Nickel is not a strong carbide former and does not generally promote the formation of intermetallic compounds, although there is evidence that its presence in the alloy may influence precipitation kinetics [1]. There is some evidence that the presence of nickel in ferritic alloys improves general corrosion resistance, particularly in reducing environments such as those containing sulfuric acid. However, nickel has been associated with a decrease in *stress corrosion cracking* (SCC) *resistance*. The work of Copson [2] clearly showed a decrease in SCC resistance in an aggressive Cl-containing environment when nickel was added to a Fe–20Cr alloy. The famous *Copson curve* shows that the lowest SCC resistance occurs in the range 8 to 12 wt% Ni and increases with either an increase or decrease in nickel outside this range. Nickel is a good solid solution strengthener but is most beneficial in terms of improving toughness in both the martensitic and ferritic grades. Additions of up to 2 wt% Ni to high-Cr ferritic stainless steel can dramatically reduce the ductile-to-brittle fracture transition temperature (DBTT) [3].

3.1.3 Manganese

Manganese is added to virtually every steel. In austenitic stainless steels, it is normally present in the range 1 to 2 wt%. In ferritic and martensitic stainless steels, it is more commonly present at less than 1 wt%. Historically, it was added to prevent *hot shortness* during casting. This is a form of solidification cracking that is associated with the formation of low melting point iron–sulfide eutectic constituents. Since manganese combines much more readily with sulfur than does iron, the addition of sufficient manganese and the formation of stable manganese sulfide (MnS) effectively eliminated the hot shortness problem.

Manganese is generally considered to be an austenite-promoting element, although the degree of promotion is dependent on the amount present and the level of nickel, as discussed later in this chapter. It is very effective in stabilizing austenite at low temperature to prevent transformation to martensite. Its potency in promoting austenite at elevated temperature is dependent on the overall composition of the alloy. In austenitic stainless steels, such as Type 304, it appears to have little effect on promoting austenite versus ferrite.

Manganese is sometimes added to specialty alloys to increase the solubility of nitrogen in the austenite phase. For example, the addition of 15 wt% manganese to a Fe–20Cr alloy raises the nitrogen solubility from 0.25 to approximately 0.4 wt% [4]. The manganese effect on mechanical properties is minimal. It provides some solid solution strengthening and appears to have little effect on embrittlement.

3.1.4 Silicon

Silicon is also present in virtually all stainless steels and is added primarily for deoxidation during melting. In most alloys it is present in the range 0.3 to 0.6 wt%. In some cases aluminum may be substituted as a deoxidizer, but this is rarely the case in stainless steels. It has been found to improve corrosion resistance when present at levels of 4 to 5 wt% and is added to some heat-resisting alloys in the range 1 to 3 wt% to improve oxide scaling resistance at elevated temperature. The role of silicon in promoting ferrite or austenite is not entirely clear. In austenitic stainless steels, up to 1 wt% seems to have no effect on the phase balance, but higher levels appear to promote ferrite. In ferritic and martensitic stainless steels, silicon appears to help promote ferrite.

Silicon forms a number of iron silicides (FeSi, Fe_2Si, Fe_3Si, Fe_5Si_3) and a Cr_3Si intermetallic, all of which tend to embrittle the structure. It also expands the composition range over which sigma phase forms [5]. Silicon is known to segregate during solidification, resulting in the formation of low melting eutectic constituents, particularly in combination with nickel [6]. For these reasons, it is usually held below 1 wt%.

Silicon is well known to improve the fluidity of molten steel. For this reason, it may be added in somewhat higher than normal amounts to weld filler metals. Some stainless steels, particularly the austenitic grades, tend to be quite sluggish in the molten state, and the addition of silicon can greatly improve the fluidity.

3.1.5 Molybdenum

Molybdenum is added to a number of stainless steels and has different functions depending on the particular grade. For the ferritic, austenitic, and duplex grades, Mo is added in amounts up to 6 % or more in super austenitics in order to improve corrosion resistance, particularly with respect to pitting and crevice corrosion. In austenitic stainless steels, Mo also improves elevated temperature strength. For example, the addition of 2 wt% Mo to a standard 18Cr–8Ni alloy results in a 40% increases in tensile strength at 760°C (1400°F) [1]. This can also have a negative effect, since alloys containing Mo will be more difficult to hot work. Some of the martensitic stainless steels contain Mo as a carbide former. The addition of as little as 0.5 wt% Mo increases the secondary hardening characteristics of the steel, resulting in higher room-temperature yield and tensile strength and improved elevated-temperature properties. Molybdenum is a ferrite-promoting element, and its presence will promote ferrite formation and retention in the microstructure. This can be a potential problem in the martensitic grades, where residual room-temperature ferrite can reduce toughness and ductility.

3.1.6 Carbide-Forming Elements

In addition to Cr and Mo, a number of other elements added to stainless steels also promote carbide formation. These include *niobium, titanium, tungsten, tantalum*, and *vanadium*. Niobium and titanium are added to austenitic stainless steels to provide stabilization of the carbon to avoid intergranular corrosion. Both elements form an MC-type carbide that resists dissolution during welding and heat treatment, thereby preventing the formation of Cr-rich $M_{23}C_6$ carbides that are associated with the onset of intergranular corrosion. This phenomenon is discussed in greater detail in Chapter 6. Tungsten, tantalum, and vanadium are added to some specialty stainless steels, primarily to provide elevated temperature strength by the formation of a fine dispersion of carbides. These elements tend to promote ferrite in the microstructure since they tie up carbon and effectively neutralize this powerful austenite promoter. In addition, these elements in solid solution promote ferrite formation in their own right.

3.1.7 Precipitation-Hardening Elements

Aluminum, titanium, copper, and *molybdenum* can be added to stainless steels to promote a precipitation reaction that hardens the alloy. The precipitation-hardenable (PH) martensitic alloys contain Cu, Al, and Mo and can be heat-treated to produce room-temperature yield strengths in excess of 1375 MPa (200 ksi). Austenitic PH stainless steels usually contain titanium and aluminum and form Ni_3Ti and Ni_3Al [gamma-prime (γ')] precipitates that are similar to the hardening precipitates in Ni-base superalloys. Aluminum in solid solution is a potent ferrite-promoting element. Copper, on the other hand, is a weak austenite-promoting element. Nearly pure copper precipitates can be used for hardening in martensitic stainless steels such as 17-4PH.

3.1.8 Interstitial Elements: Carbon and Nitrogen

Carbon is present in all steels, but unlike C–Mn and low alloy structural steels, it is usually desirable to control carbon below 0.1 wt%. The exception is the martensitic grades, where carbon is critical for the transformation strengthening of these alloys. In solution, carbon provides an interstitial strengthening effect, particularly at elevated temperatures. In most alloys, carbon combines with other elements to form carbides, as discussed previously. In the case of Cr-rich $M_{23}C_6$ carbides, a degradation in corrosion resistance can result, and for this reason low-carbon (L-grade) alloys are produced where carbon is kept below 0.04 wt%. It should be noted that $M_{23}C_6$ carbides contain nearly four times as many metal atoms (mostly chromium) as carbon atoms, and a chromium atom is a bit over four times as heavy as a carbon atom. So, on a weight percent basis, $M_{23}C_6$ carbide formation can remove as high as 16 times as much chromium as carbon from solid solution.

Nitrogen is usually present as an impurity in many of the stainless steels but is an intentional addition to some of the austenitic and almost all the duplex grades. Similar to carbon, nitrogen is a powerful solid solution strengthening agent, and additions of as little as 0.15 wt% can dramatically increase the strength of austenitic alloys [7]. The strengthening effect of nitrogen in austenite is especially pronounced at cryogenic temperatures. For the duplex stainless steels, nitrogen is added to improve strength, but more important, to increase resistance to pitting and crevice corrosion. Some duplex grades contain up to 0.3 wt% nitrogen. As noted previously, the solubility of nitrogen in stainless steels is relatively low, particularly in the ferrite phase. The addition of manganese to the austenitic stainless steels increases nitrogen solubility. In the ferritic and duplex grades, Cr_2N will precipitate in the ferrite phase if the solubility limit is exceeded, as can be observed in the weld metals and HAZs of these alloys if appreciable austenite fails to form during cooling from temperatures above about 1100°C (2010°F).

Carbon and nitrogen are the most potent of the austenite-promoting elements, and thus the levels of these elements must be controlled carefully if precise microstructure balance is required. As noted previously, this can be controlled by the level of each element in the alloy or by the addition of elements that will form carbides (Nb, Ti) or nitrides (Ti, Al) and effectively neutralize their effect in the matrix. Nitrogen pickup from the atmosphere can produce a departure from the desired microstructure if shielding during arc welding is not adequate. In high-nitrogen austenitic and duplex alloys, nitrogen loss during welding can be a problem. For duplex stainless steels, nitrogen is sometimes added to the shielding gas to maintain weld metal nitrogen levels.

3.1.9 Other Elements

There are a number of other intentional alloying additions made to various stainless steels for very specific applications. *Sulfur, selenium*, and *lead* are added to *free-machining grades* to improve machinability by allowing higher machining speeds and improved tool life. These additions reduce the corrosion resistance and typically make the alloys unweldable, although control of solidification behavior (primary ferrite versus primary austenite) may negate the effect of sulfur. This is discussed in Chapter 6. *Tungsten* is added to some duplex stainless steels to improve pitting corrosion resistance, and it appears to promote ferrite formation. *Aluminum* is used in some of the lower chromium ferritic grades to improve general corrosion resistance. *Cobalt* is an effective solid-solution strengthener, and in the martensitic stainless steels it can be added to increase the martensite start (M_S) temperature. Cobalt promotes austenite formation.

3.2 FERRITE-PROMOTING VERSUS AUSTENITE-PROMOTING ELEMENTS

Stainless steels are iron-based alloys containing from 12 to over 50 wt% alloying additions. Alloying elements affect the equilibrium phase relationships relative to the stability of the austenite, ferrite, and martensite phases. Elements added to stainless steels can be divided into those that promote, or stabilize, either the ferrite or the austenite phase. Remember that martensite is a transformation product that forms from austenite upon cooling from elevated temperature. If austenite does not form at high temperatures, martensite cannot form at low temperatures.

Austenitic stainless steels contain high levels of nickel and other austenite-forming elements which promote the formation of the austenite phase, so it is stable to room temperature and below. Ferritic stainless steels contain a balance of elements, such as high chromium contents, such that ferrite is the predominant metallurgical phase present. Martensitic stainless steels are austenitic at elevated temperature, but this austenite is unstable and transforms upon cooling. By balancing austenite- and ferrite-promoting elements, the microstructure of the stainless steel can be controlled. This balance has important implications with respect to mechanical properties, corrosion resistance, and weldability.

Ferrite-Promoting Elements

- Chromium
- Molybdenum
- Silicon
- Niobium
- Titanium
- Aluminum
- Vanadium
- Tungsten

Austenite-Promoting Elements

- Nickel
- Manganese
- Carbon
- Nitrogen
- Copper
- Cobalt

3.3 CONSTITUTION DIAGRAMS

Considerable effort has been directed toward the prediction of stainless steel weld metal constitution over the past 75 years. Most of this research has dealt with the compositional effects on the weld microstructure of these alloys. Various predictive diagrams and equations have been developed, which are based on the chemical compositions of the alloys of interest. This section provides a detailed chronological history of these research activities. Much of the information presented here has been summarized by Olson [8] and Balmforth [9].

3.3.1 Austenitic–Ferritic Alloy Systems: Early Diagrams and Equivalency Relationships

By far the most interest in predicting stainless steel weld metal constitution has been related to the austenitic and austenitic–ferritic alloy systems. The roots of this interest began in 1920, when Strauss and Maurer [10] introduced a nickel–chromium diagram that allowed for the prediction of various phases in the microstructure of wrought, slowly cooled steels. This diagram included only phase stability lines for austenite, martensite, troostosorbite (an archaic term referring to tempered martensite and bainite), and pearlite. The design of this diagram served as a model for many diagrams to follow. With percent nickel and chromium as the *y*-axis and *x*-axis, respectively, the diagram illustrated the effect of each element and their interactions on the microstructure of the wrought alloy.

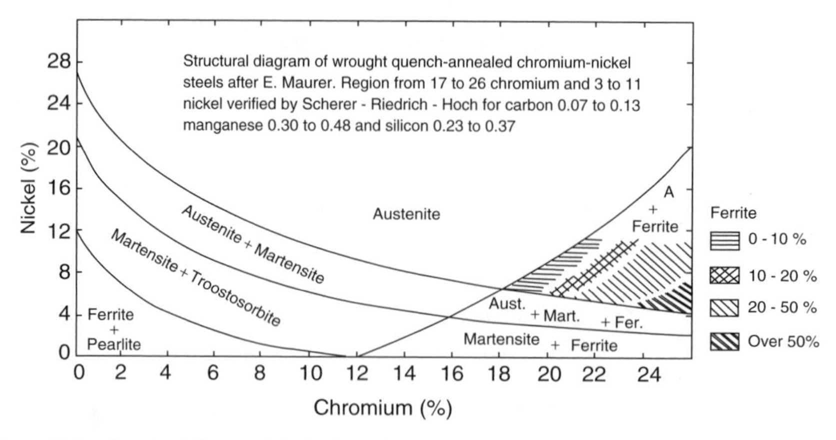

Figure 3.1 Strauss–Maurer nickel–chromium microstructure diagram as modified by Scherer et al. (From Scherer et al. [11].)

The *Strauss–Maurer diagram* was modified in 1939 by Scherer et al. [11], who added austenite–ferrite stability lines. This revised diagram, shown in Figure 3.1, uses the Strauss–Maurer axes that represented the actual chromium and nickel content. The left side of the diagram (where the lines are concave toward the upper right) contains the lines proposed by Strauss and Maurer; the right side of the diagram is the contribution of Scherer et al. The use of curved lines is significant, and it became the pattern for researchers for the next 30 years. The composition ranges of the diagram spanned 0 to 28 wt% nickel and 0 to 26 wt% chromium. Within the nominal ranges of carbon, silicon, and manganese in stainless steel plate and welds, this diagram was useful for predicting the presence of various phases. The authors included phase regions for austenite, ferrite, martensite, pearlite, troostosorbite, and mixtures of these phases. Although the diagram was developed for wrought materials, the austenite–ferrite relationships are also fairly accurate for welds, because the scale of the diagram is relatively coarse and the effect of cooling rate (processing procedure) is masked by the uncertainty of the diagram [12].

Newell and Fleischman [13] developed an expression for austenite stability on the previous diagram based on alloy content. They recognized that other elements besides chromium and nickel had an effect on the microstructure. Their formula was developed to predict whether certain chromium–nickel alloys could be fabricated into seamless tubing. They claimed that the alloy must be free from the formation of ferrite to allow the mill to pierce the billet successfully. The *Newell–Fleischman equation* for the austenite/austenite-plus-ferrite boundary is as follows:

$$\mathrm{Ni} = \frac{(\mathrm{Cr} + 2\mathrm{Mo} - 16)^2}{12} - \frac{\mathrm{Mn}}{2} + 30(0.10 - \mathrm{C}) + 8 \tag{3.1}$$

In equation (3.1) and those that are presented subsequently, the chemical symbols indicate the weight percentage of the element present. According to the equation, molybdenum is twice as effective in promoting ferrite as chromium, and manganese is one-half and carbon is 30 times as effective as nickel in promoting austenite. Much of the research on the development of constitution diagrams and the prediction of weld metal microstructures has centered on determining the coefficients for these formulas, which have been termed *chromium-equivalent* and *nickel-equivalent equations*.

During World War II, the use of stainless steel electrodes for welding armor became a topic of great interest. In 1943, Field et al. [14], in their study of armor welding, determined that the Newell–Fleischman equation could not be applied directly to weld deposits. To accommodate the higher cooling rates associated with welding, as compared to wrought products, the constant term "8" was changed to "11" to account for the higher nickel level at which some ferrite can be found in the largely austenitic as-welded microstructure, as is shown by the following equation for the austenite/austenite-plus-ferrite boundary:

$$\mathrm{Ni} = \frac{(\mathrm{Cr} + 2\mathrm{Mo} - 16)^2}{12} - \frac{\mathrm{Mn}}{2} + 30(0.10 - \mathrm{C}) + 11 \qquad (3.2)$$

This equation was later rewritten in the following form to separate the ferrite-promoting elements on the right from the austenite-promoting elements on the left:

$$\mathrm{Ni} + 0.5\mathrm{Mn} + 30\mathrm{C} = \frac{(\mathrm{Cr} + 2\mathrm{Mo} - 16)^2}{12} + 14 \qquad (3.3)$$

Other investigators also had begun to realize that the various alloying elements could be grouped together into equivalency relationships. In their investigation of 25Cr–20Ni weld metals, Campbell and Thomas [15], showed how the combination of chromium, molybdenum, and niobium influenced the microstructure and properties of the alloys. They reported a chromium-equivalent expression for this relationship:

$$\text{chromium equivalent} = \mathrm{Cr} + 1.5\mathrm{Mo} + 2\mathrm{Nb} \qquad (3.4)$$

Based on equilibrium diagrams, Thielemann [16] showed the effectiveness of the various alloying elements in expanding the gamma loop compared to chromium. He stated that it was convenient to express the relative effects of the various alloying elements in terms of the effectiveness of chromium because of its common presence in practically all alloys intended for service at elevated temperatures, and because it is the least effective of the common alloying elements in eliminating the austenite phase (i.e., promoting ferrite) in iron. The chromium equivalent values according to Thielemann are given in Table 3.1. In a study of alloying element effects on intergranular corrosion of low carbon austenitic chromium–nickel steels, Binder et al. [17] reported the following equation for the austenite stability boundary relative to delta ferrite:

$$30\mathrm{C} + 26\mathrm{N} + \mathrm{Ni} - 1.3\mathrm{Cr} + 11.1 = 0 \qquad (3.5)$$

Thomas [18] reported a linear equation for predicting austenite stability relative to delta ferrite, which included terms for additional elements:

$$\mathrm{Ni} + 0.5\mathrm{Mn} + 30\mathrm{C} = 1.1(\mathrm{Cr} + \mathrm{Mo} + 1.5\mathrm{Si} + 0.5\mathrm{Nb}) - 8.2 \qquad (3.6)$$

TABLE 3.1 Coefficients for Elements in Chromium and Nickel Equivalents

		Ferrite-Promoting Elements										Austenite-Promoting Elements						
Year	Investigators	Cr	Si	Nb	Mo	Ti	Al	V	W	Ta	Mn	Ni	Mn	C	N	Cu	Co	Mn^2
1940	Thielemann [16]	1.0	5.2	4.5	4.2	7.2	12.0	11.0	2.1	2.8	—	3.0[a]	2.0[a]	40.0[a]	—	1.0[a]	—	—
1943	Field et al. [14]	1.0	—	—	2.0	—	—	—	—	—	—	1.0	0.5	30.0	—	—	—	—
1946	Campbell and Thomas [15]	1.0	—	2.0	1.5	—	—	—	—	—	—	—	—	—	—	—	—	—
1947	Schaeffler [19]	1.0	2.5	2.0	1.8	—	—	—	—	—	—	1.0	0.5	30.0	—	—	—	—
1949	Avery	1.0	1.6	2.8	—	—	—	—	—	—	—	1.0	—	17.0	11.0	—	—	—
	Henry et al.	1.0	1.0	2.0	2.0	5.0	—	—	—	—	—	1.0	0.5	30.0	—	—	—	—
	Schaeffler [22]	1.0	1.5	0.5	1.0	—	—	—	—	—	—	1.0	0.5	30.0	—	—	—	—
	Thomas [18]	1.0	1.5	0.5	1.0	—	—	—	—	—	—	1.0	0.5	30.0	—	—	—	—
1956	DeLong [26,28]	1.0	1.5	0.5	1.0	—	—	—	—	—	—	1.0	0.5	30.0	30.0	—	—	—
1960	Schneider [24]	1.0	2.0	—	1.5	—	—	5.0	—	—	—	1.0	0.5	30.0	—	—	1.0	—
1967	Guiraldenq	1.0	1.5	—	2.0	4.0	3.0	—	—	—	0.45	1.0	—	30.0	20.0	—	—	—
	Runov	1.0	—	—	—	3.5	—	—	—	—	—	—	—	—	—	—	—	—
1969	Ferree [51]	1.0	1.5	—	—	—	—	—	—	—	—	1.0	0.5	30.0	30.0	0.3	—	—
1971	Kaltenhauser [63]	1.0	6.0	—	4.0	8.0	2.0	—	—	—	—	4.0[a]	2.0[a]	40.0[a]	40.0[a]	—	—	—
1972	Potak and Sagalevich [31]	1.0	1.0	2.0	0.9	1.0	4.0	4.0	1.5	0.5	—	1.0	0.5	27.0	27.0	0.33	0.4	—
1973	Hull [33]	1.0	0.48	0.14	1.21	2.20	2.48	2.27	0.72	0.21	—	1.0	0.11	24.5	18.4	0.44	0.41	− 0.0086
	Lefevre et al. [65]	1.0	—	—	—	8.0	—	—	—	—	—	1.0	—	10.0	—	—	—	—
1974	Castro and de Cadenet [52]	1.0	1.5	0.5	1.0	2–5	—	—	0.5	—	—	1.0	0.5	30.0	10–25	0.6	—	—
	Schoefer [29,30]	1.0	1.5	1.0	1.0	—	—	—	—	—	—	1.0	0.5	30.0	26.0	—	—	—
1976	Patriarca et al. [67]	1.0	6.0	5.0	4.0	8.0	12.0	11.0	1.5	—	—	4.0[a]	2.0[a]	40.0[a]	30.0[a]	1.0[a]	2.0[a]	—
1977	Wright and Wood [64]	1.0	5.0	—	4.0	7.0	12.0	—	—	—	—	3.0[a]	2.0[a]	40.0[a]	40.0[a]	1.0[a]	—	—
1978	Novozhilov et al. [41]	1.0	1.5	0.5	1.5	3.5	—	—	—	—	—	1.0	0.5	30.0	8–45	—	—	—
1979	Hammar and Svensson [38]	1.0	—	—	1.37	—	—	—	—	—	—	1.0	0.31	22.0	14.2	—	—	—
1980	Kakhovskii et al. [25]	1.0	1.5	—	1.0	3.5	—	1.0	—	—	—	1.0	0.5	30.0	30.0	—	—	—
	Suutala [46]	1.0	1.5	2.0	1.37	3.0	—	—	—	—	—	1.0	0.31	22.0	14.2	1.0	—	—
1982	Espy [34]	1.0	1.5	0.5	1.0	—	3.0	5.0	—	—	—	1.0	0[b]	30	20–30	0.33	—	—
1983	Kotecki [42]	1.0	1.5	0.5	0.7	—	—	—	—	—	—	—	—	—	—	—	—	—
1988	Siewert et al. [45]	1.0	—	0.7	1.0	—	—	—	—	—	—	1.0	—	35.0	20.0	—	—	—
1991	Panton-Kent [68]	1.0	6.0	4.0	4.0	8.0	2.0	—	—	—	—	—	—	—	—	—	—	—
1992	Kotecki and Siewert [54]	1.0	—	0.7	1.0	—	—	—	—	—	—	1.0	0[c]	35.0	20.0	0.25	—	—
1999	Gooch et al. [69]	3.0	—	—	1.0	16.0	2.0	—	—	—	—	4.0[a]	—	40.0[a]	40.0[a]	4.0[a]	—	—
2000	Balmforth and Lippold [71]	1.0	—	—	2.0	10	10	—	—	—	—	1.0	—	35.0	20.0	—	—	—

[a]Factors not in terms of nickel equivalent. Divide by Ni factor to determine Ni equivalency. [b]Mn factor replaced by a constant of 0.87. [c]Mn factor replaced by a constant of 0.35.

These linear equations became the predecessors to the linear constitution diagrams that are in use today.

3.3.2 Schaeffler Diagram

Anton Schaeffler [19] recognized that the preceding research could be put to practical use when combined and applied to welding. He saw that the Strauss–Maurer diagram provided a map for predicting the microstructures of the wrought chromium–nickel alloys, while the Newell–Fleischman [13], and Field et al. [14] equations could be applied almost directly to welding. His research then focused on combining this information to produce a constitution diagram for weld metals that would allow the prediction of weld metal microstructure based on chemical composition. His diagram contained chromium- and nickel-equivalent formulas for the axes, with ranges for the specific weld metal microstructural phases plotted in the diagram. Ferrite-promoting elements are included in the chromium equivalent, while austenite-promoting elements are included in the nickel equivalent. One of the original Schaeffler diagrams is shown in Figure 3.2. This was a major advancement for the prediction of weld microstructure and led to the development of other diagrams.

Schaeffler based his multiplying factors for computation of the equivalency formulas on previous research and his own experience. He assumed values of 2.5 for silicon, 1.8 for molybdenum, and 2 for niobium based on his own research, and stated that these were in relative agreement with the values suggested by Thielemann and by Campbell and Thomas. The original Schaeffler nickel-equivalent equation [19] is given by

$$Ni_{eq} = Ni + 0.5Mn + 30C \tag{3.7}$$

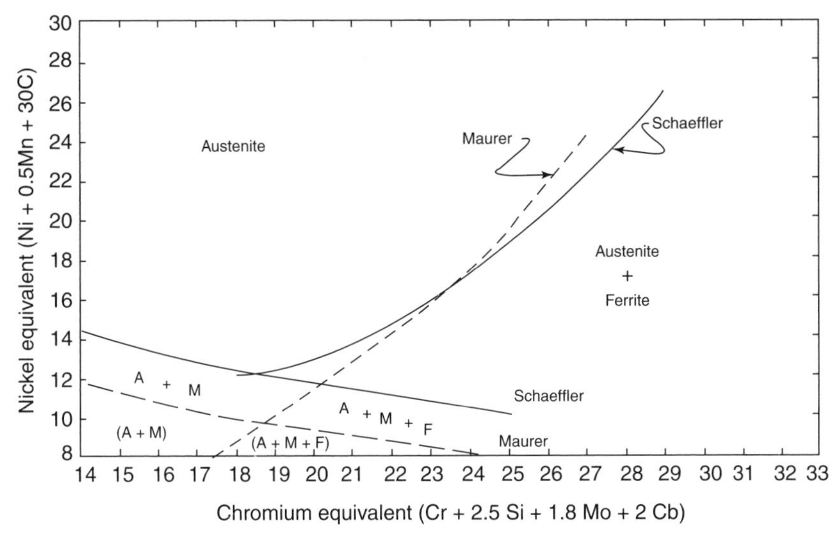

Figure 3.2 Schaeffler diagram of 1947, with the curve from the Strauss–Maurer diagram included for comparison. (From Schaeffler [19]. Courtesy of the American Welding Society.)

When the Newell–Fleischman equation is rearranged to separate positive from negative values, one side forms a nickel-equivalent formula that is the same as Schaeffler's. The other side is a chromium-equivalent formula that has been modified by Schaeffler. The original Schaeffler chromium-equivalent equation [19] is given by

$$Cr_{eq} = Cr + 2.5Si + 1.8Mo + 2Nb \tag{3.8}$$

It is interesting to note that Schaeffler did not include a nitrogen term in his nickel-equivalent formula, despite its behavior as a strong austenite promoter. This is probably due to the difficulty in determining the nitrogen content of steels in his time. The diagram was developed using the shielded metal arc welding (SMAW) process, where the nominal nitrogen content was estimated to be about 0.06 wt%. Because of this low value, nitrogen was not considered as an alloying element by Schaeffler; rather, it was simply incorporated into the diagram at a constant value of 0.06 wt%. The diagram proved to be reasonably accurate for most 300 series alloys of his time, using conventional arc welding processes.

Schaeffler also reported a new equation for the phase boundary between fully austenitic alloys and alloys composed of austenite and ferrite. His austenite stability equation was reported [19] as

$$Ni_{eq} = \frac{(Cr_{eq} - 16)^2}{12} + 12 \tag{3.9}$$

where Ni_{eq} represents a nickel equivalent and Cr_{eq} a chromium equivalent. This equation differs from the Newell–Fleischman and Field et al. equations in the final constant term. It is interesting to note that this equation implies curvature and that the lines on the original Schaeffler diagram, shown in Figure 3.2, are curved. The quadratic nature of the austenite stability line showed up as late as 1969, when Griffith and Wright [20] reported the following equation for the austenite/austenite-plus-ferrite boundary:

$$Ni + 0.5Mn + Cu + 35C + 27N = \frac{1}{12}(Cr + 1.5Mo - 20)^2 + 15 \tag{3.10}$$

Schaeffler further modified his diagram [21] in 1948 (Figure 3.3). Note that the austenite/austenite-plus-ferrite boundary became a straight line. The diagram increased the ability to quantitatively predict weld metal microstructure, adding additional isoferrite lines in the two-phase austenite + ferrite region, while retaining the original equivalency formulas. In 1949, Schaeffler introduced the final version of his diagram [22], shown in Figure 3.4, which is the version still in use. This diagram was the result of many additional examinations of weld metals, resulting in a revision of the coefficients for silicon, molybdenum, and niobium and a slight relocation of the phase boundaries. The new chromium-equivalent equation was changed to

$$Cr_{eq} = Cr + Mo + 1.5Si + 0.5Nb \tag{3.11}$$

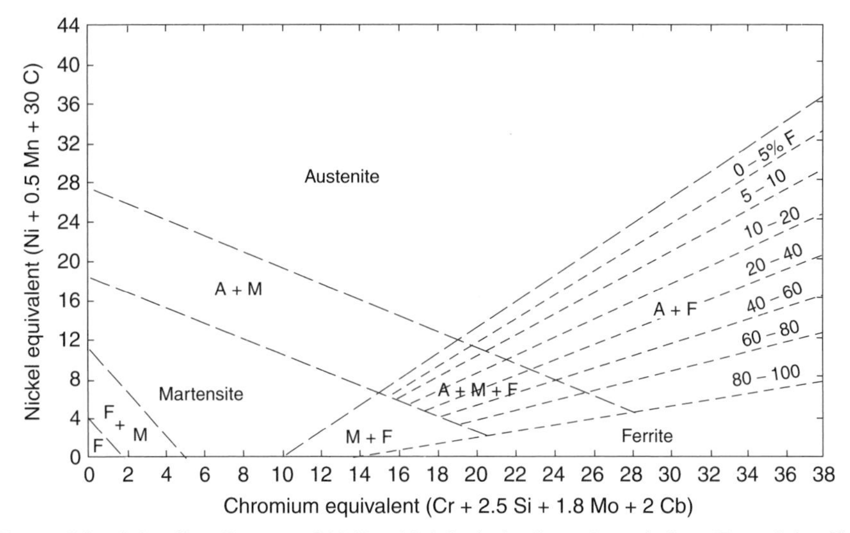

Figure 3.3 Schaeffler diagram of 1948, which includes linear boundaries. (From Schaeffler [21].)

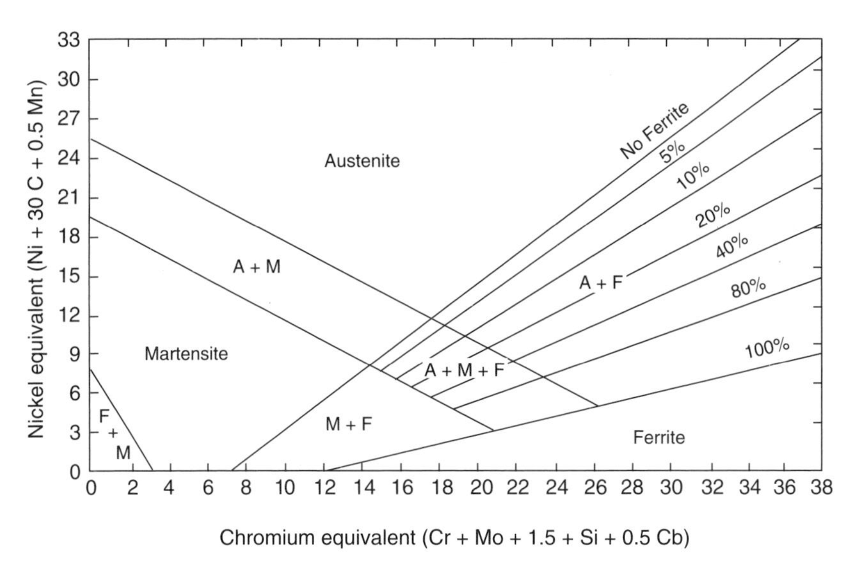

Figure 3.4 Schaeffler diagram of 1949, which is still in use. (From Schaeffler [22]. Courtesy of ASM International.)

Using the Schaeffler equivalency relationships (Cr_{eq} and Ni_{eq}), Seferian [24] developed an expression for calculating the amount of delta ferrite present in austenitic weld metal:

$$\text{delta ferrite} = 3(Cr_{eq} - 0.93Ni_{eq} - 6.7) \tag{3.12}$$

Other researchers have introduced diagrams similar to the Schaeffler diagram. In 1960, Schneider [24] developed a diagram for the prediction of cast microstructures, shown in Figure 3.5. This diagram adds the influence of cobalt to the nickel equivalent and removes niobium and adds vanadium to the chromium-equivalent formula. Kakhovskii et al. [25] introduced a modified Schaeffler diagram, shown in Figure 3.6. This diagram quantifies the influence of nitrogen in the nickel equivalent and the influence of titanium and vanadium in the chromium equivalent.

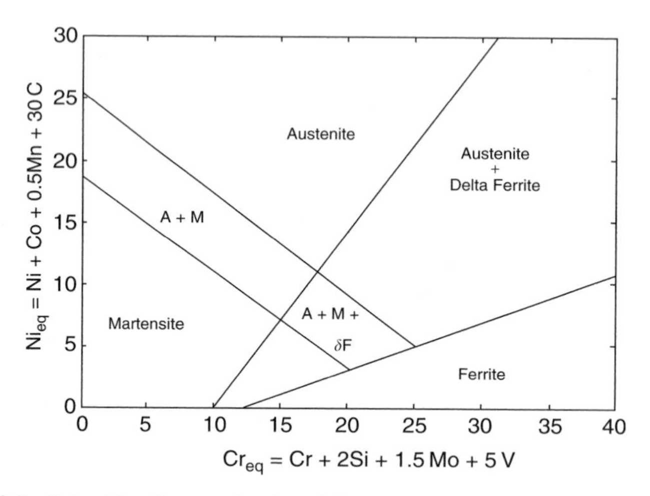

Figure 3.5 Schneider diagram developed for cast materials. (From Schneider [24].)

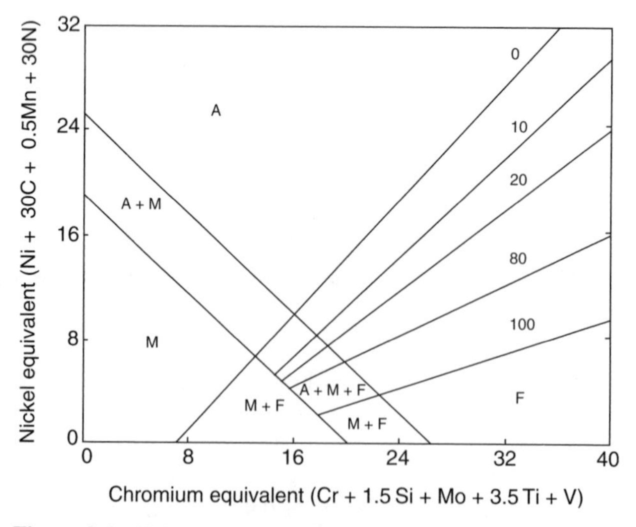

Figure 3.6 Kakhovskii diagram. (From Kakhovskii et al. [25].)

3.3.3 DeLong Diagram

In 1956, DeLong et al. [26] introduced what was to become the next major trend in the development of constitution diagrams. Rather than predicting weld metal constitution for the entire composition range of stainless steels, these investigators focused on a particular region of interest, that of the 300 series austenitic stainless steels. The enlarged scale and more precise line positions enabled a more detailed prediction of the ferrite content in austenitic stainless steel weld metal. They also investigated the influence of nitrogen on the weld metal microstructure, showing that it has a powerful influence on ferrite content. The original DeLong diagram, shown in Figure 3.7, differs from the Schaeffler diagram for the same region in a number of ways. First, a term for nitrogen was added to the nickel equivalent, which affected the location of the lines on the diagram. The new nickel equivalent was given as

$$Ni_{eq} = Ni + 0.5Mn + 30C + 30N \tag{3.13}$$

Second, the slope of the isoferrite lines was increased to account for the discrepancies the investigators discovered between the measured and calculated ferrite content on high-alloyed stainless steel types, such as 316, 316L, and 309. A third difference is that the spacing between isoferrite lines is relatively constant, whereas on the Schaeffler diagram the spacing between isoferrite lines varies, being noticeably greater at about 10% ferrite than in the 0 to 5% or 15 to 20% area.

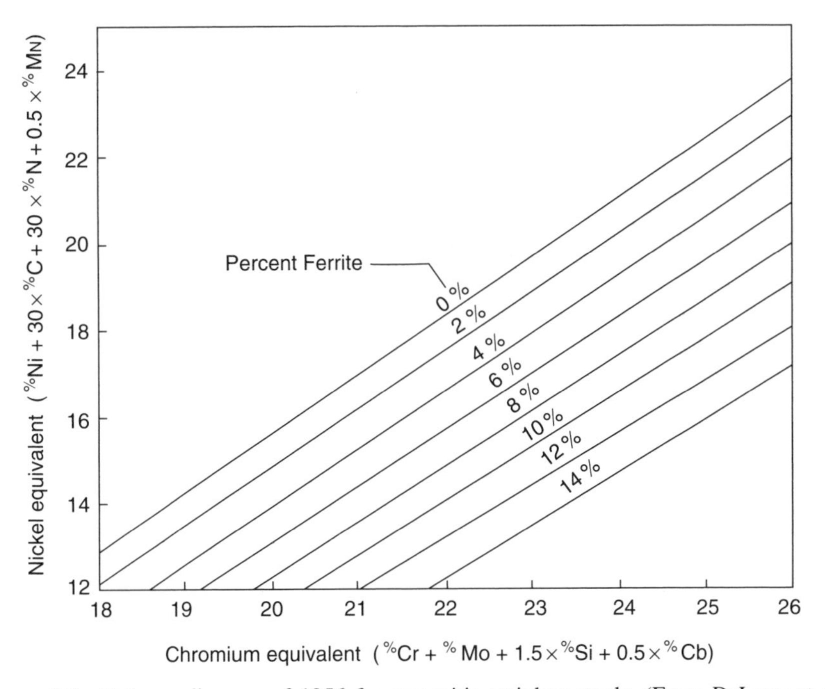

Figure 3.7 DeLong diagram of 1956 for austenitic stainless steels. (From DeLong et al. [26]. Courtesy of the American Welding Society.)

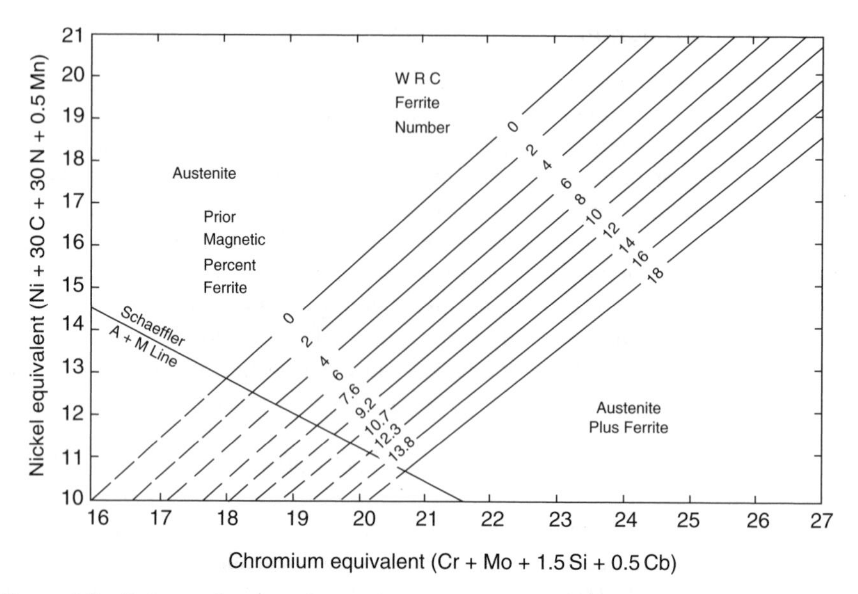

Figure 3.8 DeLong diagram of 1973, which introduced the concept of Ferrite Number. (From Long and DeLong [27]. Courtesy of the American Welding Society.)

Long and DeLong [27] proposed further modifications to this diagram in 1973. The revised diagram is shown in Figure 3.8. After further research, some changes were made to the lines on the diagram to improve its ability to predict delta ferrite. The major change at this point was the addition of a *Ferrite Number* (FN) *scale* to the diagram. This term resulted from the difficulty of measuring the ferrite content quantitatively by volume in stainless steel welds. FN values are based on magnetic measurements, which are possible because the BCC delta ferrite is ferromagnetic, while the FCC austenite is not. FN units are not intended to relate directly to percent ferrite, although at values below 10 they are considered to be similar.

The Welding Research Council Subcommittee on Welding Stainless Steels adopted FN as its value for measuring ferrite in 1973 [28], and its method for calibration is specified in the AWS A4.2 and ISO 8249 standards. Long and DeLong [27] also reported that their diagram, which has been termed the *DeLong–WRC diagram*, is fairly insensitive to the normal range of heat input variations associated with arc welding. Thus, it could be applied with a reasonable degree of accuracy to processes such as SMAW, GTAW, GMAW, and SAW.

3.3.4 Other Diagrams

There are a number of other constitution diagrams of other forms that have been proposed by various investigators. The *Schoefer diagram* [29,30] for the estimation of ferrite content in cast austenitic alloys was derived from the Schaeffler diagram, although its final form is quite different. Schoefer modified the Schaeffler equivalents for use with his diagram:

$$Cr_{eq} = Cr + 1.5Si + Mo + Nb - 4.99 \tag{3.14}$$

and

$$Ni_{eq} = Ni + 30C + 0.5Mn + 26(N - 0.02) + 2.77 \tag{3.15}$$

The effect of these modifications is to transform the coordinates of the Schaeffler diagram so that the origin is at the point where the isoferrite lines converge. Thus, the ratio of the modified chromium equivalent to the modified nickel equivalent is the reciprocal of the slope of the isoferrite lines. The reciprocal was chosen to make the composition ratio increase directly with increasing ferrite content. The relationship of composition ratio and ferrite content is therefore, expressed on the Schoefer diagram as a single curve, as shown in Figure 3.9. The original Schoefer diagram was given in percent ferrite; the one shown here has been updated to include FN values.

Potak and Sagalevich [31] proposed a new form of structural diagram for stainless steel castings and weld metal. The diagram contains a ferrite formation chromium equivalent, Cr^F_{eq}, which is plotted along the horizontal axis (Figure 3.10). The martensite formation chromium equivalent, Cr^M_{eq}, is plotted on the vertical axis. These two equivalents consider the effects (relative to chromium) of all the alloying elements on the formation of delta ferrite and martensite, respectively, and are very comprehensive in considering potential alloying additions. The two equations are given by

$$\begin{aligned} Cr^F_{eq} = {} & Cr - 1.5Ni + 2Si - 0.75Mn - K_f(C + N) + Mo + 4Al \\ & + 4Ti + 1.5V + 0.5W + 0.9Nb - 0.6Co - 0.5Cu \end{aligned} \tag{3.16}$$

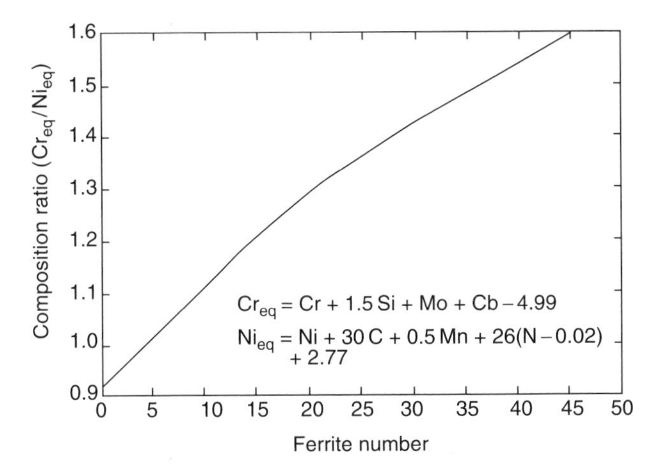

Figure 3.9 Schoefer diagram developed for cast materials. (From Beck et al. [29].)

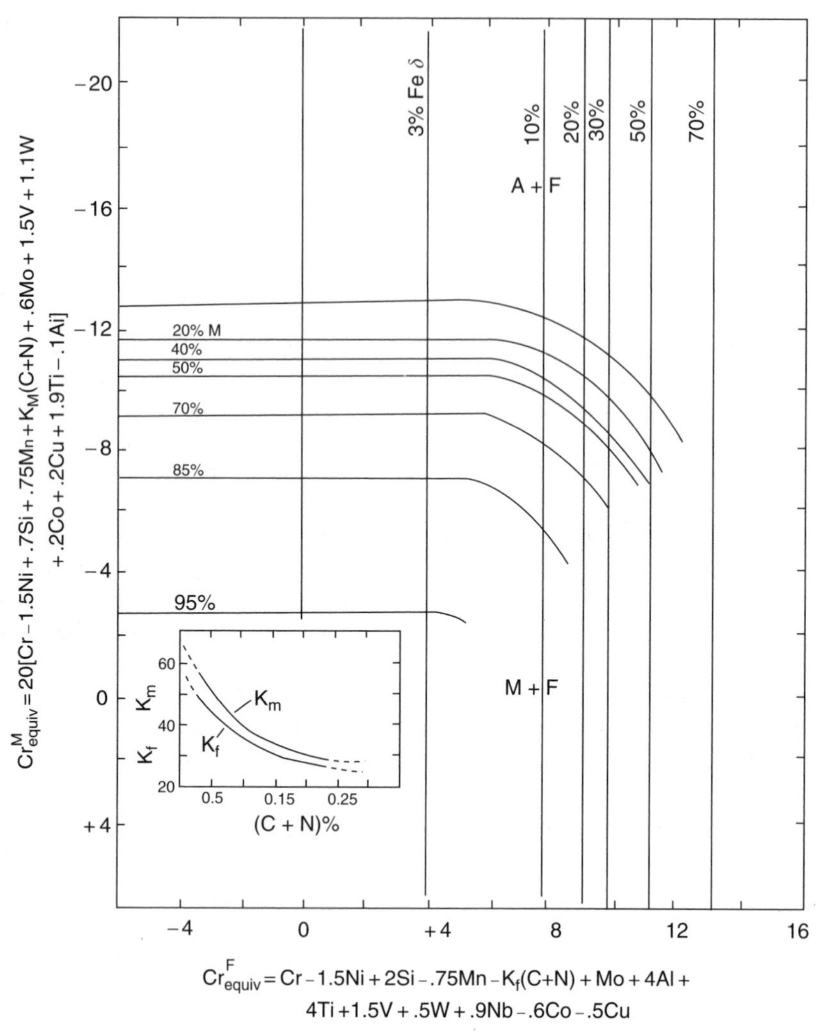

Figure 3.10 Potak and Sagalevich diagram. (From Potak and Sagalevich [31].)

and

$$\text{Cr}_{\text{eq}}^{\text{M}} = 20[\text{Cr} - 1.5\text{Ni} + 0.7\text{Si} + 0.75\text{Mn} + K_m(\text{C} + \text{N}) + 0.6\text{Mo} + 1.5\text{V} + 1.1\text{W} + 0.2\text{Co} + 0.2\text{Cu} + 1.9\text{Ti} - 0.1\text{Al}] \tag{3.17}$$

The diagram is quite complex, requiring the determination of K_f and K_m values from a plot incorporated within the complete diagram. Several notes to the diagram were also included: (1) allow for about 0.02% N for steels not alloyed with nitrogen (except for grades alloyed with titanium or aluminum); (2) for steels containing >5% Ni, the ferrite formation equivalent for nickel is calculated by the formula 2.5 + % Ni; (3) for steels alloyed with titanium or niobium, the calculation must include the amount of the element present in solid solution (about 0.8 times the total content), and the carbon must be reduced by 1/4 and 1/7.5 relative to the titanium or niobium contents of the carbides, respectively (for steels containing about 0.1% C). The Potak and Sagalevich diagram has been reported to predict microstructures more accurately in the triplex martensite–ferrite–austenite constitution region [8].

Another form of constitution diagram has been proposed by Carpenter et al. [32] for the Fe–Mn–Ni–Al weld metal alloy system. Their diagram (Figure 3.11) uses weight percent aluminum for the *x*-axis and the following nickel-equivalent formula for the *y*-axis:

$$\mathrm{Ni_{eq}} = \mathrm{Ni} + 2\mathrm{Mn} + 30\mathrm{C} \tag{3.18}$$

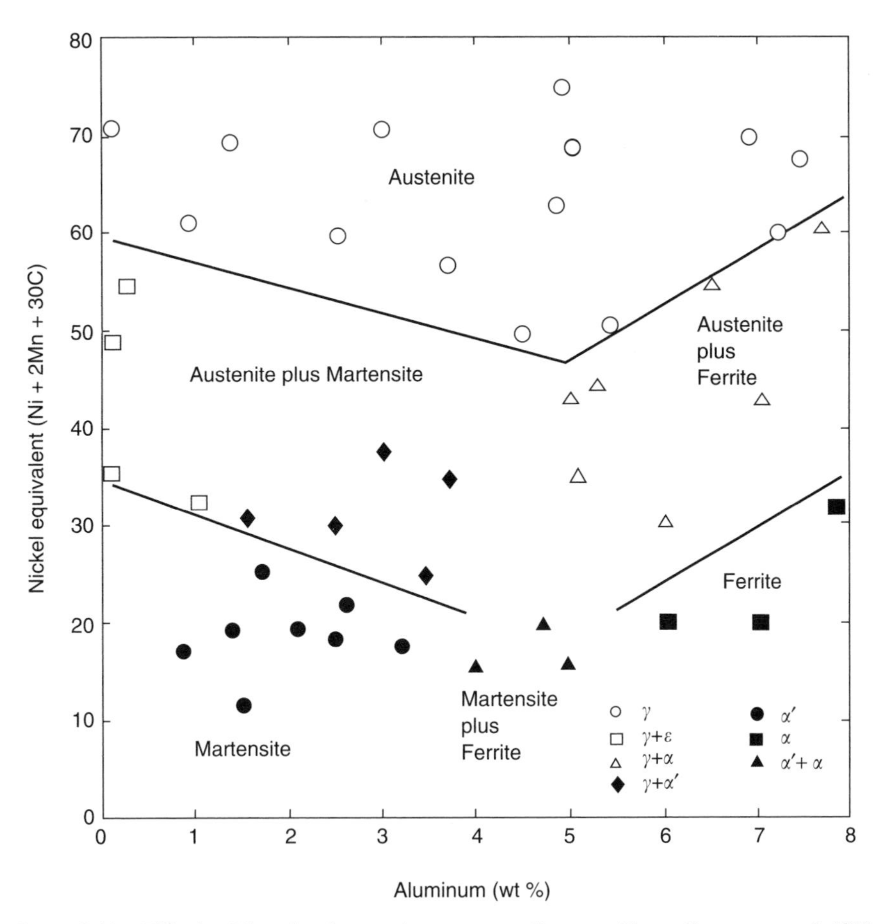

Figure 3.11 Effective $\mathrm{Ni_{eq}}$–aluminum microstructure diagram. (From Carpenter et al. [32]. Courtesy of the American Welding Society.)

The authors note that both α and ε martensite can form from austenite in this alloy system, and that the microstructure predicted by the upper right corner of the diagram is very similar to a Type 308 stainless steel weld deposit, which is characterized by an austenitic matrix and an aluminum-rich delta ferrite phase. The diagram has similarities to the Schaeffler diagram (Figure 3.4), although there is a large difference in the coefficient for manganese. This seems to be due to the fact that Carpenter et al. were concentrating on austenite stability relative to martensite formation at low temperature rather than austenite stability relative to ferrite formation at high temperature.

The DeLong–WRC diagram was widely accepted and used for a number of years as the standard for predicting FN in austenitic stainless steel welds, as evidenced by its inclusion in the ASME Boiler and Pressure Vessel Code and other standards. However, researchers continued to investigate the effects of specific elements and to suggest modifications to the coefficients, equations, and diagrams to improve FN prediction. The coefficient for manganese has been in question from the time that Schaeffler introduced his diagram, and has been much studied.

Hull [33] used stainless steel chill castings of high manganese content to simulate weld metal and produced a revised portion of the Schaeffler diagram (Figure 3.12) which shows the austenite-to-delta ferrite boundary. The equivalency formulas were quite comprehensive, including a large number of potential alloying elements for

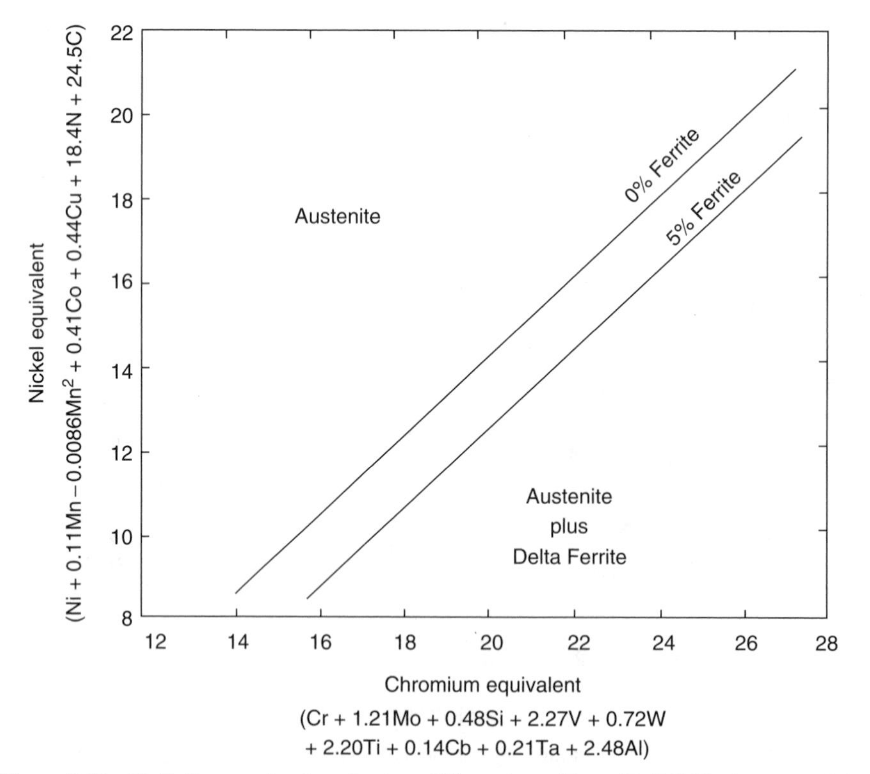

Figure 3.12 Hull diagram developed using chill castings. (From Hull [33]. Courtesy of the American Welding Society.)

stainless steels. The nickel equivalent contained a quadratic term for manganese ($0.11\text{Mn} - 0.0086\text{Mn}^2$), which indicates that manganese promotes austenite to a lesser degree as its content is increased, and at very high levels, it actually promotes the formation of ferrite. Espy [34] studied nitrogen-strengthened austenitic stainless steels that contained manganese at levels up to 15 wt% and found that manganese had little effect on weld microstructure as an austenite promoter. He concluded that the austenitizing potential of manganese could be better represented as a constant term of 0.87 rather than by a term proportional to the manganese content.

Szumachowski and Kotecki [35] found that the DeLong diagram underestimated FN for weld metal above 2.5 wt% manganese. They also determined that the austenitizing potential of manganese could be represented by a constant term, but their value of 0.35 was less than that proposed by Espy. Better agreement between measured and predicted FNs for an extended manganese range, up to 12.5 wt%, was found by using a modified nickel-equivalent formula:

$$\text{Ni}_{\text{eq}} = \text{Ni} + 30(\text{C} + \text{N}) - 0.35 \tag{3.19}$$

Espy [34] determined a modified set of equivalents for use with the Schaeffler diagram:

$$\text{Cr}_{\text{eq}} = \text{Cr} + \text{Mo} + 1.5\text{Si} + 0.5\text{Nb} + 5\text{V} + 3\text{Al} \tag{3.20}$$

and

$$\text{Ni}_{\text{eq}} = \text{Ni} + 30\text{C} + 0.87 \text{ for Mn} + 0.33\text{Cu} + k_n(\text{N} - 0.045) \tag{3.21}$$

with k_n as a variable coefficient equal to 30 with 0.0 to 0.2 wt% nitrogen, 22 with 0.21 to 0.25 wt% nitrogen, and 20 with 0.26 to 0.35 wt% nitrogen. This indicates that nitrogen is less effective as an austenite former as its content is increased.

McCowan et al. [36] studied 18Cr–9Ni stainless steels with high manganese and nitrogen levels and concluded that manganese and nitrogen interacted in a complex manner that could be expressed by

$$\begin{aligned}\text{Ni}_{\text{eq}} = {} & \text{Ni} + 29(\text{C} + \text{N}) + 0.53(\text{Mn}) - 0.05(\text{Mn})^2 \\ & - 2.37(\text{MnN}) + 0.94(\text{MnN})^2 - 0.71\end{aligned} \tag{3.22}$$

This is a modified Ni-equivalent expression, which was intended to be used with the DeLong diagram.

The potency factor for nitrogen has been the focus of many investigations. Constant coefficients of 13.4, 14.2, 18.4, and 20 for nitrogen have been reported by various researchers [34,37–39]. Ogawa and Koseki [40] reported a coefficient of 30 for predicting FN but a coefficient of 18 for predicting the effect of nitrogen on the solidification mode. Novozhilov et al. [41] reported that the coefficient for nitrogen in the nickel equivalent depends to a large extent on the composition of the metal. In their studies, the coefficient for nitrogen varied from 8 to 45.

Novozhilov et al. also reported that the value of the coefficient for molybdenum should be 1.5. This is consistent with most of the values reported for this factor, which have varied between 1 and 2. In 1983, Kotecki [42] reexamined the effect of molybdenum in determining ferrite content. He found that the DeLong diagram, which considers molybdenum equal to chromium in promoting ferrite, overestimated the Ferrite Number in weld metals. He suggested that a molybdenum coefficient of 0.7 be substituted in the chromium equivalent as a way to correct the overestimation.

Kotecki [43] also investigated the coefficient for silicon in the chromium equivalent and found that the value of 1.5, which is used in both the Schaeffler and DeLong diagrams, overstates the effect. From a study of welds containing from 0.34 to 1.38 wt% silicon, he determined that a coefficient of 0.1 for silicon was more predictive. Takemoto et al. [44] reported a nonlinear effect of silicon on the formation of ferrite. They concluded that the effect was very small below 2 wt% silicon, but large at higher silicon contents.

3.3.5 WRC-1988 and WRC-1992 Diagrams

In the mid-1980s, the Subcommittee on Welding Stainless Steel of the Welding Research Council initiated activity to revise and expand the Schaeffler and DeLong diagrams in an effort to improve the accuracy of ferrite prediction for stainless steel weld metal. In 1988, in a study funded by WRC, Siewert et al. [45] proposed a new predictive diagram, which covered an expanded range of compositions, from 0 to 100 FN, compared to the 0 to 18 FN range of the DeLong diagram. This diagram (Figure 3.13) also included boundaries, based on studies of other researchers [38,40,46,47], that defined solidification mode and became known as the *WRC-1988 diagram*. This diagram was developed from a database of approximately 950 welds, gathered from electrode manufacturers, research institutes, and the literature. Multiple linear regression analysis and other statistical techniques used in the development of this diagram can be found in the detailed version of the study [48]. An important feature of this diagram is that since the chemical analysis data came from numerous sources, versus largely or completely single-source data for the preceding diagrams, the WRC-1988 diagram should be relatively free of any effects of biases in chemical analysis.

New equivalency formulas were developed which removed the manganese coefficient from the nickel equivalent, thereby eliminating the systematic overestimation of FN for highly alloyed weld metals. The data were also used to evaluate and compare the WRC-1988 diagram to the DeLong diagram, and it was determined that the new diagram was significantly more accurate. The WRC-1988 equivalency formulas are given as

$$Cr_{eq} = Cr + Mo + 0.7Nb \tag{3.23}$$

and

$$Ni_{eq} = Ni + 35C + 20N \tag{3.24}$$

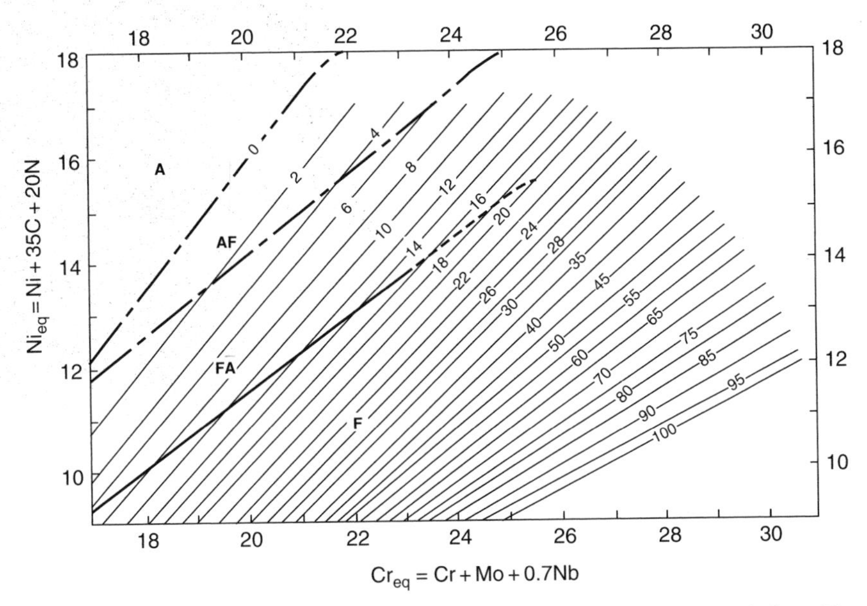

Figure 3.13 WRC-1988 diagram that includes solidification mode boundaries. (From Siewert et al. [45]. Courtesy of the American Welding Society.)

After its introduction, the WRC-1988 diagram was evaluated and reviewed. Kotecki [49] used independent data from over 200 welds to confirm the improved predictive accuracy of the diagram compared to the DeLong diagram. After introduction of the WRC-1988 diagram, the effect of copper on ferrite content became a topic of interest because of the increased use of duplex stainless steels, which may contain up to 2% copper. Lake [50] showed that the addition of a coefficient for copper in the WRC-1988 nickel equivalent would improve the accuracy of FN prediction when copper is an important alloying element. He proposed a value of from 0.25 to 0.30 for the copper coefficient.

Various investigators have proposed a coefficient for copper in the Schaeffler and DeLong nickel equivalents, suggesting that copper is significant in the determination of ferrite content. A number of investigators have proposed coefficients for Cu, including Ferree [51] of 0.3, Hull [33] of 0.44, Potak and Sagalevich [31] of 0.5, and Castro and deCadenet [52] of 0.6. Using Lake's data as a basis, Kotecki [53] proposed a coefficient of 0.25 for copper in the Ni-equivalent formula.

In 1992, Kotecki and Siewert [54] proposed a new diagram, which was exactly the same as the WRC-1988 diagram except that it included a coefficient of 0.25 for copper in the nickel equivalent:

$$Ni_{eq} = Ni + 35C + 20N + 0.25Cu \tag{3.25}$$

The WRC-1992 diagram is presented in Figure 3.14. Recognizing that the WRC-1992 diagram was limited in the range of ferrite prediction compared to the

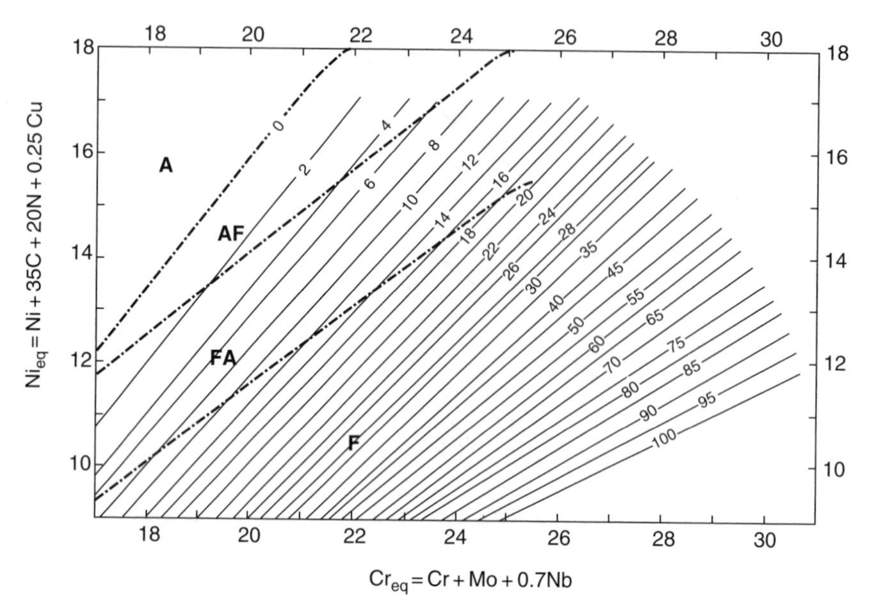

Figure 3.14 WRC-1992 diagram. (From Kotecki and Siewert [54]. Courtesy of the American Welding Society.)

Schaeffler diagram, they also proposed an extended version of the diagram so that FN predictions could be made for dissimilar metal welds (see Chapter 9). Whereas the extended axes allow a wide range of base and filler metal compositions to be plotted, the FN prediction is valid only when the weld metal composition falls within the original iso-FN lines (0 to 100 FN) of the diagram. Extension of the iso-FN lines into the surrounding regions could result in erroneous predictions since the limits of the FN lines were determined by the extent of the original database.

At the present time, the WRC-1992 diagram is the most reliable and accurate diagram available for prediction of the Ferrite Number in austenitic and duplex stainless steel weld metal. It has been widely accepted worldwide and has been included in a number of international codes and standards, replacing the DeLong diagram in the ASME code. One potential shortcoming of the WRC-1992 diagram and of many of those that preceded it is the absence of a factor for titanium. Similar to niobium, titanium is a potent carbide former and can influence the phase balance by removing carbon from the matrix. It is also a ferrite-promoting element in the absence of carbon. Hull [33] assigned a value of 2.2 and Suutala [46] a value of 3.0 to titanium in their Cr-equivalent relationships. Although few of the austenitic stainless steels contain titanium, when it is present at levels exceed 0.2 wt%, including a factor of 2 to 3 for titanium in the WRC-1992 Cr-equivalent formula may improve the accuracy of FN prediction.

3.4 AUSTENITIC–MARTENSITIC ALLOY SYSTEMS

The boundary between austenite and martensite has also received some attention, although not nearly as extensively as the austenite/austenite + ferrite boundary. Much of the research has involved Fe–Mn–Ni austenitic alloys with lower chromium, which places them in the left-hand portion of the Schaeffler diagram (Figure 3.4). The weld metal microstructures for these alloys containing less than 18 wt% Cr consist of various fractions of austenite and martensite. Eichelman and Hull [55] were the first to study the effect of alloying on the martensite start temperature, M_S, of stainless steel alloys. They determined that the M_S temperature in degrees Fahrenheit could be calculated approximately for alloys containing 10 to 18 wt% Cr with the equation

$$M_S(°F) = 75(14.6 - Cr) + 110(8.9 - Ni) + 60(1.33 - Mn) + 50(0.47 - Si) + 3000[0.068 - (C + N)] \quad (3.26)$$

By assuming that the martensite start temperature of interest is 20°C (68°F) and solving in terms of manganese equivalency, this equation reduces to

$$0 = 38.55 - 1.25Cr - 1.83Ni - Mn - 0.83Si - 50(C + N) \quad (3.27)$$

Of note is the fact that the manganese/nickel ratio for this equation is near the ratio of 0.5 originally proposed by Schaeffler. This was not the case for the Andrews [56] equation, which has been found to predict the austenite–martensite constitution boundary in iron–manganese–nickel weld metal with less than 5 wt% chromium:

$$M_S(°C) = 539 - 423C - 30.4\,Mn - 17.7Ni - 12.1\,Cr - 7.5Mo \quad (3.28)$$

Using a martensite start temperature of 20°C (68°F), the austenite–martensite boundary is described as

$$0 = 17.07 - 13.9C - Mn - 0.58Ni - 0.4Cr - 0.25Mo \quad (3.29)$$

In this case, the manganese was found to be twice as effective an austenite-stabilizing element as nickel, a factor of 4 greater than the Schaeffler ratio. Self et al. [57] found that for chromium levels around 9 wt%, manganese and nickel exhibit approximately equal efficiencies in stabilizing austenite.

Self et al. [57] also reported a best-fit general criteria in order for a fully austenitic weld metal to occur in the range 0 to 16 wt% chromium. This was determined to be

$$Mn + (0.0833Cr + 0.5)Ni + 0.0742(Cr)^2 - 1.2Cr > 14.00 \quad (3.30)$$

These investigators also produced a diagram by simplifying their austenite stability criteria, rearranging the variables and plotting data from their own study and some of Schaeffler's original data [19]. This diagram is presented in Figure 3.15.

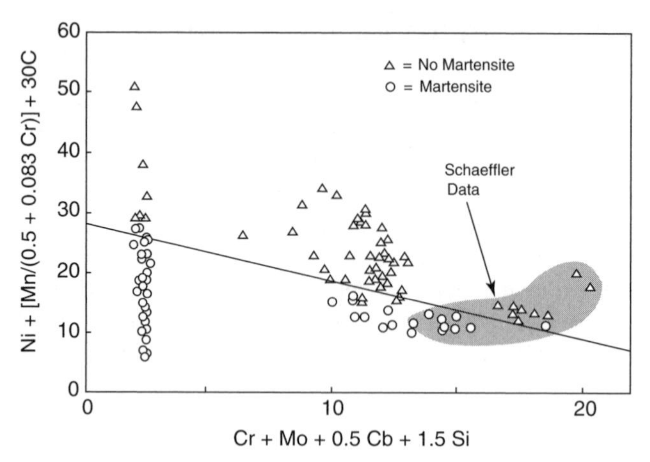

Figure 3.15 Martensite start composition for room temperature as a function of modified chromium and nickel equivalents. (From Self et al. [57]. Courtesy of the American Welding Society.)

Self et al. [58], using a statistical regression analysis of the data from 16 different investigations, obtained an expression for the martensite start temperature as a function of alloy composition. Their equation, which incorporates the switch in the relative austenite stability experienced for manganese relative to nickel with variations in chromium content, is given as

$$M_S(°C) = 526 - 12.5Cr - 17.4Ni - 29.7Mn - 31.7Si - 354C - 20.8Mo - 1.34(CrNi) + 22.4(Cr + Mo)C \quad (3.31)$$

The austenite–martensite start boundary lines calculated as a function of service temperature according to Self et al. are presented in Figure 3.16. Notice the large shift in this boundary as the temperature approaches cryogenic service.

In a study of the phase compositions of a Fe–Ni–Co–Mo–Ti–Si deposited weld metal, Barmin et al. [59] extended the concept of the Potak and Sagalevich [31] diagram construction to prediction of the amounts of martensite, austenite, and Laves phase. Figure 3.17 illustrates a diagram that has a vertical axis parameter given by

$$\beta_{\gamma}^{20} = 18 - 1.5Ni + 0.1Co - 0.7\,Mo - 0.5Ti - 0.2Si \quad (3.32)$$

where β_{γ}^{20} represents the austenite stability expression. When $\beta_{\gamma}^{20} = 0$, this equation describes the boundary between martensite and martensite–austenite at about 20°C (68°F). The horizontal axis parameter is described by

$$\beta_{\varepsilon}^{1400} = 2.5 - 0.01Ni - 0.06Co - 0.12Mo - 0.50Ti - 1.00Si \quad (3.33)$$

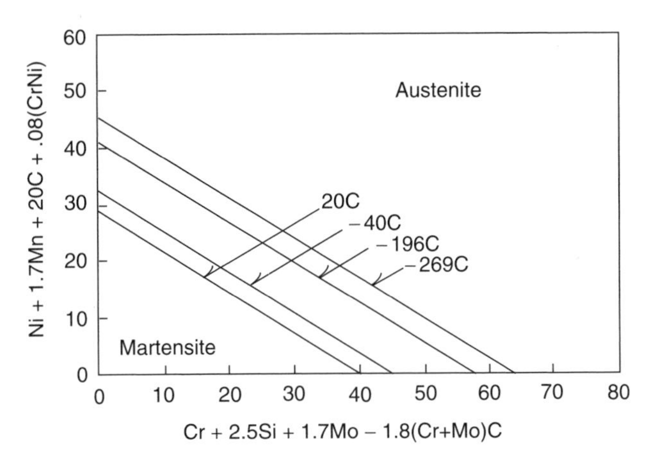

Figure 3.16 Martensite start composition for various temperatures. The (Cr + Mo)C term is used only if extensive carbide precipitation is probable. (From Self et al. [58].)

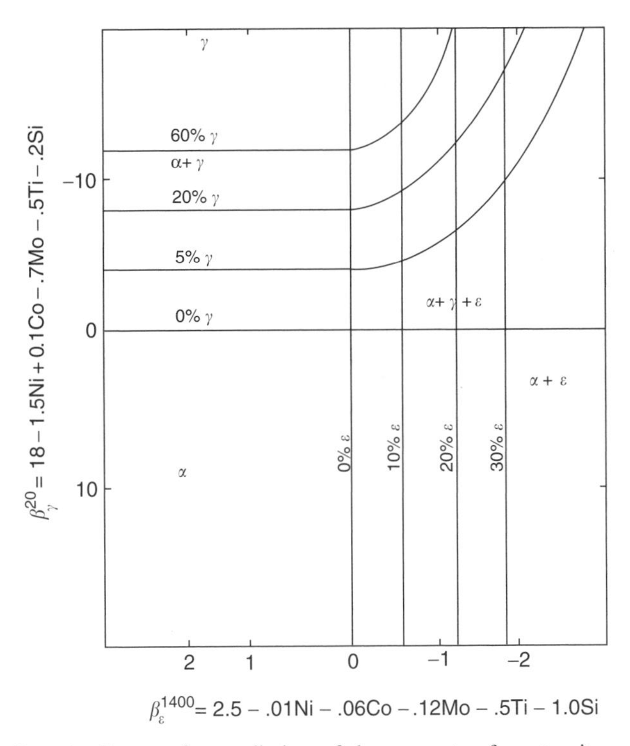

Figure 3.17 Barmin diagram for prediction of the amounts of martensite, austenite, and Laves phase. (From Barmin et al. [59].)

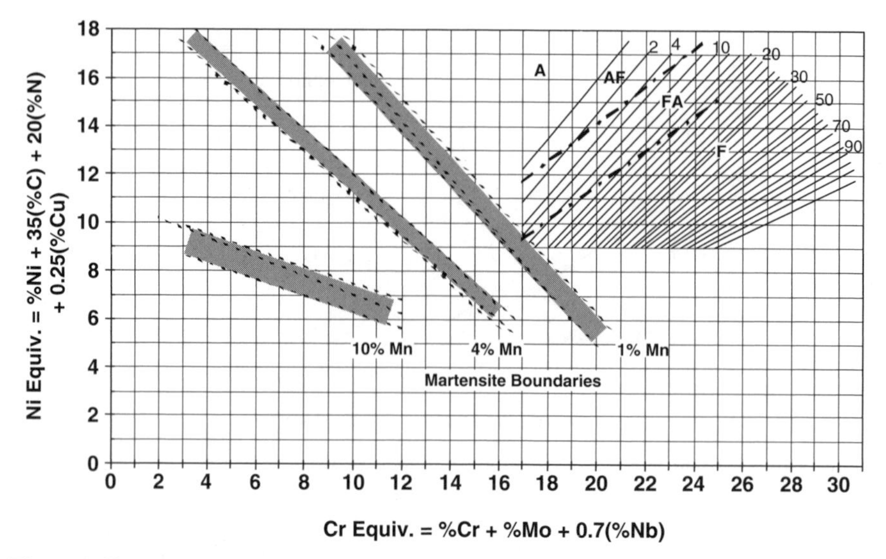

Figure 3.18 WRC-1992 diagram, with martensite boundaries for 1, 4, and 10 wt% Mn. The boundaries are shown as shaded bands to indicate a degree of uncertainty in their positions. At a given Mn content, compositions above and to the right of the appropriate boundary are predicted to be free of martensite. (From Kotecki [61]. Courtesy of the American Welding Society.)

The numerical value of $\beta_{\varepsilon}^{1400}$ can be regarded as a measure of the approximation of an alloy to saturation at about 1400°C (2552°F), and for steels quenched from that temperature it can be regarded as the degree of supersaturation of the solid solution. The degree of saturation is related to the amount of Laves phase formation.

Kotecki [60–62] examined the formation of martensite in compositions on the WRC-1992 diagram. He developed a series of boundaries, for 1 wt% Mn, 4 wt% Mn, and 10 wt% Mn, which separated austenitic and austenitic–ferritic compositions that are free of martensite, and pass an ASME 2T bend test, from those that contain appreciable martensite and fail a 2T bend test. Since Mn does not appear in the nickel equivalent of the WRC-1992 diagram, a separate boundary is required for each Mn level. At a given Mn level, the chromium and nickel equivalents do not require adjustment for Mo or N concentration. However, C above about 0.1 wt% needs to be treated as being at 0.1% in calculating the nickel equivalent for accurate prediction of the presence or absence of martensite. This modification of the WRC-1992 diagram is shown in Figure 3.18.

3.5 FERRITIC–MARTENSITIC ALLOY SYSTEMS

Equations and diagrams for the prediction of phase balance in ferritic, martensitic, and ferritic–martensitic stainless steels have also been developed. Recognizing that the Thielemann [16] chromium equivalency values for the various alloying elements were based on binary-phase diagrams and were developed for wrought base materials, Kaltenhauser [63] modified these constants and proposed a ferrite factor for determining the tendency to form martensite in ferritic stainless steel welds. In particular, Kaltenhauser was interested in low-chromium ferritic grades such as Type 409. This equation has come to be known as the *Kaltenhauser factor* or *K-factor*:

$$\text{Kaltenhauser factor} = Cr + 6Si + 8Ti + 4Mo + 2Al + 40(C + N) - 2Mn - 4Ni \quad (3.34)$$

When the K-factor falls below a certain level, a fully ferritic microstructure is ensured. Kaltenhauser proposed that there is not a single value for the K-factor that ensures a fully ferritic microstructure for the entire family of ferritic stainless steel alloys. Rather, a unique value has been suggested for individual alloys based on metallographic examination of numerous welds of various compositions. For the low-chromium grades, the critical K-factor is 13.5, while for the medium-chromium grades, the critical value is 17. This K-factor level and resulting microstructural effect are normally achieved by reducing carbon, by adding titanium, or by a combination of the two.

Wright and Wood [64] also presented a chromium-equivalent equation based on a modified Thielemann [16] system, where

$$Cr_{eq} = Cr + 5Si + 7Ti + 4Mo + 12Al - 40(C + N) - 2Mn - 3Ni - Cu \quad (3.35)$$

Generally, chromium-equivalent values greater than 12 are required to ensure fully ferritic microstructures within the compositional limits of common ferritic stainless steels. This relationship varies only slightly from that of Kaltenhauser, except for the addition of a factor for copper and a large difference in the potency of aluminum.

Lefevre et al. [65] studied a much wider range of low-chromium alloys than is commercially available and were able to develop a constitution diagram for these alloys based on modified chromium and nickel equivalents. Much like the DeLong diagram, this diagram (Figure 3.19) was a subset of the composition space in the Schaeffler diagram. It has axis values that are calculated by determining the chromium and nickel equivalents and subtracting a respective constant. These formulas are

$$Cr_{eq} = Cr + 8Ti - 11 \quad (3.36)$$

and

$$Ni_{eq} = Ni + 10C - 0.4 \quad (3.37)$$

In a 1990 review of ferritic stainless steel weldability, Lippold [66] extended this idea and replotted the Lefevre diagram using the K-factor equivalents, shown in Figure 3.20. The shape of the ferrite + martensite region approximates that of the Schaeffler diagram (Figure 3.4). However, based on the work of both Lefevre et al. and Kaltenhauser, this diagram provides a more accurate means for predicting the

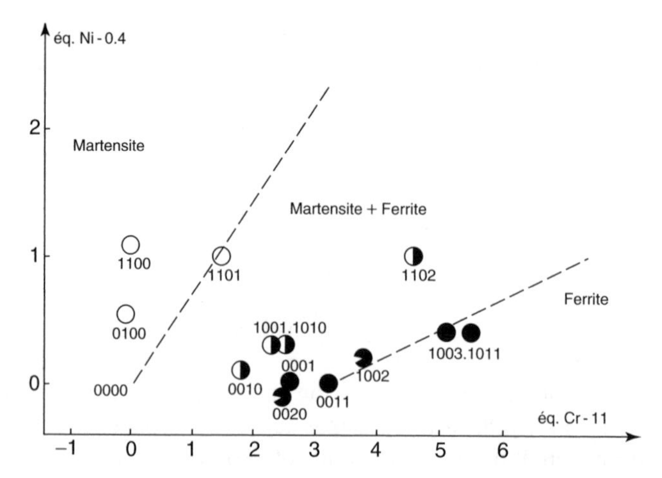

Figure 3.19 Lefevre diagram for ferritic and martensitic stainless steels. (From Lefevre et al. [65].)

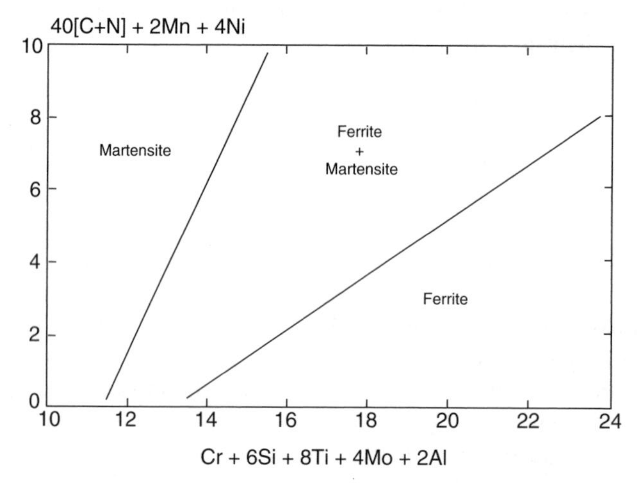

Figure 3.20 Ferritic stainless steel constitution diagram developed by Lippold using Kaltenhauser factors. (From Lippold [66]. Courtesy of the Edison Welding Institute.)

microstructure in low-chromium stainless steels, since it takes into account the effects of titanium and aluminum. Since the microstructural data for the diagram was developed by quenching low-chromium alloys from 1100°C (2021°F), it can be used, as a first approximation, to predict the weld microstructure of these alloys. Note that the line between the fully ferritic and ferrite + martensite regions corresponds to the critical K-factor proposed by Kaltenhauser for low-chromium alloys (KF = 13.5) and approximates that recommended for medium-chromium alloys

(KF = 17). This diagram also has utility in predicting whether delta ferrite exists in martensitic stainless steel welds. Isoferrite lines, which would allow the prediction of volume percent martensite or ferrite in the weld metal, are absent since insufficient data were available in the literature for their development.

In a study of 9Cr–1Mo steels for potential use in steam generators, Patriarca et al. [67] used some of the Thielemann [16] constants to modify a chromium-equivalent equation developed by General Electric researchers for predicting the amount of delta ferrite that would form in the martensitic weld structure. The equation used was

$$Cr_{eq} = Cr + 6Si + 4Mo + 1.5W + 11V + 5Nb + 12Al + 8Ti - 40C - 2Mn - 4Ni - 2Co - 30N - Cu \qquad (3.38)$$

This equation was more comprehensive in predicting alloy potency, including factors for tungsten, vanadium, niobium, and cobalt in addition to those commonly reported. In his study of the phase balance in 9Cr–1Mo steels, Panton-Kent [68] used the chromium equivalent from Patriarca et al. and modified the K-factor for use with martensitic alloys by adding a term for niobium. He found that the following modified *ferrite factor* (FF) was more accurate for the more restricted range of weld metal compositions that he produced using commercially available SMAW electrodes:

$$FF = Cr + 6Si + 8Ti + 4Mo + 2Al + 4Nb - 2Mn - 4Ni - 40(C + N) \qquad (3.39)$$

When applying this relationship, weld metal with a ferrite factor of approximately 7.5 or less would be expected to have a fully martensitic microstructure.

Gooch et al. [69] at TWI developed a *preliminary* diagram (Figure 3.21) to predict the ferrite content in the HAZ of low-C 13% Cr steels, called *supermartensitic stainless steels*. The number of alloys evaluated was small, particularly those with high Ni contents, so this diagram should be considered qualitative rather than being a rigorous predictor of microstructure. Also, since this diagram was based on HAZ microstructure, its applicability to weld metals is questionable.

Most recently, Balmforth and Lippold [70,71] have developed a ferritic–martensitic stainless steel constitution diagram that covers a range of composition encompassing most of the commercial ferritic and martensitic stainless steel grades. They developed this diagram by producing over 200 button melts of varying composition and then determining the fraction of phases present using quantitative metallography. The resulting diagram is shown in Figure 3.22.

The nickel equivalent for this diagram is essentially identical to that of the WRC-1992, but the chromium equivalent has a large factor for Al and Ti. This diagram has proven to be very accurate within the composition range from which it was developed (up to $Ni_{eq} = 6$ and $Cr_{eq} = 24$). The axes have been extended slightly beyond these values to accommodate a wider range of alloys. The composition ranges within which this diagram is considered valid are shown below the diagram. This range

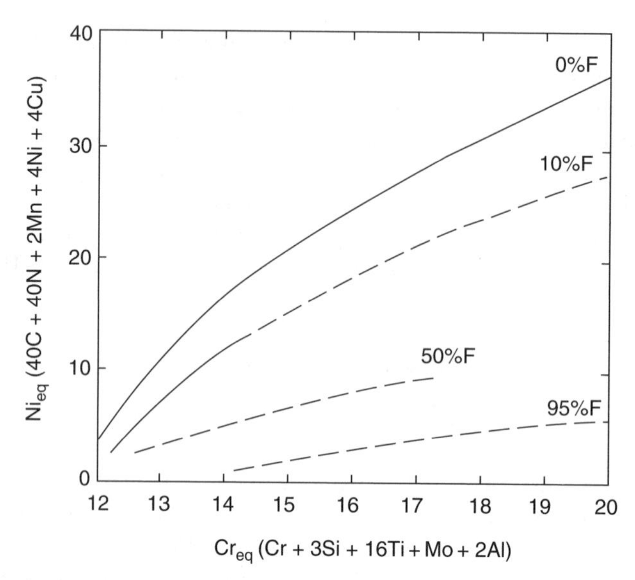

Figure 3.21 Diagram for predicting the ferrite content in the HAZ of low-C 13% Cr martensitic stainless steels. (From Gooch et al. [69].)

covers most ferritic and martensitic alloys, although the use of this diagram for low-carbon alloys (<0.03 wt%) may be questionable. These constraints are discussed in Chapters 4 and 5. The Balmforth diagram also contains a region in the upper portion of the diagram where austenite is stable. This is consistent with the Schaeffler diagram (Figure 3.4).

3.6 NEURAL NETWORK FERRITE PREDICTION

Neural networks are based on a simple architecture whereby input data are related to an output result (or results) by a system of interconnected nodes. These networks are termed *neural* because they mimic the function of the human brain. They normally consist of layers of three types of nodes: an input layer, a hidden layer, and an output layer. Recently, a neural network approach has been applied to the prediction of the Ferrite Number using an input layer consisting of the composition of the alloy [72–74], as shown in Figure 3.23.

By using the composition and FN data used to construct the WRC-1992 diagram and data from other sources, the neural network was "trained." It is reported to be more accurate than the WRC-1992 diagram. Since it employs nonlinear regression techniques, it cannot simply be displayed as a diagram with fixed coefficients for the various elements. More recently, the effect of cooling rate has also been incorporated into the network [75]. These networks are now accessible through the Internet [76].

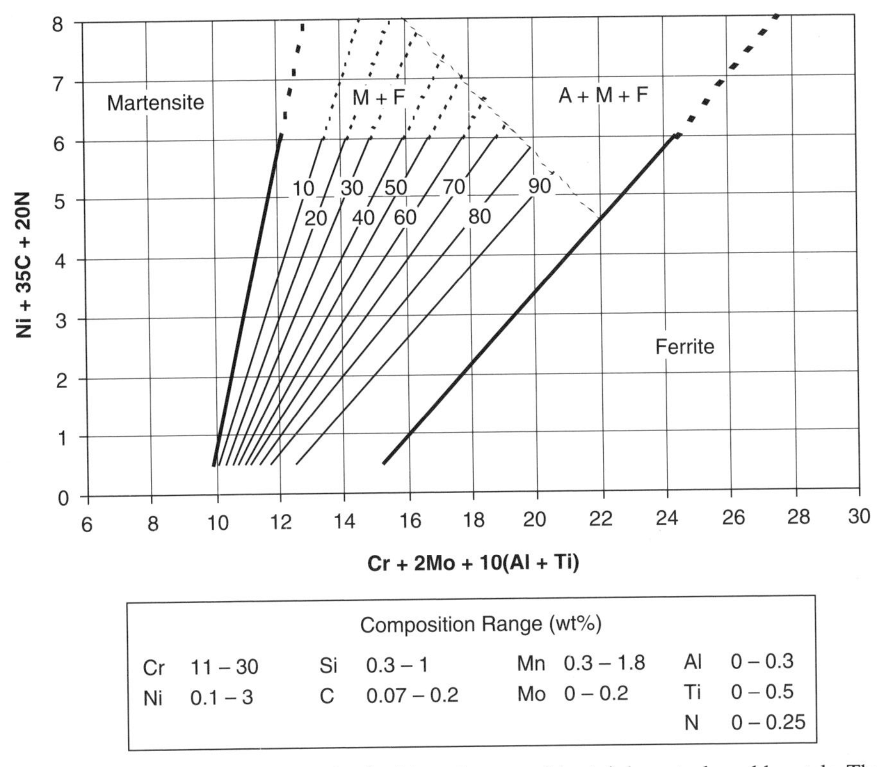

Composition Range (wt%)

Cr	11 – 30	Si	0.3 – 1	Mn	0.3 – 1.8	Al	0 – 0.3
Ni	0.1 – 3	C	0.07 – 0.2	Mo	0 – 0.2	Ti	0 – 0.5
						N	0 – 0.25

Figure 3.22 Balmforth diagram for ferritic and martensitic stainless steels weld metals. The composition range over which this diagram was developed is shown below the diagram. (From Balmforth and Lippold [71]. Courtesy of the American Welding Society.)

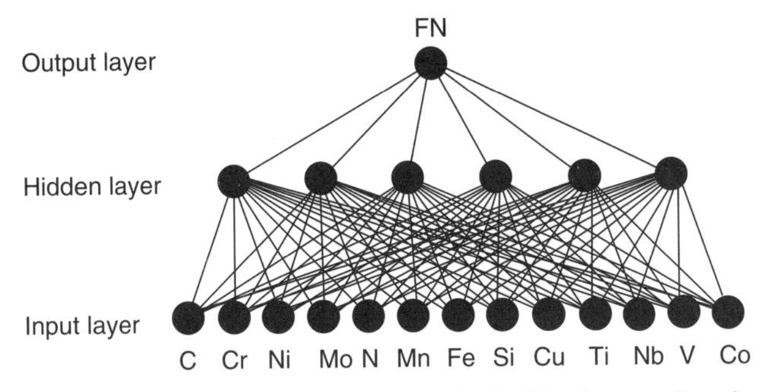

Figure 3.23 Neural network for prediction of the Ferrite Number as a function of composition. (From Vitek et al. [73]. Courtesy of the American Welding Society.)

Table 3.1 summarizes the various coefficients proposed by researchers for the elements making up chromium and nickel equivalents. It cannot include coefficients from the neural network approach, however, because there are no simple relationships among the various elements included in the neural network.

REFERENCES

1. Peckner, D., and Bernstein, I. M. 1977. *Handbook of Stainless Steels*, McGraw-Hill, New York.
2. Copson, H. R. 1959. *Physical Metallurgy of Stress-Corrosion Fracture*, Interscience, New York, p. 126.
3. Floreen, S., and Hayden, H. W. 1968. *Transactions of the American Society for Metals*, 61:489–499.
4. Wentrup, H., and Reif, O. 1949. *Archiv fuer das Eisenhuettenwesen*, 20:359–362.
5. Kubaschewski, O. 1982. *Iron: Binary Phase Diagrams*, Springer-Verlag, New York.
6. Dahl, W., Duren, C., and Musch, H. 1973. *Stahl und Eisen*, 93:813–822.
7. Irvine, J. J., et al. 1961. *Journal of the Iron and Steel Institute, London*, 199:153–169.
8. Olson, D. L. 1985. Prediction of austenitic weld metal microstructure and properties, *Welding Journal*, 64(10):281s–295s.
9. Balmforth, M. 1998. M.S. thesis, Ohio State University.
10. Strauss, B., and Maurer, E. 1920. Die hochlegierten Chromnickelstahle als nichtrostende Stahle, *Kruppsche Monatshefte*, 1(8):129–146.
11. Scherer, R., Riedrich, G., and Hoch, G. 1939. Einfluss eines Gahalts an Ferrit in austenitischer Chrom–Nickel–Stahlen auf den Kornzerfall, *Archiv fuer das Eisenhuettenwesen*, 13:52–57, July.
12. Siewert, T. A., McCowan, C. N., and Olson, D. L. 1992. Ferrite number prediction for stainless steel welds, in *Key Engineering Materials*, Vol. 69/70, Trans Tech Publications, Zurich, Switzerland, pp. 149–166.
13. Newell, H. D., and Fleischmann, M. 1938. Hot rolled metal article and method of making same, U.S. patent 2,118,683.
14. Field, A. L., Bloom, F. K., and Linnert, G. E. 1943. *Development of Armor Welding Electrodes: Relation to the Composition of Austenitic (20Cr–10Ni) Electrodes to the Physical and Ballistic Properties of Armor Weldments*, OSRD Report 1636, July 20.
15. Campbell, H. C., and Thomas, R. D., Jr. 1946. The effect of alloying elements on the tensile properties of 25–20 weld metal, *Welding Journal*, 25(11):760s–768s.
16. Thielemann, R. H. 1940. Some effects of composition and heat treatment on the high-temperature rupture properties of ferrous alloys, *Transactions of the American Society for Metals*, 40:788–804.
17. Binder, W. O., Brown, C. M., and Franks, R. 1949. Resistance to sensitization of austenitic chromium–nickel steels of 0.03% max. carbon content, *Transactions of the American Society for Metals*, 41:1301–1346.
18. Thomas, R. D., Jr. 1949. A constitution diagram application to stainless weld metal, *Schweizer Archiv fuer Angewandte Wissenschaft und Technik*, 1:3–24.
19. Schaeffler, A. L. 1947. Selection of austenitic electrodes for welding dissimilar metals, *Welding Journal*, 26(10):601s–620s.
20. Griffith, A. J., and Wright, J. C. 1969. *Mechanical Properties of Austenitic and Metastable Stainless Steel Sheet and Their Relations with Press Forming Behaviour*, Publication 117, Iron and Steel Institute, London, p. 52.
21. Schaeffler, A. L. 1948. Welding dissimilar metals with stainless electrodes, *Iron Age*, 162:72, July.

22. Schaeffler, A. L. 1949. Constitution diagram for stainless steel weld metal, *Metal Progress*, 56(11):680–680B.
23. Seferian, D. 1959. *Metallurgie de la Soudure*, Dunod, Paris.
24. Schneider, H. 1960. Investment casting of high-hot strength 12% chrome steel, *Foundry Trade Journal*, 108:562–563.
25. Kakhovskii, N. I., Lipodaev, V. N., and Fadeeva, G. V. 1980. The arc welding of stable austenitic corrosion-resisting steels and alloys, *Avtomaticheskaya Svarka*, 33(5):55–57.
26. DeLong, W. T., Ostrom, G. A., and Szumachowski, E. R. 1956. Measurement and calculation of ferrite in stainless-steel weld metal, *Welding Journal*, 35(11):521s–528s.
27. Long, C. J., and DeLong, W. T. 1973. The ferrite content of austenitic stainless steel weld metal, *Welding Journal*, 52(7):281s–297s.
28. DeLong, W. T. 1973. Calibration procedure for instruments to measure the delta ferrite content of austenitic stainless steel weld metal, *Welding Journal*, 52(2):69s.
29. Beck, F. H., Schoefer, E. A., Flowers, J. W., and Fontana, M. G. 1965. *New Cast High-Strength Alloy Grades by Structure Control*, ASTM Special Technical Publication 369, American Society for Testing and Materials, West Conshokocken, PA.
30. Schwartzendruber, L. J., Bennett, L. H., Schoefer, E. A., DeLong, W. T., and Campbell, H. C. 1974. Mossbauer-effect examination of ferrite in stainless steel welds and castings, *Welding Journal*, 53(1):1s–12s.
31. Potak, Y. M., and Sagalevich, E. A. 1972. Structural diagram for stainless steels as applied to cast metal and metal deposited during welding, *Avtomaticheskaya Svarka*, 25(5): 10–13.
32. Carpenter, B., Olson, D. L., and Matlock, D. K. 1985. A diagram to predict aluminum passivated stainless steel weld metal microstructure, paper presented at AWS Annual Convention.
33. Hull, F. C. 1973. Delta ferrite and martensite formation in stainless steels, *Welding Journal*, 52(5):193s–203s.
34. Espy, R. H. 1982. Weldability of nitrogen-strengthened stainless steels, *Welding Journal*, 61(5):149s–156s.
35. Szumachowski, E. R., and Kotecki, D. J. 1984. Effect of manganese on stainless steel weld metal ferrite, *Welding Journal*, 63(5):156s–161s.
36. McCowan, C. N., Siewert, T. A., Reed, R. P., and Lake, F. B. 1987. Manganese and nitrogen in stainless steel SMA welds for cryogenic service, *Welding Journal*, 66(3):84s–92s.
37. Okagawa, R. K., Dixon, R. D., and Olson, D. L. 1983. The influence of nitrogen from welding on stainless steel weld metal microstructure, *Welding Journal*, 62(8):204s–209s.
38. Hammar, O., and Svensson, U. 1979. Influence of steel composition on segregation and microstructure during solidification of austenitic stainless steels, in *Solidification and Casting of Metals*, Metals Society, London.
39. Mel'Kumor, N., and Topilin, V. V. 1969. Alloying austenitic stainless steel with nitrogen, *Obrabotka Metallov*, 8:47–51.
40. Ogawa, T., and Koseki, T. 1988. Weldability of newly developed austenitic alloys for cryogenic service, II: high-nitrogen stainless steel weld metal, *Welding Journal*, 67(1): 8s–17s.
41. Novozhilov, N. M., et al. 1978. On the austenitising and ferritising effect of elements in austenitic ferritic weld metals, *Welding Production*, 25(6):12–13.

42. Kotecki, D. J. 1983. *Molybdenum Effect on Stainless Steel Weld Metal Ferrite*, IIW Document II-C-707-83, American Council of the International Institute of Welding, Miami, FL.

43. Kotecki, D. J. 1986. *Silicon Effect on Stainless Steel Weld Metal Ferrite*, IIW Document II-C-779-86, American Council of the International Institute of Welding, Miami, FL.

44. Takemoto, T., Murata, Y., and Tanaka, T. 1987. Effect of manganese on phase stability of Cr–Ni nonmagnetic stainless steel, in *High Manganese Austenitic Stainless Steels*, R. A. Lula, ed., ASM International, Materials Park, OH, pp. 23–32.

45. Siewert, T. A., McCowan, C. N., and Olson, D. L. 1988. Ferrite Number prediction to 100 FN in stainless steel weld metal, *Welding Journal*, 67(12):289s–298s.

46. Suutala, N., Takalo, T., and Moisio, T. 1980. Ferritic–austenitic solidification mode in austenitic stainless steel welds, *Metallurgical Transactions*, 11A(5):717–725.

47. Kujanpää, V., Suutala, N., Takalo, T., and Moisio, T. 1979. Correlation between solidification cracking and microstructure in austenitic and austenitic–ferritic stainless steel welds, *Welding Research International*, 9(2):55.

48. McCowan, C. N., Siewert, T. A., and Olson, D. L. 1989. *Stainless Steel Weld Metal: Prediction of Ferrite Content*, WRC Bulletin 342, Welding Research Council, New York, April.

49. Kotecki, D. J. 1988. *Verification of the NBS-CSM Ferrite Diagram*, IIW Document II-C-834-88, American Council of the International Institute of Welding, Miami, FL.

50. Lake, F. B. 1990. Effect of Cu on stainless steel weld metal ferrite content, paper presented at AWS Annual Convention.

51. Ferree, J. A. 1969. Free machining austenitic stainless steel, U.S. patent 3,460,939.

52. Castro, R. J., and de Cadenet, J. J. 1968. *Welding Metallurgy of Stainless and Heat-Resisting Steels*, Cambridge University Press, Cambridge.

53. Kotecki, D. J. 1990. Ferrite measurement and control in duplex stainless steel welds, in *Weldability of Materials: Proceedings of the Materials Weldability Symposium*, ASM International, Materials Park, OH, October.

54. Kotecki, D. J., and Siewert, T. A. 1992. WRC-1992 constitution diagram for stainless steel weld metals: a modification of the WRC-1988 diagram, *Welding Journal*, 71(5): 171s–178s.

55. Eichelman, G. H., and Hull, F. C. 1953. The effect of composition on the temperature of spontaneous transformation of austenite to martensite in 18–8-type stainless steel, *Transactions of the American Society for Metals*, 45:77–104.

56. Andrews, K. 1965. Empirical formulae for the calculation of some transformation temperatures, *Journal of the Iron and Steel Institute*, 203:721–727.

57. Self, J. A., Matlock, D. K., and Olson, D. L. 1984. An evaluation of austenitic Fe–Mn–Ni weld metal for dissimilar metal welding, *Welding Journal*, 63(9):282s–288s.

58. Self, J. A., Olson, D. L., and Edwards, G. R. 1984. The stability of austenitic weld metal, in *Proceedings of IMCC*, Kiev, Ukraine, July.

59. Barmin, L. N., Korolev, N. V., Grigor'ev, S. L., Logakina, I. S., and Manakova, N. A. 1980. The phase composition of iron–nickel–cobalt–molybdenum–titanium–silicon system deposited metal, *Avtomaticheskaya Svarka*, 33(10):22–24.

60. Kotecki, D. J. 1999. A martensite boundary on the WRC-1992 diagram, *Welding Journal*, 78(5):180s–192s.

61. Kotecki, D. J. 2000. A martensite boundary on the WRC-1992 diagram, 2: the effect of manganese, *Welding Journal*, 79(12):346s–354s.
62. Kotecki, D. J. 2001. *Weld Dilution and Martensite Appearance in Dissimilar Metal Joining*, IIW Document II-1438-01, American Council of the International Institute of Welding, Miami, FL.
63. Kaltenhauser, R. H. 1971. Improving the engineering properties of ferritic stainless steels, *Metals Engineering Quarterly*, 11(2):41–47.
64. Wright, R. N., and Wood, J. R. 1977. Fe–Cr–Mn microduplex ferritic–martensitic stainless steels, *Metallurgical Transactions A*, 8A(12):2007–2011.
65. Lefevre, J., Tricot, R., and Castro, R. 1973. Noveaux aciers inoxydables á 12% de chrome, *Revue de Metallurgie*, 70(4):259.
66. Lippold, J. C. 1991. *A Review of the Welding Metallurgy and Weldability of Ferritic Stainless Steels*, EWI Research Brief B9101, Columbus, OH.
67. Patriarca, P., Harkness, S. D., Duke, J. M., and Cooper, L. R. 1976. U.S. advanced materials development program for steam generators, *Nuclear Technology*, 28(3):516–536.
68. Panton-Kent, R. 1991. Phase balance in 9%Cr1%Mo steel welds, *Welding Institute Research Bulletin*, January/February.
69. Gooch, T. G., Woolin, P., and Haynes, A. G. 1999. Welding metallurgy of low carbon 13% chromium martensitic steels, in *Proceedings of Supermartensitic Stainless Steels, 1999*, Belgian Welding Institute, Ghent, Belgium, pp. 188–195.
70. Balmforth, M. C., and Lippold, J. C. 1998. A preliminary ferritic–martensitic stainless steel constitution diagram, *Welding Journal*, 77(1):1s–7s.
71. Balmforth, M. C., and Lippold, J. C. 2000. A new ferritic–martensitic stainless steel constitution diagram, *Welding Journal,* 79(12):339s–345s
72. Vitek, J. M., Iskander,Y. S., Oblow, E. M., Babu, S. S., and David, S. A. 1999. Neural network model for predicting Ferrite Number in stainless steel welds, in *Proceedings of the 5th International Trends in Welding Research*, ASM International, Materials Park, OH, pp. 119–124.
73. Vitek, J. M., Iskander, Y. S., and Oblow, E. M. 2000. Improved Ferrite Number predication in stainless steel arc welds using artificial neural networks, 1: neural network development, *Welding Journal,* 79(2):33s–40s.
74. Vitek, J. M., Iskander, Y. S., and Oblow, E. M. 2000. Improved Ferrite Number predication in stainless steel arc welds using artificial neural networks, 2: neural network results, *Welding Journal,* 79(2):41s–50s.
75. Vitek, J. M., David, S. A., and Hihman, C. R. 2003. Improved Ferrite Number predication model that accounts for cooling rate effects, 1: model development, *Welding Journal,* 82(1):10s–17s.
76. Oak Ridge National Laboratory, http://engm01.ms.ornl.gov, courtesy of J. M. Vitek.

CHAPTER 4

MARTENSITIC STAINLESS STEELS

Martensitic stainless steels are based on the Fe–Cr–C ternary system. They undergo an allotropic transformation and form martensite from austenite under most thermo-mechanical processing situations, except when cooling is very slow, such as during furnace cooling. These steels are generally termed *air hardening*, because when withdrawn from a furnace as austenite, cooling in still air is sufficiently rapid to produce extensive martensite. Normal weld cooling rates are also sufficiently rapid to produce weld metal and HAZ microstructures that are predominantly martensitic.

A wide range of strengths is achievable with martensitic stainless steels. Yield strengths ranging from 275 MPa (40 ksi) in an annealed condition to 1900 MPa (280 ksi) in the quenched and tempered condition (for high-carbon grades) are possible. Normally, tempering of the quenched steel is required to achieve acceptable toughness and ductility for most engineering applications. High hardness levels are also achievable, promoting metal-to-metal wear resistance and abrasion resistance.

In general, corrosion resistance of the martensitic stainless steels is not as good as that of the other grades, due to the relatively low chromium content (12 to 14 wt%) and high carbon content (compared to more corrosion-resistant stainless steels). These alloys are generally selected for applications where a combination of high strength and corrosion resistance under ambient atmospheric conditions is required. The low-chromium and low-alloying element content of the martensitic stainless steels also makes them less costly than the other types.

Welding Metallurgy and Weldability of Stainless Steels, by John C. Lippold and Damian J. Kotecki
ISBN 0-471-47379-0

Common applications of martensitic stainless steels include steam, gas, and jet engine turbine blades that operate at relatively low temperatures, steam piping, large hydroturbines, freshwater canal locks, piping and valves for petroleum gathering and refining, and cladding for continuous caster rolls. Low-carbon *supermartensitic grades* are being used increasingly for oil and gas pipelines. The high-chromium, high-carbon grades are used to make items such as surgical instruments, cutlery, gears, and shafts.

Martensitic stainless steels are generally not used above 650°C (1200°F), due to degradation of both mechanical properties and corrosion resistance. Because of the formation of untempered martensite during cooling after welding, the martensitic alloys are considered the least weldable of the stainless steels. Similar to many structural steels, special precautions must be taken when welding the martensitic grades, particularly those with over 0.1 wt% carbon.

4.1 STANDARD ALLOYS AND CONSUMABLES

A list of standard wrought and cast martensitic stainless steels is provided in Table 4.1. A more comprehensive list of alloys is provided in Appendix 1. Martensitic stainless steels can be subdivided into three groups based on their susceptibility to *hydrogen-induced cracking* or *cold cracking*. The groupings are based on carbon content, as this largely determines the hardness of the martensite in the as-welded condition, which directly influences cold-cracking susceptibility. The least troublesome group consists of those steels with about 0.06 wt% C or less, which limits the maximum hardness to about 35 Rockwell C (HRC). For these, successful welding procedures are similar to the procedures for welding low-alloy high-strength steels. The second group consists of those steels with carbon content from 0.06 up to about 0.30 wt%.

TABLE 4.1 Composition of Standard Wrought and Cast Martensitic Stainless Steels

		Composition (wt%)[a]					
Type	UNS No.	C	Cr	Mn	Si	Ni	Other
403	S40300	0.15	11.5–13.0	1.00	0.50	—	—
410	S41000	0.15	11.5–13.5	1.00	1.00	—	—
410NiMo	S41500	0.05	11.4–14.0	0.50–1.00	0.60	3.5–5.5	0.50–1.00 Mo
414	S41400	0.15	11.5–13.5	1.00	1.00	1.25–2.50	—
416	S41600	0.15	12.0–14.0	1.25	1.00	—	0.15 S min., 0.6 Mo
420	S42000	0.15 min	12.0–14.0	1.00	1.00	—	—
422	S42200	0.20–0.25	11.5–13.5	1.00	0.75	0.5–1.0	0.75–1.25Mo, 0.75–1.25W, 0.15–0.3V
431	S43100	0.20	15.0–17.0	1.00	1.00	1.25–2.50	—
440A	S44002	0.60–0.75	16.0–18.0	1.00	1.00	—	0.75 Mo
440B	S44003	0.75–0.95	16.0–18.0	1.00	1.00	—	0.75 Mo
440C	S44004	0.95–1.20	16.0–18.0	1.00	1.00	—	0.75 Mo
CA-15	—	0.15	11.5–14.0	1.00	1.50	1.00	0.50 Mo
CA-6NM	—	0.06	11.5–14.0	1.00	1.00	3.5–4.5	0.40–1.0 Mo

[a]A single value is a maximum.

TABLE 4.2 Martensitic Stainless Steel AWS Filler Metal Classifications

AWS Classification	UNS No.	Composition (wt%)[a]						Base Metal
		C	Cr	Mn	Si	Ni	Mo	
E410-XX	W41010	0.12	11.0–13.5	1.0	0.90	0.7	0.75	410, CA-15
ER410	S41080	0.12	11.5–13.5	0.6	0.5	0.6	0.75	410, CA-15
E410TX-X	W41031	0.12	11.0–13.5	0.60	1.0	0.60	0.5	410, CA-15
E410NiMo-XX	W41016	0.06	11.0–12.5	1.0	0.90	4.0–5.0	0.40–0.70	410NiMo, CA-6NM
ER410NiMo	S41086	0.06	11.0–12.5	0.6	0.5	4.0–5.0	0.4–0.7	410NiMo, CA-6NM
E410NiMoTX-X	W41036	0.06	11.0–12.5	1.0	1.0	4.0–5.0	0.40–0.70	410NiMo, CA-6NM
ER420	S42080	0.25–0.40	12.0–14.0	0.6	0.5	0.6	0.75	420

[a]A single value is a maximum.

The as-welded hardness of this second group ranges above 35 HRC up to about 55 HRC, with greater cracking risk and correspondingly higher preheat requirements up to 315°C (600°F). The third group consists of those steels with more than 0.30 wt% C and an as-welded hardness of 55 to 65 HRC. This group requires very specialized procedures for welding without cracking.

A large number of martensitic alloys are available, with chromium contents ranging from 11.5 to 18 wt%. Most of the structural grades contain carbon in the range 0.1 to 0.25 wt%. Some alloys contain small additions of Mo, V, and W to provide high-temperature strength via the formation of stable carbides. The addition of nickel improves toughness. The 440 grades contain high carbon levels and are used in applications where high hardness, wear resistance, and corrosion resistance are required.

It is often desirable to weld martensitic stainless steels with matching or near-matching filler metals in order to match the strength of the base metal. These weld metals also respond to postweld heat treatment in a manner similar to the base metal. However, not all base metals have matching filler metals in the AWS filler metal specifications. Table 4.2 lists AWS filler metal classifications that deposit martensitic stainless steel weld metal, and the base metals for which these filler metals are a match. There are also many proprietary unclassified filler metals, mostly in the form of tubular wires for submerged arc welding that produce weld metal in the medium-carbon range, intended primarily for overlay of rolls in continuous casters for steel mills. In some applications it is desirable to select austenitic filler metals for martensitic base metals, particularly when weld metal hydrogen cracking is a concern, since the austenite has higher solubility for hydrogen and is effectively immune to hydrogen-induced cracking. Another situation in which it is often desirable to use austenitic filler metals is when the martensitic base metal is in the annealed condition and the weldment is to be used in the as-welded condition.

Dilution of the austenitic stainless steel filler metal by the martensitic base metal often results in a two-phase austenite + ferrite microstructure that will be far softer than the base metal unless the base metal is in the annealed condition. Some precautions should be exercised during postweld heat treatment of weld metals containing ferrite, since sigma phase embrittlement is possible within the standard tempering range.

Ni-base filler metals are also metallurgically compatible with the martensitic alloys and will result in a fully austenitic weld deposit if the base metal dilution is kept low. Generally, dissimilar filler metals are not used, due to both performance (strength) and cost considerations. They may be useful, however, in transition joints between martensitic stainless steels and austenitic alloys, where it is desirable to have a transition in the coefficient of thermal expansion or where the weld metal strength is required to be stronger than the weakest component.

4.2 PHYSICAL AND MECHANICAL METALLURGY

In the broadest sense, stainless steels are based on the iron–chromium binary system. As can be seen in the Fe–Cr system in Figure 2.1, the body-centered cubic (BCC)

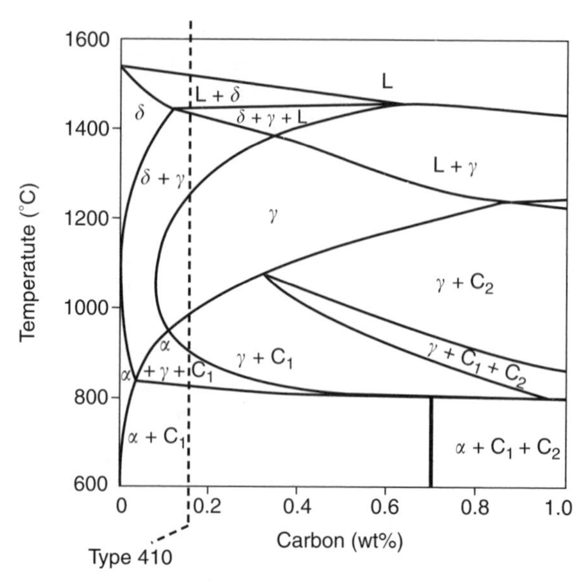

Figure 4.1 Fe–13Cr pseudobinary diagram with a nominal carbon content of Type 410 superimposed. (From Castro and Tricot [1].)

ferrite phase is stable over a wide range of composition and temperature. At low Cr concentrations, less than about 12 wt%, austenite (FCC) forms from the ferrite upon cooling in the solid state. The region of the phase diagram in which austenite is stable is often referred to as the *gamma loop* because of its shape. Under equilibrium cooling conditions, austenite that forms within the gamma loop will retransform to ferrite, but under more rapid cooling conditions the austenite will form martensite. Most martensitic stainless steels contain other alloying elements, usually carbon, that expand the gamma loop, thereby promoting austenite and facilitating the transformation to martensite. As such, ternary Fe–Cr–C diagrams are more appropriate for describing the phase equilibrium in martensitic stainless steels.

As described in Chapter 2, the Fe–Cr–C ternary system best describes the phase transformations that occur on-heating and on-cooling in martensitic stainless steel weldments. A pseudobinary section through this ternary system at a constant 13 wt% Cr (Figure 4.1) can be used to determine phase stability from the solidification temperature range to room temperature [1]. Note that in the range 0.1 to 0.25 wt% carbon, these steels solidify as ferrite, but form some austenite, or a mixture of ferrite and austenite, at the end of solidification. Upon cooling from the solidification range, the diagram predicts that within this carbon range all the ferrite transforms to austenite. At temperatures below 800°C (1470°F), the equilibrium phases are ferrite and carbide ($Cr_{23}C_6$).

Under normal weld cooling conditions, the austenite that is present at elevated temperatures will transform to martensite. Many of the martensitic stainless steels will retain some high-temperature ferrite in the martensite matrix. The retention of ferrite is discussed later, but its presence is a function of the balance of ferrite-promoting to austenite-promoting elements. At higher carbon contents, the austenite phase field expands, promoting fully martensitic structures. A higher carbon content results in a harder and more brittle martensite that is more prone to hydrogen-induced cracking and possible brittle fracture.

An isothermal transformation diagram [2] such as one for Type 410 shown in Figure 4.2 can be used to predict the microstructure that forms in the weld metal and HAZ during cooling from the austenite phase field. For Type 410, the "nose" of the ferrite formation curve occurs at times exceeding 100 seconds. For most fusion welds, this results in the formation of a predominantly martensitic structure with high hardness (about 45 HRC) in the weld metal and portions of the HAZ.

It is noteworthy in Figure 4.2 that if the high-temperature austenite can be cooled to below 700°C (1290°F) in less than 200 seconds, or about 3 minutes, followed by holding at a slightly higher temperature than the M_S temperature, the metal can remain austenite for periods up to a week or more. This feature can be very useful in selecting preheat and interpass temperatures for martensitic stainless steels. Multiple weld passes can be made under this condition, and the weld deposit will remain austenite, which easily accommodates weld shrinkage strains. At the same time, hydrogen can escape, albeit slowly, because its diffusion coefficient is much lower in austenite than in ferrite or martensite. When welding is finished and the weldment is finally allowed to cool, the mass of austenitic weld metal can transform to martensite. This transformation is accompanied by a volume expansion because martensite is less dense than austenite, so the weld metal may arrive at ambient temperature under compressive stresses, which do not promote cracking. This approach

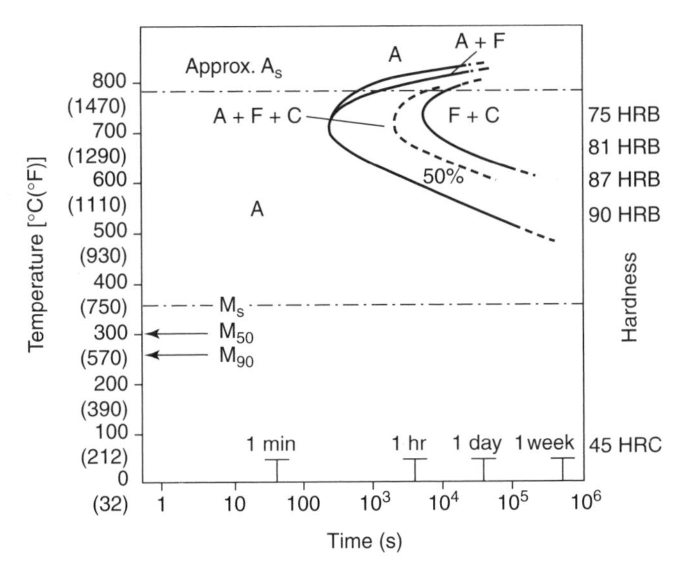

Figure 4.2 Isothermal transformation diagram for a Type 410 stainless steel. (From McGannon [2]. Courtesy of U.S. Steel Corporation.)

is commonly used in overlaying low-alloy steel roll bodies with martensitic stainless steel weld metal for service as continuous caster rolls.

The temperature range over which martensite forms is primarily a function of composition. Almost all alloying additions (cobalt is one exception) tend to lower the martensite start temperature (M_S), with carbon having the most potent influence. A number of relationships have been developed to predict the martensite start temperature in stainless steels [3–9]. These are listed in Table 4.3 and include the various coefficients for alloying additions that are present in martensitic stainless steels. The constant value represents the baseline from which these weighted factors are subtracted in order to estimate the M_S temperature in °C. For example, the predictive equation from Gooch [3] is

$$M_S(°C) = 540 - (497C + 6.3Mn + 36.3Ni + 10.8Cr + 46.6Mo) \quad (4.1)$$

Some caution should be exercised in the use of these relationships since prediction differences among these equations can be greater than 100°C (180°F) for a given composition. Generally, choosing the safer prediction among the several possibilities is the approach recommended. For example, for selection of a preheat temperature, the highest M_S prediction should be selected. If one is concerned about complete martensite transformation upon cooling from the austenite phase field, the lowest M_S value should be selected (the martensite finish temperature, M_F, is approximately 100°C (180°F) below the M_S value).

TABLE 4.3 Predicting Equations for Martensite Start Temperatures (°C)

Reference	Constant	Coefficients								
		C	Mn	Si	Cr	Ni	Mo	W	Co	Other
Payson and Savage [4]	499	−317	−33	−11	−28	−17	−11	−11	—	—
Irvine et al. [5]	551	−474	−33	−11	−17	−17	−21	−11	—	—
Steven and Haynes [6]	561	−474	−33	—	−17	−17	−21	—	—	—
Kung and Rayment [8] (modification of Steven and Haynes [6])	561	−474	−33	−7.5	−17	−17	−21	—	+10	—
Andrews [7]	539	−423	−30.4	—	−12.1	−17.7	−7.5	—	—	—
Kung and Rayment [8] (modification of Andrews [7])	539	−423	−30.4	−7.5	−12.1	−17.7	−7.5	—	+10	—
Gooch [3]	540	−497	−6.3	—	−10.8	−36	−46.6	—	—	—
Self et al. [9]	526	−354	−29.7	−31.7	−12.5	−17.4	−20.8	—	—	[a]

[a] −1.34(%Ni × %Cr) + 22.4(%Cr + %Mo) × %C.

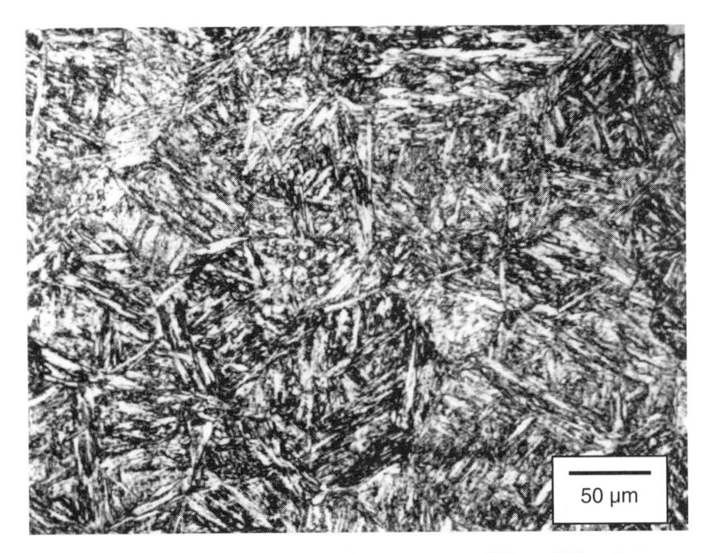

Figure 4.3 Quench and tempered microstructure of Type 410 stainless steel.

In most martensitic stainless steels containing from 0.1 to 0.25 wt% C, the M_S value is relatively high, typically in the range 200 to 400°C (390 to 750°F). Since the completion of the martensite transformation or M_F temperature is normally about 100°C (180°F) below M_S, the transformation is complete at room temperature. In more highly alloyed steels, particularly those containing 4% Ni or more, where the M_F value may be below room temperature, some austenite can be retained in the microstructure. This can be potentially beneficial to toughness.

A representative microstructure of a tempered martensitic stainless steel wrought base metal is shown in Figure 4.3. This structure consists essentially of ferrite and carbides, where tempering promotes the formation of Cr-rich and/or alloy carbides and a transition from the BCT (body-centered tetragonal) crystal structure of the martensite to the BCC structure of the ferrite.

The minimum mechanical properties required for several martensitic stainless steels in the annealed and/or tempered conditions are provided in Table 4.4. A more comprehensive list of mechanical properties in various product forms and thermo-mechanical conditions may be found elsewhere [10].

4.3 WELDING METALLURGY

4.3.1 Fusion Zone

The fusion zones of martensitic stainless steels with nominally 11 to 14 wt% Cr and 0.1 to 0.25 wt% C solidify as delta ferrite. Segregation of C and other alloying elements during solidification can in some cases result in the formation of austenite, or a mixture of ferrite and austenite, at the end of solidification. As the

TABLE 4.4 Minimum Mechanical Properties of Martensitic Stainless Steels

Type	Condition	Tensile Strength		Yield Strength		Elongation (%)
		MPa	ksi	MPa	ksi	
403	Annealed	485	70	275	40	20
	Intermediate temper	690	100	550	80	15
	Hard temper	825	120	620	90	12
410	Annealed	485	70	275	40	20
	Intermediate temper	690	100	550	80	15
	Hard temper	825	120	620	90	12
420	Annealed	690	100	—	—	15
	Tempered at 204°C (400°F)	1720	250	1480	215	8
431	Annealed	760	110	—	—	—
	Intermediate temper	795	115	620	90	15
	Hard temper	1210	175	930	135	13
440C	Annealed	760	110	450	65	14
	Tempered at 315°C (600°F)	1970	285	1900	275	2

weld metal cools in the solid state, the austenite will consume the ferrite, resulting in a fully austenitic structure below about 1100°C (2012°F). The austenite will transform to martensite upon further cooling. This transformation is represented by the following sequence and is shown schematically in Figure 4.4. A representative microstructure of this sequence is provided in Figure 4.6*a*.

Transformation path 1: fully martensitic microstructure

$$L \rightarrow L + F_P \rightarrow F_P \rightarrow F_P + A \rightarrow A \rightarrow \text{martensite}$$

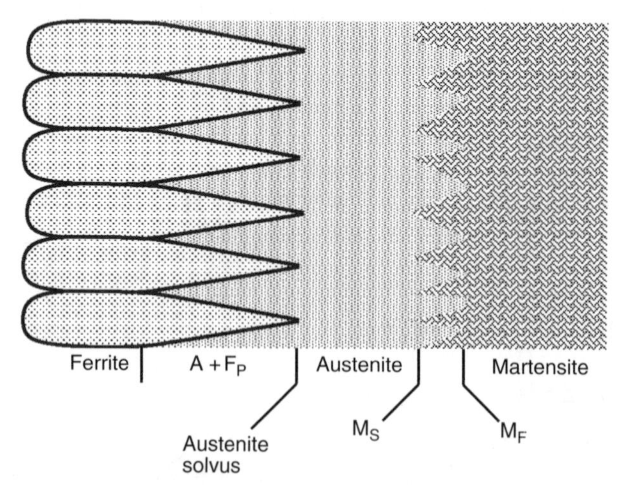

Figure 4.4 Transformation behavior of a fully martensitic fusion zone. F_P, primary ferrite; M_S, martensite start temperature; M_F, martensite finish temperature.

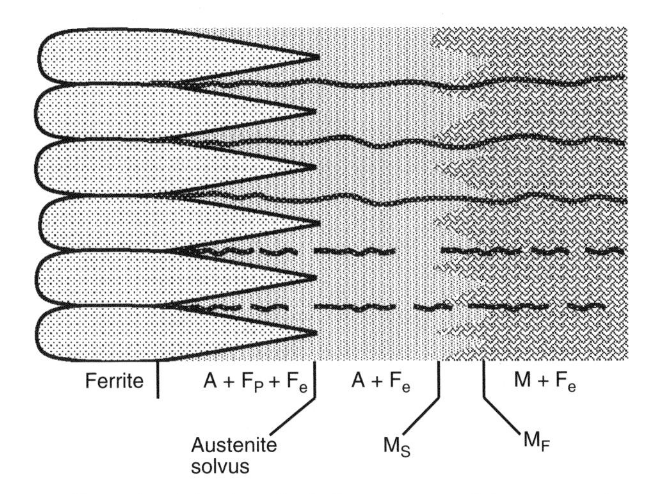

Figure 4.5 Transformation behavior of a fusion zone containing martensite and ferrite. F_P, primary ferrite; F_e, eutectic ferrite; M_S, martensite start temperature; M_F, martensite finish temperature.

If some ferrite forms at the end of solidification, this ferrite may be sufficiently enriched in ferrite-promoting elements (particularly Cr, and Mo if present) such that it does not transform to austenite on cooling below the solidification temperature range. This ferrite resides along the solidification grain and subgrain boundaries, and the final weld microstructure will consist of a mixture of martensite and *eutectic ferrite*, since it is presumed that this ferrite forms via a eutectic reaction at the end of solidification. The amount of ferrite will depend on the ratio of ferrite- to austenite-promoting elements and the solidification conditions. This transformation sequence is represented by the following sequence and is shown schematically in Figure 4.5. A representative microstructure resulting from this transformation path is shown in Figure 4.6*c*.

Transformation path 2: two-phase martensite + eutectic ferrite microstructure

$$L \rightarrow L + F_P + (A + F_e) \rightarrow F_P + A + F_e \rightarrow A + F_e \rightarrow M + F_e$$

It is also possible that some of the original primary ferrite does not transform completely to austenite at elevated temperatures and remains in the structure upon cooling to room temperature. This transformation path is represented by the following:

Transformation path 3: two-phase martensite + primary ferrite microstructure

$$L \rightarrow L + F_P \rightarrow F_P \rightarrow A + F_P \rightarrow M + F_P$$

A microstructure resulting from this transformation sequence is shown in Figure 4.6*b*.

Castro and de Cadenet [11] have also suggested that ferrite can be retained along the cores of the original ferrite dendrites due to incomplete transformation of the

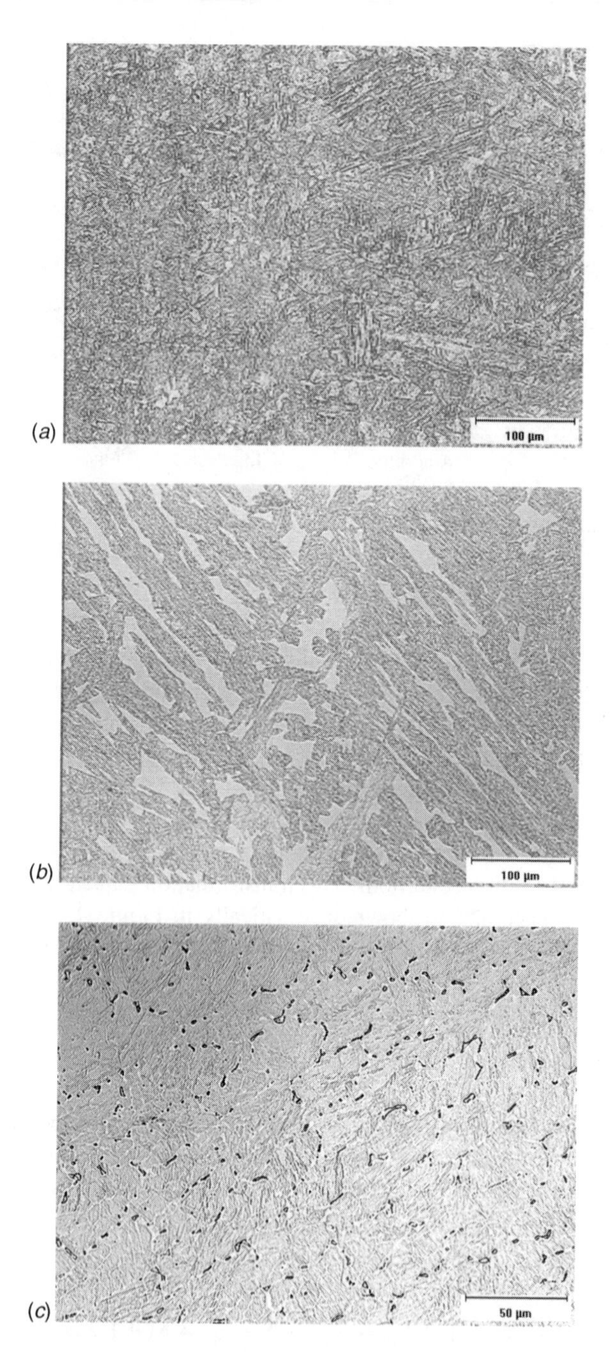

Figure 4.6 Fusion zone microstructure of martensitic stainless steels: (*a*) fully martensitic Type 410; (*b*) martensite with some retained ferrite in Type 410 (dark etching phase is martensite, light etching phase is ferrite); (*c*) martensite with ferrite along solidification subgrain boundaries from a 12Cr–1Mo alloy (HT9) (dark etching phase is ferrite).

austenite. This is similar to the "skeletal" ferrite formation mechanism that occurs in austenitic stainless steels. Although retention of ferrite at the dendrite cores is theoretically possible, when ferrite is present in these weld metals, it is usually along the solidification grain and subgrain boundaries, or distributed in and around the martensite, as shown in Figure 4.6.

Some carbide precipitation may also occur on cooling, depending on cooling rate. These carbides are normally of the type $M_{23}C_6$ or M_7C_3, where the "M" is predominantly Cr and Fe. The M_7C_3 carbides are usually restricted to the higher-carbon alloys (greater than 0.3 wt% C).

4.3.2 Heat-Affected Zone

In the as-welded condition, the HAZ of martensitic stainless steel welds can exhibit a number of distinct microstructural regions. A macrograph of an autogenous weld in a 12Cr–1Mo stainless steel is shown in Figure 4.7. This figure shows a narrow dark band in the HAZ just adjacent to the fusion boundary and a wider, light-etching region outside this. Figure 4.8 is the same diagram as Figure 4.1, now with the carbon content indicated to describe microstructure evolution in a low-carbon (0.15 wt%) alloy. This diagram is used to describe the regions of the HAZ in Figure 4.7.

In a low-carbon alloy, four distinct regions can be identified by metallurgical examination in an optical microscope and by microhardness traverses, such as the one shown in Figure 4.9 for the 12Cr–1Mo alloy of Figure 4.7. Region 1 represents the portion of the HAZ just adjacent to the fusion boundary. In this region, the bulk of the microstructure at elevated temperature consists of austenite, but some ferrite may be present at the austenite grain boundaries (Figure 4.10*a*). Since carbon is an austenite promoter, increasing carbon above 0.15 wt% will expand the austenite phase field and reduce the amount of ferrite in the elevated-temperature microstructure. Upon cooling to room temperature, the austenite transforms to martensite and some of the ferrite remains in the microstructure. The amount of ferrite that is present at room temperature will be a function of the amount that was present initially and the rate of

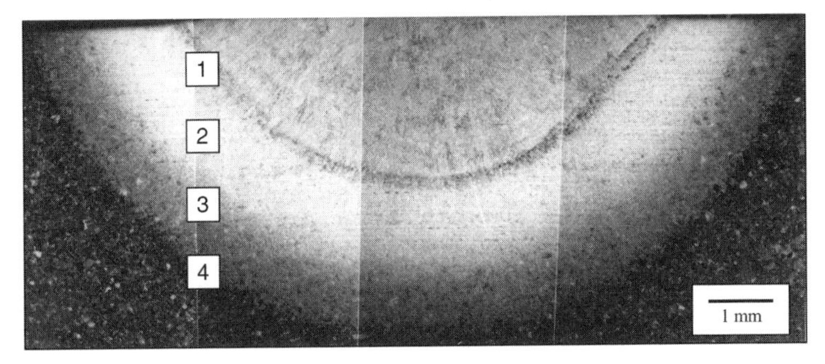

Figure 4.7 Macrograph of an autogenous GTA weld in a 12Cr–1Mo stainless steel showing distinct regions in the HAZ. (From Lippold [12]. Courtesy of TMS/AIME.)

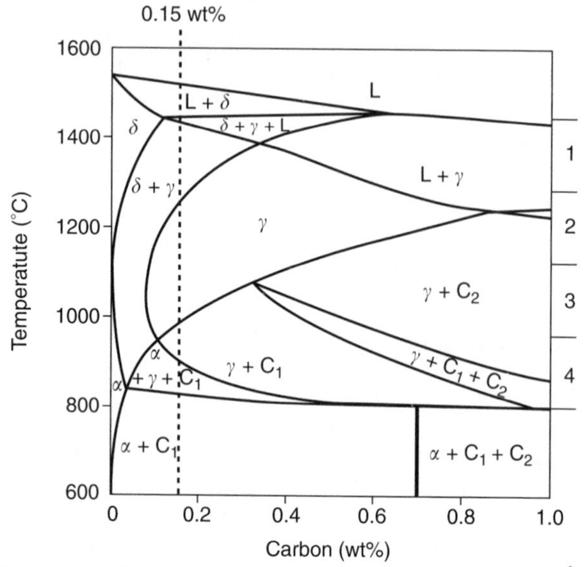

Figure 4.8 13% Cr pseudobinary phase diagram showing the four HAZ microstructure regions that can exist in a 0.15 wt% carbon martensitic stainless steel. (From Castro and Tricot [1].)

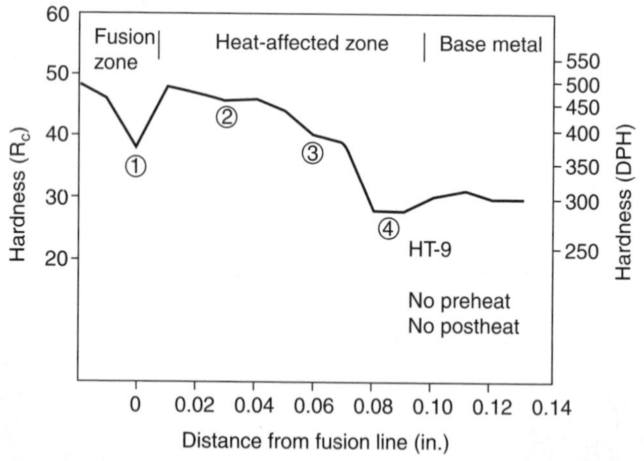

Figure 4.9 Microhardness traverse across the HAZ of a GTA weld in a 12Cr–1Mo–0.5W–0.3V–0.2C alloy, as-welded condition (no preheat or PWHT). (From Lippold [12]. Courtesy of TMS/AIME.)

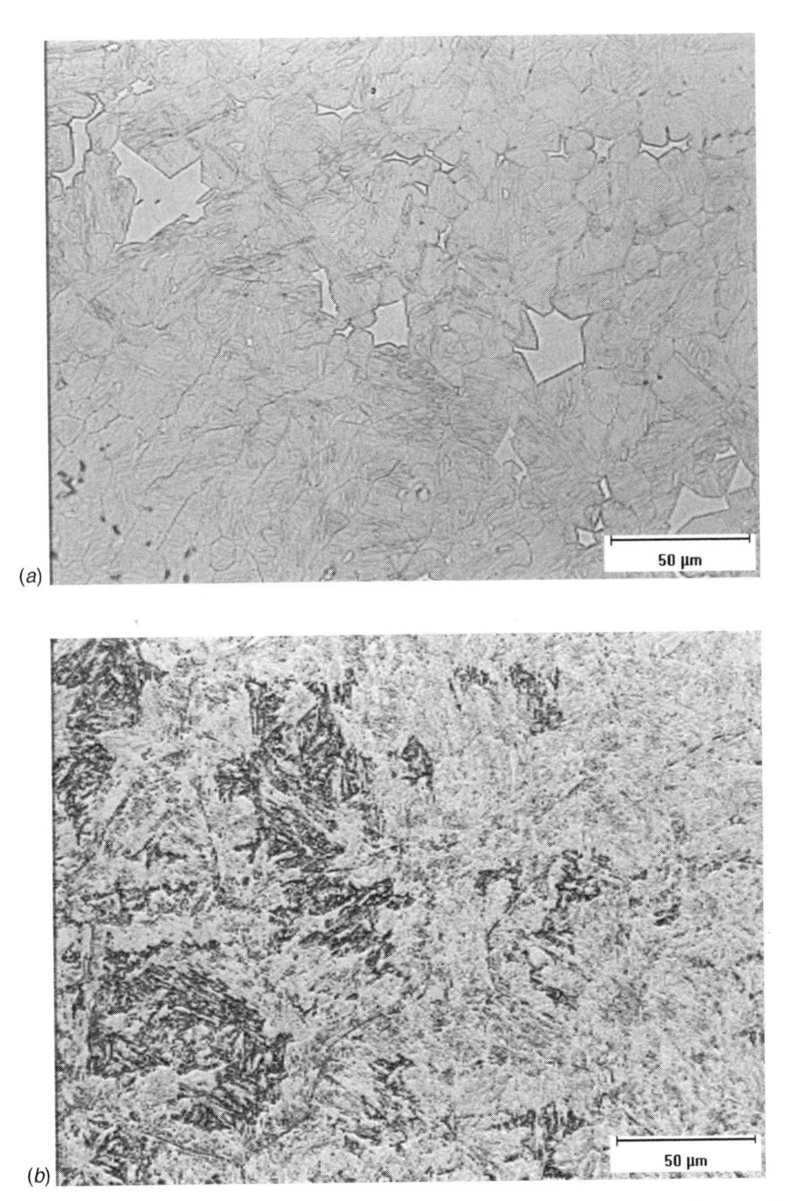

Figure 4.10 Photomicrographs of the HAZ in 12Cr–1Mo steel (HT9): (*a*) region near fusion boundary showing untempered martensite and ferrite; (*b*) region at edge of HAZ showing mixture of tempered and untempered martensite.

dissolution of this ferrite as it cools through the austenite phase field. The presence of ferrite can promote local softening relative to the adjacent fusion zone and HAZ, as shown in Figure 4.9.

In region 2 of the HAZ, the microstructure will be fully austenitic at elevated temperature. The temperatures in this region of the HAZ are sufficiently high that the base metal carbides will dissolve and austenite grain growth will occur. Upon cooling, this region will be fully martensitic. Because all or most of the carbon will have gone back into solution in the austenite, the peak HAZ hardness will generally occur in this region.

Region 3 of the HAZ is also heated into the austenite phase field during welding, but because the temperature is lower than in region 2, carbide dissolution will be incomplete and austenite grain growth will not be so pronounced. This reduction in grain growth is due to both the lower temperature experienced and the pinning effect of the undissolved base metal carbides. Failure to dissolve the carbides results in a lower austenite carbon concentration and a subsequent reduction in the hardness of the martensite that forms upon cooling.

In region 4, little or no transformation to austenite occurs and the microstructure under optical metallography appears virtually identical to the quench and tempered base metal. Within this temperature regime [800 to 950°C (1470 to 1740°F)] carbide coarsening can occur, resulting in some local softening relative to the base metal (Figure 4.9).

If alloys with higher carbon content are considered, the two-phase austenite + ferrite region will shrink and eventually disappear (Figure 4.8). This will result in elimination of the softened region at the fusion boundary, since untempered martensite will extend all the way to the fusion boundary. An alloy of 0.4 wt% C, for example, will have no ferrite in the HAZ near the fusion boundary, as predicted by Figure 4.8. At lower carbon contents, considerable ferrite may form in the HAZ near the fusion boundary, resulting in more pronounced softening.

4.3.3 Phase Transformations

The predominant phase transformation in the martensitic stainless steel welds is the austenite-to-martensite transformation that occurs in the fusion zone and regions of the HAZ that have been heated into the austenite phase field. If more than a few percent of ferrite exists in a predominantly martensitic stainless steel, the disparity in mechanical properties between the ferrite and the martensite needs to be taken into account. In a hot-worked wrought steel, this is not so important because the ferrite and martensite zones become aligned parallel to the rolling direction. Then, unless tensile strain is applied in the thickness direction (which is not very common unless thick plate sections are considered), strain will be distributed uniformly in ferrite and martensite. But in weld metal, ferrite areas tend to be oriented perpendicular to the surface, so that any strain applied parallel to the surface tends to be concentrated disproportionately in the ferrite, resulting in fracture at low stress and low elongation. Figure 4.11 shows a nominally martensitic weld metal with excessive ferrite. The softer ferrite is lighter etching and contains larger *diamond pyramid hardness* (DPH)

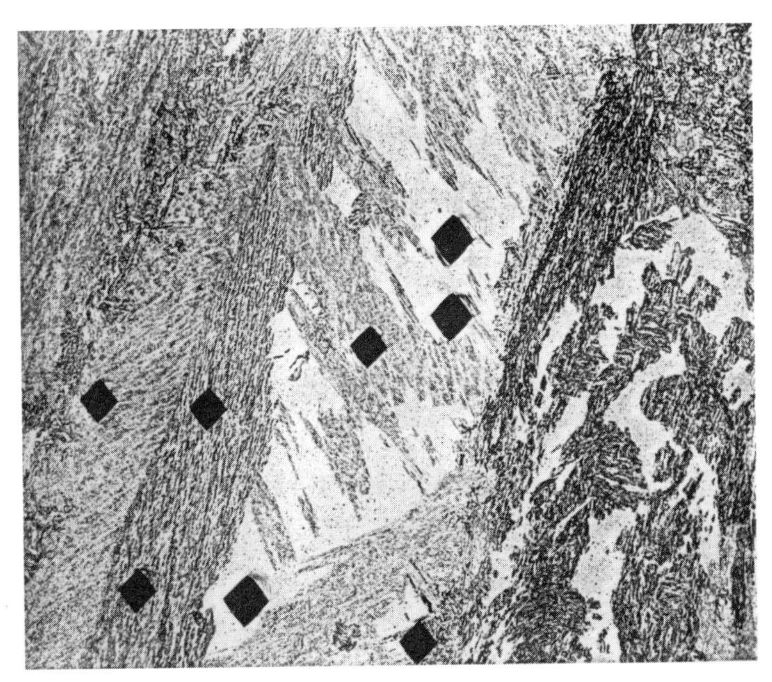

Figure 4.11 Ferrite in a martensitic stainless steel weld metal. Composition: 0.05% C, 0.9% Mn, 0.6% Si, 14.1% Cr, 2.1% Ni, 1% Mo.

indentations than those in martensite. So it is important to anticipate or predict the presence of ferrite in a martensitic stainless steel weld metal.

The Balmforth diagram [13], discussed in Chapter 3 and shown again here in Figure 4.12, can be used to predict the ferrite content in the weld metal of most martensitic stainless steels. Note that this diagram predicts *as-welded* ferrite content and does not take into account any ferrite dissolution that may occur during subsequent thermal exposure, such as during multipass welding or postweld heat treatment. In Figure 4.12 the composition limits of Types 410 and 420 have been superimposed on the diagram. This illustrates that both these alloys have the potential to form weld metal ferrite within their composition limits. In practice, the filler metal compositions tend to be biased such that ferrite retention in the weld metal is minimized.

4.3.4 Postweld Heat Treatment

Postweld heat treatment (PWHT) is almost always required for martensitic stainless steels. Even at carbon levels as low as 0.1 wt%, the as-welded hardness may exceed 30 to 35 HRC. PWHT is used primarily to temper the martensite, but it also provides some stress relief when high levels of residual stress are present. PWHT is normally performed in the range 480 to 750°C (895 to 1380°F), although PWHT at temperatures as low as 200°C (390°F) has been used. There is virtually no softening if tempering is

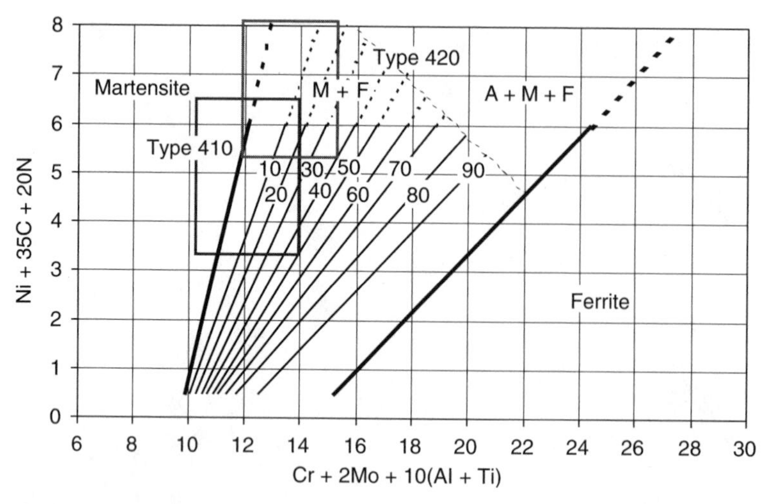

Figure 4.12 Balmforth constitution diagram with Types 410 and 420 stainless steel composition limits superimposed. (From Balmforth and Lippold [13]. Courtesy of the American Welding Society.)

done below 480°C (900°F), but tempering below this temperature can be used to improve toughness and dimensional stability after machining. Heat treatment times depend on section thickness, but normally, 30 minutes to 2 hours is sufficient.

The usual motivation for PWHT of a martensitic stainless steel weldment is to temper martensite. For PWHT to accomplish this, it is necessary for martensite to be present at the tempering temperature. Consider the case of a continuous caster roll overlaid by submerged arc welding with 420 stainless. This would normally be done using the preheat and interpass temperature range 305 to 425°C (600 to 800°F), which is above the M_S temperature. So, upon completion of welding, there is no martensite in the weld metal; it is austenite and can remain austenite for a week or more. Now, if the roll is put directly into the PWHT furnace before the roll cools off, the PWHT will be performed on the austenite. In this case, one of two things may happen. If the PWHT were done at 565°C (1050°F) to achieve a final hardness of about 30 HRC for ease of machining, and the PWHT time was only two hours, Figure 4.2 indicates that the austenite will still be present when the rolls are withdrawn from the furnace. (The isothermal transformation curve for 420 stainless is virtually the same as that for the 410 stainless shown in Figure 4.2, but the M_S temperature is lower.) Then that austenite will transform to fresh martensite upon cooling, and the resulting hardness will be over 45 HRC. The machinist who is next to finish-machine the roll will not be happy. On the other hand, if a much longer time at temperature had been chosen, such as 16 hours, the austenite could transform at that temperature into ferrite and carbides, and the result would be hardness well below 20 HRC, probably about 90 HRB. This will make the machinist happy but the caster operator will not get much life from the roll because it is too soft. So it is essential, after welding, to cool the weldment to allow for martensite transformation before undertaking PWHT. Normally, this means cooling to about 100°C (210°F) before beginning PWHT.

Metallurgically, tempering promotes transformation of the martensite to ferrite and very fine carbides. This transformation reduces strength but improves ductility and toughness. If carbides other than chromium carbides form at intermediate temperatures, there may be a degree of secondary hardening that offsets the softening of the martensite. This allows lower-carbon weld metal, containing alloying elements besides chromium, to match the hardness and strength of higher-carbon base metal when used with a properly chosen PWHT.

An example of this is shown in Figure 4.13. Note the secondary hardening at about 480°C (900°F), and the more gradual softening above that temperature, of the 423L composition as compared to the higher-carbon 420 composition. This more gradual softening is also desirable if a PWHT is being used to achieve a desired hardness for machining and/or service requirements. For continuous caster roll alloys, hardness for machining of 30 to 35 HRC is often required. The 423L alloy, with secondary hardening characteristics, will fall into this hardness range over a much wider range of tempering temperatures than will the 420 alloy. This can be important in an industrial application because a furnace load may consist of a dozen rolls, and not every roll will see exactly the same temperature during PWHT (i.e., heat-treatment furnaces have temperature gradients that result in uneven heating). Care must be taken not to temper higher-Cr alloys for excessive times since sigma phase precipitation in the ferrite is possible. This will result in embrittlement of the structure.

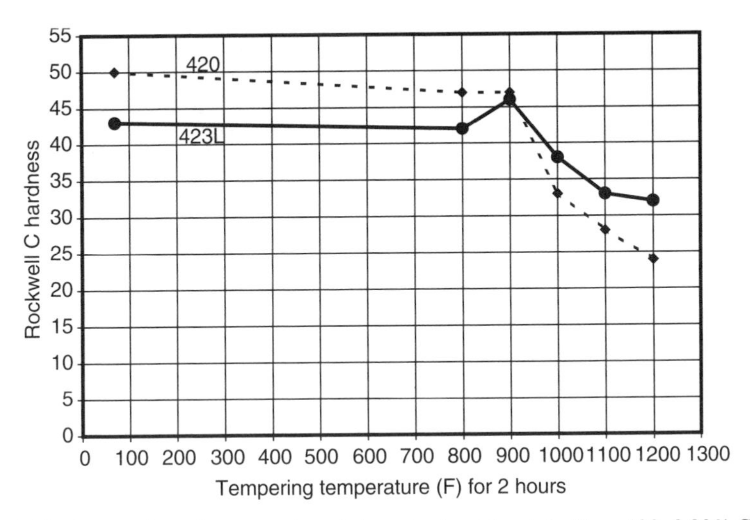

Figure 4.13 Tempering of martensitic stainless steel weld metals. Type 420: 0.20% C, 1.2% Mn, 0.5% Si, 12.0% Cr; type 423L: 0.15% C, 1.2% Mn, 0.4% Si, 11.5% Cr, 2.0% Ni, 1.0% Mo, 0.15% V.

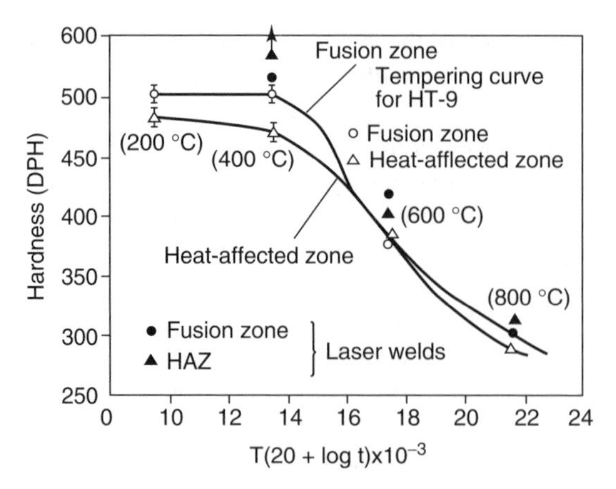

Figure 4.14 Tempering curves for the fusion zone and HAZ of a 12Cr–1Mo–0.5W–0.3V–0.2C steel. (From Lippold [14].)

To fully optimize properties relative to the base metal, the entire structure can be given a *solution heat treatment* (SHT), then quenched and tempered. The SHT will re-austenitize the entire structure, dissolve most or all of the ferrite in the weld metal and HAZ, and result in a uniform martensitic structure upon quenching. Various tempering treatments can then be used to achieve desired strength, ductility, and toughness. Unfortunately, this approach is usually not practical, due to size and/or logistical constraints. SHT can also potentially result in significant distortion of large and/or complex components.

The tempering temperature range for martensitic stainless steels is normally from 480 to 750°C (900 to 1380°F). Within this tempering range the hardness of the martensite decreases as a function of time, with more rapid tempering occurring at higher temperatures. The tempering response of 12Cr–1Mo alloy whose as-welded hardness profile is shown in Figure 4.9 is presented in Figure 4.14. This plot uses a Larson–Miller parameter to combine temperature and time into a single variable. The HAZ response corresponds to region 2 in Figure 4.9, where the highest hardness exists in the as-welded structure. Note that tempering below 600°C (1110°F) is essentially ineffective in reducing the hardness and that there is virtually no difference in behavior between the fusion zone of the autogenous GTA weld and the HAZ.

4.3.5 Preheat, Interpass, and Postweld Heat Treatment Guidelines

Similar to structural steels, careful preheat and interpass control is required to avoid hydrogen-induced cracking. Martensitic stainless steels that contain less than 0.06 wt% C (of which 410NiMo, CA-6NM, and the supermartensitics are examples) may require no special preheat or interpass control in thin sections and a 120°C (250°F) minimum in sections above 12 mm (0.5 in.) thickness. Alloys with carbon levels between 0.06 and 0.3 wt% (which constitute most of the structural alloys) require preheat and interpass temperature control. In thin sections, the preheat and interpass is usually set below M_S, to allow complete transformation to martensite and hydrogen diffusion.

In thicker sections, preheat and interpass temperature should be above M_S to prevent possible cracking during fabrication. Following welding, the assembly is then cooled slowly to room temperature to allow adequate time for hydrogen diffusion during the transformation process. Remember that hydrogen has a high solubility in austenite, and hydrogen cracking is much less likely if the transformation occurs more slowly.

Alloys with carbon contents above 0.3 wt% must use preheat and interpass temperatures above the M_S value, for the reasons described above. These alloys can then be heated directly to the austenite temperature range, cooled slowly to allow martensite formation, and tempered. This procedure will optimize weldment properties relative to the base metal. A second procedure for handling these high-carbon martensitic stainless steels is, after welding with preheat and interpass temperature as above, to heat the entire weldment immediately (without cooling) to a temperature at which austenite can transform most rapidly to ferrite and carbides [typically about 700°C (1290°F), if there is little or no nickel present] and hold for several hours to permit the equilibrium isothermal transformation to take place. In this approach, the weld metal transforms and the base metal is heavily tempered, so that the entire weldment is soft upon cooling. The weldment can then easily be machined, if necessary, to nearly final dimensions, then austenitized again and hardened.

Preheat and interpass temperature can be controlled such that the transformation to martensite is either promoted or avoided during weld cooling. Similar to structural steels, some level of preheat is usually recommended as a precaution for driving off any surface moisture, and a level of 120°C (248°F) is usually considered a safe minimum. For most martensitic stainless steels, this level of preheat and interpass temperature will also allow full transformation to martensite during weld cooling, as shown in situation A in Figure 4.15.

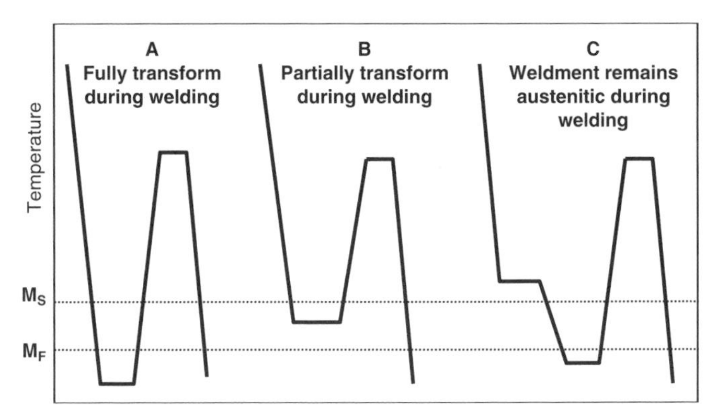

Figure 4.15 Transformation behavior during weld cooling and subsequent postweld heat treatment.

If preheat and interpass temperatures are maintained below the M_F value (situation A in Figure 4.15), the weld metal and HAZ will transform to martensite during the welding process. This will allow for tempering of the martensite by subsequent passes. If the interpass temperature is between M_S and M_F, some austenite will remain in the microstructure and be reheated to the tempering temperature (situation B in Figure 4.15). This will result in the formation of untempered martensite upon cooling from the tempering temperature. With thick-section welds or high-carbon alloys, it is sometimes advisable to maintain the interpass temperature above M_S, as shown in situation C. This will prevent transformation to martensite during welding and avoid hydrogen cracking. Following welding, the entire structure is cooled below M_F and then tempered.

It is important to know the transformation temperatures of the base and filler metals that are being used in order to properly control preheat and interpass temperatures and avoid the unexpected presence of untempered martensite, as in situation B. This can be accomplished by using relationships to calculate the M_S temperature, such as those shown in Table 4.3. Again, precautions should be taken in the use of these relationships because of the variance among them and because some were developed for low-alloy steels, not specifically for martensitic stainless steels. It is recommended that a minimum 50°C "safety factor" be used when calculating M_S.

An early formula for low-alloy steels by Payson and Savage [4] was the basis for a later formula for martensitic stainless steels by Irvine et al. [5]. Steven and Haynes [6], and Andrews [7] separately developed predicting formulas for low-alloy steels, to which formulas Kung and Rayment [8] recommended adding factors for silicon and cobalt to obtain good agreement with measured M_S temperatures of high-alloy martensitic steels. Gooch [3] recommended a formula specifically for martensitic stainless steels. Self et al. [9] applied regression analysis to data from 16 different investigations to obtain a formula including interactions among compositional factors. The list in Table 4.3 is by no means exhaustive. The eight predicting equations mentioned above are also given in Table 4.3. Each prediction is developed from the steel or weld metal composition and the constant and coefficients of Table 4.3 using the following pattern:

$$M_S(°C) = \text{constant} + \%C \times C\ \text{coefficient} + \%Mn \times Mn\ \text{coefficient} + \cdots \quad (4.2)$$

Note that most of the coefficients are negative, so the M_S temperature predicted is always much lower than the initial constant. The case study in section 4.7 compares predictions of the various equations for a number of alloys.

PWHT presents special problems when high-nickel martensitic stainless steels are welded. A case in point is CA-6NM or 410NiMo. With about 4.5% Ni in this alloy, the A_{C1} (the temperature at which austenite begins to form on heating of the martensite) is reduced to around 635°C (1175°F). This effectively eliminates final PWHT at this and higher temperatures because cooling causes fresh martensite to form from any austenite that is present. CA-6NM and 410NiMo are often used as alloys for handling crude oil before it is scrubbed of acidic impurities, including water, H_2S, CO_2, and chlorides. In the contaminated condition, the crude is often described as *sour*. This environment is sufficiently corrosive to "charge" the steel with hydrogen in service, and hard welds have experienced cracking due to the hydrogen charging. The petroleum industry normally sets a maximum hardness

TABLE 4.5 Double Tempering of 410NiMo Weld Metal

Weld Metal Composition (wt%)						
C	Mn	Si	Cr	Ni	Mo	N
0.034	0.62	0.24	12.73	3.87	0.57	0.017

Hardness, R_C		
As-Welded	Tempered 2 hours at 675°C (1250°F)	Tempered 2 hours at 675°C (1250°F), Air Cooled, and Tempered 4 hours at 615°C (1140°F)
34	27	18

Tensile Strength		Yield Strength		Elongation in 2 in. (%)	CVN Absorbed Energy	
MPa	ksi	MPa	ksi		J at −75°C	ft-lb at −103°F
757	110	584	85	20	59	44

limit of 22 HRC for the weld, which is difficult to achieve by tempering below the A_{C1} temperature. One solution found has been to use a double-tempering treatment. The first temper is done above the A_{C1} but below the A_{C3} temperature where austenite formation is complete, followed by cooling to ambient temperature, and the second temper is done below the A_{C1}. The first temper causes a fraction of the weld (and probably also of the base metal) to transform to austenite, but at the same time, it heavily tempers the martensite that does not transform. Cooling to ambient temperature is vital because it causes the fraction of the weld that was austenitized during the first temper to transform again to martensite. This new martensite is then softened by the second tempering treatment. Table 4.5 describes a 410NiMo weld metal, the double temper used, and the resulting weld properties.

4.4 MECHANICAL PROPERTIES OF WELDMENTS

The required minimum mechanical properties of martensitic stainless steel weld metals based on AWS A5.4-92 are presented in Table 4.6. Note that only two weld filler metals, E410 and E410NiMo, fall under this AWS specification and that the properties listed are representative of the tempered weld metal. The relatively low tensile strength of the E410 weld metal reflects the severe tempering treatment called out in this specification. High tensile strengths [up to 625 MPa (90 ksi), or even much higher] can be achieved with less severe tempering.

4.5 WELDABILITY

Because of the presence of untempered martensite following welding, the martensitic stainless steels may be susceptible to hydrogen-induced cracking. The use of

TABLE 4.6 Mechanical Properties of Weld Metals

AWS Classification	UNS No.	Tensile Minimum Strength MPa	ksi	Minimum Elongation in 50 mm (2 in.) (%)	Postweld Heat Treatment
E410-XX	W41010	520	75	20	[a]
E410NiMo-XX	W41016	760	110	15	[b]

Source: ANSI/AWS A5.4-92.

[a]Heat to 730 to 760°C (1350 to 1400°F), hold for 1 hour, furnace cool at a rate of 55°C (100°F) per hour to 315°C (600°F) and air-cool to ambient. (This is a very severe temper.)

[b]Heat to 595 to 620°C (1100 to 1150°F), hold for 1 hour, and air-cool to ambient.

preheat and postweld heat treatment is generally recommended when welding these alloys. These thermal treatments also reduce the residual stresses. Low hydrogen processes and practice reduce the amount of hydrogen uptake during welding and are essential when joining martensitic stainless steels.

Toughness and ductility are important material properties. Materials that have poor toughness are potentially subject to catastrophic brittle failure under dynamic loading. Materials that lack ductility are also prone to sudden catastrophic failure. PWHT can help improve the toughness and ductility of welded martensitic stainless components by tempering the martensitic structure. Table 4.7 presents tensile test results for several martensitic stainless steel weld metals produced by submerged arc welding. The alloys are all used for overlay of continuous caster rolls. The reduction in hardness and strength, and increase in ductility, with increasing PWHT temperature are readily apparent from the data. It is noteworthy that the alloys containing a small vanadium addition soften more slowly as the PWHT temperature is raised. This is undoubtedly due to the secondary hardening effect of vanadium carbides in the PWHT temperature range.

4.5.1 Solidification and Liquation Cracking

Most martensitic stainless steels solidify as ferrite, and hence their susceptibility to weld solidification cracking is low. However, certain factors are known to increase the likelihood of solidification cracking. These factors include the presence of niobium in the alloy and very low Mn levels. Very high carbon martensitic stainless steels may solidify as austenite, which renders them more sensitive to solidification cracking. Cracking has been observed in Nb-bearing 12 wt% Cr alloys, presumably due to the segregation of Nb. Liquation cracking is rare in martensitic stainless steels.

4.5.2 Reheat Cracking

Reheat cracking occurs during the heating cycle imposed on a weld by PWHT or heating of previous passes in multipass weldments. Molybdenum has been associated with reheat cracking in these steels. Impurities such as sulfur, phosphorus, antimony, tin, boron, and copper have also been linked to this cracking phenomenon. Minimizing the

TABLE 4.7 Effect of PWHT on Martensitic Stainless Steel Weld Metals

Alloy	410	410NiMo[a]	420	423L[a]	423Cr[a]	424A[a]
% C	0.08	0.05	0.23	0.15	0.15	0.09
% Mn	0.8	0.8	1.2	1.2	1.2	0.8
% Si	0.4	0.5	0.4	0.4	0.4	0.4
% Cr	12.5	13.0	13.0	11.5	13.5	13.0
% Ni	—	2.0	—	2.0	2.0	4.5
% Mo	—	1.0	—	1.0	1.0	1.0
% V	—	—	—	0.15	0.15	—
Hardness, Rockwell C						
As-welded	26	36	52	43	46	43
2 hours at 425°C (800°F)	25	39	48	42	45	41
2 hours at 480°C (900°F)	25	38	48	46	46	39
2 hours at 535°C (1000°F)	21	29	36	38	38	35
2 hours at 600°C (1100°F)	13	25	30	33	34	31
2 hours at 650°C (1200°F)	10	19	27	32	32	28
Tensile strength (ksi)						
After 425°C (800°F)	159	170	229	203	212	183
After 480°C (900°F)	164	159	198	205	205	184
After 535°C (1000°F)	118	132	151	168	172	149
After 600°C (1100°F)	111	124	141	160	156	138
After 650°C (1200°F)	104	117	128	153	153	143
Yield (ksi)						
After 425°C (800°F)	129	134	178	169	169	153
After 480°C (900°F)	121	132	125	149	163	156
After 535°C (1000°F)	97	115	118	143	142	127
After 600°C (1100°F)	94	105	117	144	125	113
After 650°C (1200°F)	86	89	106	124	123	106
Elongation (%)						
After 425°C (800°F)	3	7	2	4	6	10
After 480°C (900°F)	6	14	3	8	2	12
After 535°C (1000°F)	16	17	15	12	10	11
After 600°C (1100°F)	17	19	15	14	11	14
After 650°C (1200°F)	20	18	17	14	12	11

Source: Data from the files of the Lincoln Electric Company.

[a]Common name, not a standard ASTM composition given in Appendix 1.

impurity content of the steel, increasing heat input and eliminating stress concentrations can avoid reheat cracking. In general, the martensitic stainless steels are not susceptible to reheat, or postweld heat treatment, cracking that is common in many high-strength low-alloy (HSLA) steels that contain Cr, Mo, and V.

4.5.3 Hydrogen-Induced Cracking

Hydrogen-induced cracking (HIC) is a function of composition, hydrogen content, microstructure, and restraint. If one of these can be controlled, HIC can be avoided.

Low-hydrogen welding practice and the application of appropriate preheat and interpass temperature control will help to reduce hydrogen content. The combination of preheat and heat input control will also slow the weld cooling rate and allows some tempering of the martensite. Very slow cooling from the interpass temperature to below the M_F temperature is also helpful. For example, when a continuous caster roll is overlaid with martensitic stainless steel, the welded roll is usually removed from the welding fixture and covered with insulating material on all sides, to obtain cooling to about 100°C (212°F) in 16 to 24 hours.

In some cases, austenitic stainless steel filler metals can be used with the martensitic grades. This results in a two-phase austenite plus ferrite weld metal that has high solubility for hydrogen and improved toughness and ductility. These weld metals may have lower strength than the base metal and must be accounted for in the design. A common application of austenitic filler metal is for welding 410 stainless steel, which is in the annealed condition before welding and which is not to be given a PWHT. The annealed 410 is only required by ASTM A240 specifications to meet 450 MPa (65 ksi) minimum tensile strength and 210 MPa (30 ksi) minimum yield strength, so a Type 309L filler metal can be used for this application because its mechanical properties are higher than those required of the Type 410 base metal. GMAW is the most common way to do this, using ER309LSi filler metal, because low heat input can be used to minimize the thickness of the hard HAZ and because GMAW is normally a very low hydrogen process. Then the overall joint properties, in the as-welded condition, are sufficient for most applications.

4.6 SUPERMARTENSITIC STAINLESS STEELS

The supermartensitic stainless steels were introduced in the 1990s as low-cost alternatives to austenitic and duplex stainless steels in sub-sea pipeline applications. These steels have comparable properties and superior weldability to the standard martensitic grades. Because of their relatively low alloy content, they are also less expensive than the austenitic and duplex alloys. These steels are currently used in the petrochemical industry for oil and gas collection pipelines (or flowlines) that contain product with high CO_2 and H_2S content prior to separation and treatment.

The improvement in weldability relative to conventional martensitic grades is achieved primarily through the reduction of carbon to 0.02 wt% or less. This allows the formation of a "soft," low-carbon martensite that is more resistant than standard martensitic grades to hydrogen-induced cracking. Tempering is still required to optimize properties, although these steels can, in some instances, be used in the as-welded condition. To compensate for the reduction in carbon, nickel is added to promote austenite formation and to expand the gamma loop region of the phase diagram. This improves the hardenability of the material. Molybdenum is added to improve corrosion resistance, while titanium may be used to stabilize the carbon through the formation of TiC and also acts as a grain refiner. Marshall and Farrar [15] surveyed the supermartensitic stainless steels and subdivided them into three categories: *lean, medium*, and *high alloys*.

Table 4.8 lists approximate compositions representative of these three categories of supermartensitic stainless steels. These steels are evolving rapidly, with new proprietary

TABLE 4.8 Approximate Compositions of Lean, Medium, and High-Alloy Supermartensitic Stainless Steels

Alloy Level	Nominal Composition (wt %)								
	C	Mn	Si	Cr	Ni	Mo	Cu	N	Other
Lean	0.01	1.5	0.2	11	1.5	—	0.5	<0.01	—
Medium	0.01	0.5	0.2	13	4.5	1	0.5	0.05[a]	Ti[b]
High	0.01	0.5	0.2	12	6	2.5	0.2	0.05[a]	Ti or V[b]

Source: Marshall and Farrar [15].

[a]Some contain <0.01%, others as much as 0.08%.

[b]Trace up to 0.3%.

compositions appearing frequently. Industry standardization of supermartensitic stainless steels is lacking at the time of this writing. The mechanical properties are comparable to those of the standard martensitic stainless steel grades. Specified minimum yield strengths range from 625 to 760 MPa (90 to 110 ksi), tensile strength from 830 to 900 MPa (120 to 130 ksi), and tensile elongation (ductility) from 18 to 25%. These properties represent the quenched and tempered condition of the base metal. The supermartensitics solidify as ferrite, so they are very resistant to weld solidification cracking. Hydrogen cracking susceptibility is also low, due to the low carbon content and "soft" martensite that forms. As-welded hardness levels are generally below 30 HRC (about 300 H_V).

Matching composition consumables have been used experimentally with these alloys to match strength and corrosion resistance with the base metal, but limited application of matching consumables has been put into service to date. Duplex stainless steels, such as alloy 2209, can also be used to produce a tough, ductile weld deposit, but the strength will undermatch the base metal. The higher strength 25 wt% Cr *superduplex stainless steel* filler metals, known by trade names such as Zeron 100, can provide near-matching weld metal strength in the as-welded condition, and many miles of sub-sea pipelines have been laid in the North Sea and other places using this approach. Originally, some pipelines were put into service in the as-welded condition, but intergranular corrosion was observed in the HAZ [16]. It was found that a very short PWHT [650°C (1200°F) for 5 minutes followed by water quench] solved the problem of intergranular corrosion, without causing undesirable precipitation in the 22 wt% Cr duplex stainless steel root pass or in the 25% Cr superduplex fill passes. At the time of this writing, this approach appears to be commonly used for supermartensitic stainless steel offshore pipelines. It should be noted that duplex stainless steel filler metals generally cannot be used when more conventional longer-time PWHT is used, due to embrittlement by intermetallic compound precipitation during PWHT. This issue is described in more detail in Chapter 7.

As with the standard grades of 12 to 13% Cr martensitic stainless steels, the supermartensitic alloys are primarily martensitic as-welded but may contain some ferrite in the room-temperature microstructure. Because of the high Ni and very low carbon contents of these alloys, the Balmforth diagram (Figures 3.22 and 4.12) does not predict the ferrite content accurately and should not be used in its current form for these alloys.

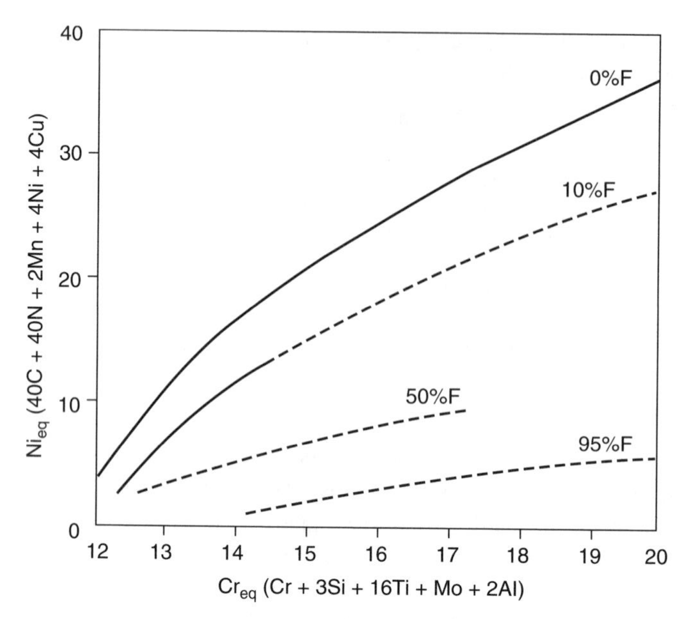

Figure 4.16 Ferrite content of the HAZ of supermartensitic stainless steels. (From Gooch et al. [17].)

Gooch and co-workers at TWI developed a *preliminary* diagram (Figure 4.16) to predict the ferrite content in the HAZ of low-C 13 wt% Cr steels. The number of alloys evaluated was small, particularly those with high Ni contents, so this diagram should be considered qualitative rather than a rigorous predictor of microstructure. Also, since this diagram was based on HAZ microstructure, its applicability to weld metals is questionable.

The high Ni content of the medium, and especially of the high-alloy, supermartensitic stainless steels, dramatically reduces the A_{C1} temperature, the temperature at which austenite starts to forming on heating. Based on the the A_{C1} relationship developed at TWI, the A_{C1} for alloys containing 4% Ni may be as low as 500°C (930°F). This can pose a problem during PWHT, since temperatures in excess of 600°C (1110°F) are normally used to temper martensitic steels. Below 600°C (1110°F), diffusion rates are relatively low and tempering times may be quite long. The following relationship was proposed by TWI to estimate the A_{C1} for 13 wt% Cr steels with less than 0.05 wt% C:

$$A_{C1}(°C) = 850 - 1500(C + N) - 50Ni - 25Mn + 25Si + 25Mo + 20(Cr - 10) \quad (4.3)$$

Marshall and Farrar [15] have indicated that the actual A_{C1} temperature for lean alloys is on the order of 650°C (1200°F) , while the high alloys are on the order of 630°C (1170°F). These temperatures are significantly higher than that the TWI relationship predicts. Based on this discrepancy, the selection of postweld tempering temperatures should be considered carefully and verified experimentally to prevent unexpected hardening of the structure due to partial transformation to austenite.

Note that Ni has a very potent effect but that Mo can be used to counteract the reduction in A_{C1}. If austenite is re-formed during PWHT, the properties of the material may be degraded. However, it should be noted that the medium- and especially the high-alloy supermartensitic stainless steels often contain considerable retained austenite as supplied, which has a beneficial effect on toughness [15].

PWHT is usually recommended when matching filler metal is used, in order to temper the martensite, thereby improving toughness and ductility while reducing strength 10 to 20%. As mentioned previously, the high Ni content reduces the A_{C1} temperature and therefore restricts the PWHT temperature to below this value. Tempering above the A_{C1} will result in a reformation of austenite and may result in subsequent loss in properties. Any austenite that forms during tempering above the A_{C1} will transform to "fresh" martensite upon cooling unless it is retained to ambient temperature. It should be noted that the 5-minute 650°C (1200°F) PWHT commonly employed in the fabrication of supermartensitic stainless steel pipelines is above the A_{C1} temperature of many of these steels.

The effect of tempering time and temperature on softening of the supermartensitics is shown in Figure 4.17. The combined effect of time and temperature is shown again using a Larson–Miller parameter. Note that the rate of softening is the same for the various carbon levels plotted and that the higher-carbon alloys actually show an aging (hardening) effect at high values of P. Figure 4.18 shows the effect of tempering temperature on austenite formation. Note that when tempered at 600°C (1110°F) as much as 30% austenite can be retained in the microstructure of a 13Cr–6Ni steel.

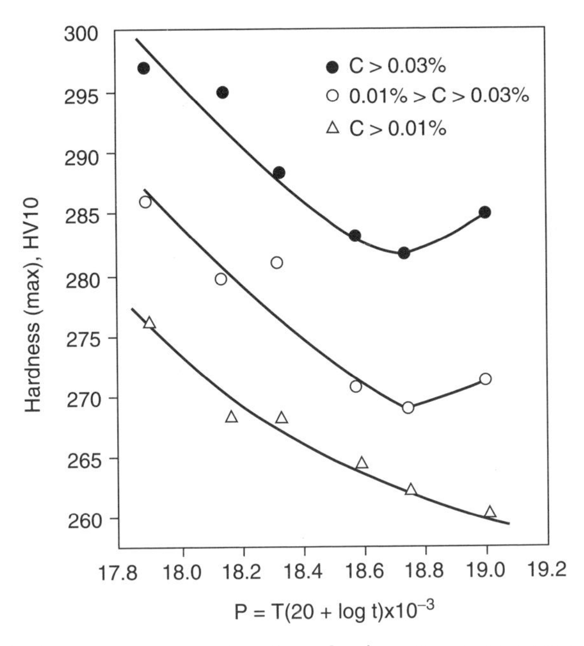

Figure 4.17 Hardness variation as a function of a time–temperature tempering parameter. (From Gooch et al. [17].)

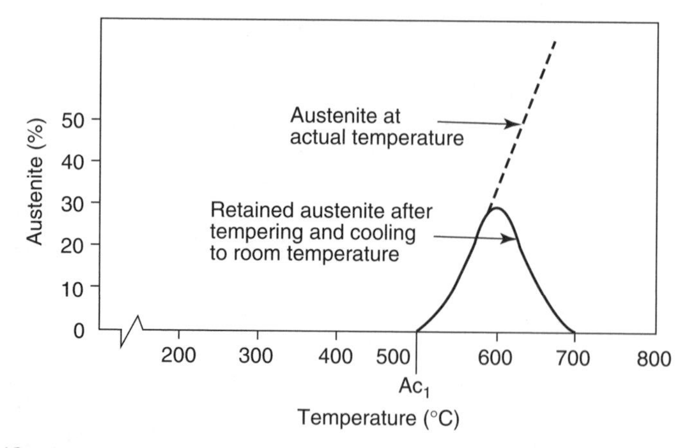

Figure 4.18 Austenite content both at elevated temperature and retained at room temperature in 13% Cr supermartensitic alloys. (From Gooch et al. [17].)

It is anticipated that the supermartensitics will continue to evolve as important engineering materials in the oil and gas transmission industry as well as in other applications. The welding metallurgy and microstructure–property relationship of these steels continues to be a subject of considerable research activity at the time of this writing. Advances in the welding metallurgy and weldability of these steels will be a key issue in their widespread implementation.

4.7 CASE STUDY: CALCULATION OF M_S TEMPERATURES OF MARTENSITIC STAINLESS STEELS

The predicting equations given in Table 4.3 produce a variety of results. Table 4.9 lists nominal compositions for several martensitic stainless steels and for one special alloy used for resurfacing continuous caster rolls. Then it shows predicted M_S temperatures according to each of the equations of Table 4.3 for these alloys. It is probably safe to say that the Payson equation should be discounted for martensitic stainless steels because its predictions seem clearly out of line for all of these alloys. For example, the M_S temperature for 410 stainless has been determined experimentally to be about 330°C (630°F), and for 420 stainless, about 300°C (570°F) [18]. This illustrates the risks in applying predicting equations without knowing their origin—the Payson relation, as noted above, was developed for low-alloy steels, not for stainless steels.

Given the differing results predicted above, a strategy is needed for making a selection of which prediction of M_S temperature should be used in deciding on minimum preheat and interpass temperature. Opting for the most conservative temperature—that least likely to allow martensite to form before welding is finished—probably makes sense. That would be the highest M_S temperature predicted for any given alloy.

Another use for the prediction of M_S temperatures concerns cooling the weldment before PWHT. It should be remembered that PWHT does not accomplish tempering of martensite if the martensite has not yet formed. So, in general, the weldment

Table 4.9 Predicted Martensite Start Temperatures for Martensitic Stainless Steels

Type:	410	414	410NiMo	420	422	431	440A	Caster Roll Alloy
UNS No.:	S41000	S41400	S41500	S42000	S42200	S43100	S44002	—
% C	0.11	0.08	0.03	0.20	0.22	0.10	0.70	0.18
% Mn	0.50	0.50	0.75	0.50	0.75	0.50	0.50	1.10
% Si	0.50	0.50	0.30	0.50	0.25	0.50	0.50	0.40
% Cr	12.5	12.5	12.75	13.0	11.8	16.0	17.0	13.5
% Ni	—	2.0	4.5	—	0.75	2.0	—	2.7
% Mo	—	—	0.75	—	1.1	—	—	1.0
% V	—	—	—	—	0.25	—	—	0.20
% W	—	—	—	—	1.1	—	—	—
% Co	—	—	—	—	—	—	—	2.0
Reference	Predicted Martensite Start Temperature (°C)							
Payson and Savage [4]	92	68	28	50	47	−37	−221	−23
Irvine et al. [5]	264	245	200	213	171	176	−92	129
Steven and Haynes [6]	280	260	213	229	196	191	−76	143
Kung and Rayment [8] (modification of Haynes [6])	276	256	211	225	194	187	−80	160
Andrews [7]	326	303	264	282	259	253	22	211
Kung and Rayment [8] (modification of Andrews [7])	322	300	262	278	257	249	18	228
Gooch [3]	347	289	219	297	271	242	5	200
Self et al. [9]	331	265	163	320	286	218	302	190

should be cooled to or below the martensite finish (M_F) temperature before PWHT. The M_F temperature is generally considered to be about 100°C (212°F) below the M_S temperature, or even lower. So, in selecting the temperature to which a weldment should be cooled before undertaking PWHT, an appropriate choice is at least 100°C (180°F) below the M_S temperature. If a conservative choice (least risky) for a maximum temperature, to which a weldment is to be cooled before PWHT, is to be based on the equations for prediction of the M_S temperature given in Table 4.3, that choice would be 100°C (180°F) below the *lowest* of the M_S temperatures predicted. Note, however, in Table 4.9 that all of the equations, except that of Self et al. [9], seem to produce unbelievably low M_S temperature predictions for the very high carbon 440A composition. Knowing the M_S temperature, or predicting it by these various equations, can be quite useful in selecting preheat and interpass temperature for martensitic stainless steels. It is also useful for determining the temperature to which a martensitic stainless steel weldment should be cooled before PWHT.

REFERENCES

1. Castro, R., and Tricot, R. 1962. Études des transformations isothermes dans les aciers inoxydables semi-ferritiques á 17% de chrome, *Memoires Scientifiques de la Revue de Metallurgie*, Part 1, 59:571–586; Part 2, 59:587–596.
2. McGannon, H. E. 1971. *The Making, Shaping, and Treating of Steel*, 9th ed., U.S. Steel Corporation, Pittsburgh, PA, p. 1176.
3. Gooch, T. G. 1977. Welding martensitic stainless steels, *Welding Institute Research Bulletin*, 18(12):343–349.
4. Payson, P., and Savage, C. H. 1944. Martensite reactions in low alloy steels, *Transactions of the American Society for Metals*, 33:261–275.
5. Irvine, K. J., Crowe, D. J., and Pickering, F. B. 1960. The physical metallurgy of 12% chromium steels, *Journal of the Iron and Steel Institute*, 195(8):386–405.
6. Steven, W., and Haynes, A. G. 1956. The temperature of formation of martensite and bainite in low-alloy steels, *Journal of the Iron and Steel Institute*, 183(8):349–359.
7. Andrews, K. 1965. Empirical formulae for the calculation of some transformation temperatures, *Journal of the Iron and Steel Institute*, 203:721–727.
8. Kung, C. Y., and Rayment, J. J. 1982. An examination of the validity of existing empirical formulae for the calculation of MS temperature, *Metallurgical Transactions A*, 13A(2): 328–331.
9. Self, J. A., Olson, D. L., and Edwards, G. R. 1984. The stability of austenitic weld metal, in *Proceedings of IMCC*, Kiev, Ukraine.
10. ASM. 1982. *Engineering Properties of Steels*, AMS International, Materials Park, OH.
11. Castro, R. J., and de Cadenet, J. J. 1974. *Welding Metallurgy of Stainless and Heat-Resisting Steels*, Cambridge University Press, Cambridge.
12. Lippold, J. C. 1984. The effect of postweld heat treatment on the microstructure and properties of the HAZ in 12Cr–1Mo–0.3V weldments, in *Proceedings of the Topical Conference on Ferritic Alloys for Use in Nuclear Energy Technologies*, Metallurgical Society of AIME, Warrendale, PA, pp. 497–506.
13. Balmforth, M. C., and Lippold, J. C. 2000. A new ferritic–martensitic stainless steel constitution diagram, *Welding Journal*, 79(12):339s–345s.
14. Lippold, J. C. 1981. Transformation and tempering behavior of 12Cr–1Mo–0.3V martensitic stainless steel weldments, *Journal of Nuclear Materials*, 104(3):1127–1131.
15. Marshall, A. W., and Farrar, J. C. M. 2001. Welding of ferritic and martensitic 11–14% Cr steels, *Welding in the World*, 45(5/6): 32–55.
16. Howard, R. D., Martin, J. W., Evans, T. N., and Fairhurst, D. 2003. Experience with 13% Cr martensitic stainless steels in the oil and gas industry, Paper PO358, *Stainless Steel World, 2003*, KCI Publishing, Zutphen, The Netherlands.
17. Gooch, T. G., Woolin, P., and Haynes, A. G. 1999. Welding metallurgy of low carbon 13% chromium martensitic steels, in *Proceedings of Supermartensitic Stainless Steels, 1999*, Belgian Welding Institute, Ghent, Belgium, pp. 188–195.
18. Unterweiser, P. M., Boyer, H. E., and Kubbs, J. J., eds. 1982. *Heat Treater's Guide: Standard Practices and Procedures for Steel*, ASM International, Materials Park, OH, pp. 424, 432.

CHAPTER 5

FERRITIC STAINLESS STEELS

Ferritic stainless steels are classified as such because the predominant metallurgical phase present is ferrite. These alloys possess good resistance to stress corrosion cracking, pitting corrosion, and crevice corrosion (particularly in chloride environments). They are used in a variety of applications where corrosion resistance, rather than mechanical properties (strength, toughness and ductility), is the primary service requirement. Low-chromium (10.5 to 12.5 wt%) grades are used for applications such as automotive exhaust systems, where resistance to general corrosion is superior to carbon steels. Medium- and high-chromium grades are used in more aggressive corrosion environments. Superferritic alloys are used in the chemical processing and pulp and paper industries, where resistance to corrosion in severe oxidizing media is required. High-chromium grades are also used in high-efficiency furnaces.

Historically, ferritic stainless steels had been used in the greatest tonnage in applications that do not require welding. For example, the medium-Cr grades are used extensively for automotive trim and other decorative/architectural applications. Since the early 1980s, the use of low- and medium-Cr grades for automotive exhaust systems has increased dramatically. Since exhaust tubes and connections are welded, the weldability of ferritic stainless steels has received increased attention.

A number of high-Cr grades have been developed over the years for use in demanding environments, such as chemical plants, pulp and paper mills, and refineries. These alloys possess superior corrosion resistance relative to the austenitic and martensitic

Welding Metallurgy and Weldability of Stainless Steels, by John C. Lippold and Damian J. Kotecki
ISBN 0-471-47379-0

grades. However, they are relatively expensive and difficult to fabricate. The weldability of alloys with medium (16 to 18%) and high (>25%) chromium content has been the subject of considerable research. Ferritic stainless steels are generally limited to service temperatures below 400°C (750°F), due to the formation of embrittling phases. High-Cr grades are particularly susceptible to 475°C (885°F) embrittlement.

Metallurgically, welds in these alloys are primarily ferritic, although martensite may be present under certain conditions, and precipitation of carbides and nitrides is common. The principal weldability issue with the ferritic grades is maintaining adequate toughness and ductility in the as-welded condition. This chapter provides insight into microstructure evolution in the weld metal and HAZ of these alloys and relates this microstructure to their mechanical and corrosion properties.

5.1 STANDARD ALLOYS AND CONSUMABLES

Over the years ferritic stainless steels have evolved in three generations relative to their general composition ranges. Early (first-generation) alloys were mainly the medium-chromium types with relatively high carbon contents. These steels are not 100% ferritic because they form some austenite in their structures during solidification and cooling or when heated to elevated temperature. The austenite that is present at elevated temperature transforms to martensite on cooling to room temperature. Second-generation ferritic alloys were developed to minimize the formation of martensite in the ferrite structure and improve weldability. They have lower carbon contents and often contain a stabilizing element (Nb or Ti) that “ties up” carbon and nitrogen, thereby promoting ferrite stability.

Third-generation ferritic stainless steels have high chromium, low interstitial (carbon + nitrogen) levels, and low impurity levels. These grades are often developed for specific applications under various trade names. The high-purity grades of stainless steels have superior corrosion resistance with moderate toughness and ductility. When welding these high-purity grades, extreme care must be taken to avoid pickup of undesirable elements, particularly nitrogen and oxygen, and to minimize grain growth. The compositions of several ferritic stainless steels are provided in Table 5.1. A more complete list of alloys is provided in Appendix 1. Note that many of these alloys have a 4XX designation, similar to that of the martensitic stainless steels, even though their microstructure and properties are different.

Castings of ferritic stainless steels have rather limited availability, and the composition ranges specified by ASTM A743 or ASTM A297 have very wide carbon ranges, so that it is possible to develop a predominately ferritic microstructure or a predominately martensitic microstructure in the as-cast condition. In the lower portion of the ASTM A743 carbon range, the cast alloy Grade CB-30 is similar to Type 442 wrought stainless steel. In the lower portion of the carbon range, the cast alloys ASTM A743 Grade CC-50 and ASTM A297 Grade HC are similar to wrought Type 446 stainless steel.

Most of the consumables used with the ferritic stainless steels are matching or near-matching compositions. A list of ferritic stainless steel consumables is provided in Table 5.2. These alloys can also be welded using austenitic stainless steel

TABLE 5.1 Composition of Standard Wrought Ferritic Stainless Steels[a]

Type	UNS No.	Composition (wt%)												
		C	Mn	P	S	Si	Cr	Ni	Mo	N	Cu	Al	Ti	Nb
		First Generation (Steels with Free Carbon)												
405	S40500	0.08	1.00	0.040	0.030	1.00	11.5–14.5	0.60	—	—	—	0.10–0.30	—	—
430	S43000	0.12	1.00	0.040	0.030	1.00	16.0–18.0	0.75	—	—	—	—	—	—
434	S43400	0.12	1.00	0.040	0.030	1.00	16.0–18.0	—	0.75–1.25	—	—	—	—	—
442	S44200	0.20	1.00	0.040	0.040	1.00	18.0–23.0	0.60	—	—	—	—	—	—
446	S44600	0.20	1.50	0.040	0.030	1.00	23.0–27.0	0.75	—	0.25	—	—	—	—
		Second Generation (Steels with Strong Carbide Formers)												
409[b]	S40900	0.08	1.00	0.045	0.030	1.00	10.50–11.75	0.50	—	—	—	—	6 × C − 0.75	—
409[c]	S40910	0.030	1.00	0.040	0.020	1.00	10.5–11.7	0.50	—	0.030	—	—	6 × C − 0.50	0.17
409[c]	S40920	0.030	1.00	0.040	0.020	1.00	10.5–11.7	0.50	—	0.030	—	—	8 × (C + N) min. 0.15–0.50	0.10 —
409[c]	S40930	0.030	1.00	0.040	0.020	1.00	10.5–11.7	0.50	—	0.030	—	—	0.05 min. Ti + Nb = [0.08 + 8 × (C + N)] − 0.75	—
436	S43600	0.12	1.00	0.040	0.030	1.00	16.0–18.0	—	0.75–1.25	—	—	—	—	5 × C − 0.80
439[d]	S43035	0.030	1.00	0.040	0.030	1.00	17.0–19.0	0.50	—	0.030	—	—	0.15[0.20 + 4(C + N)] − 1.10	—
468[e]	S46900	0.030	1.00	0.040	0.030	1.00	18.0–20.0	0.50	—	0.030	—	—	0.07–0.3	0.10–0.60

(*continued*)

TABLE 5.1 *(Continued)*

Type	UNS No.	Composition (wt%)												
		C	Mn	P	S	Si	Cr	Ni	Mo	N	Cu	Al	Ti	Nb
Third Generation (Steels with Very Low Carbon and/or Strong Carbide Formers)														
444	S44400	0.025	1.00	0.040	0.030	1.00	17.5–19.5	1.00	1.75–2.00	0.035	—	—	Ti + Nb = [0.20 + 4(C + N)] − 0.80	—
XM-27	S44627	0.010	0.40	0.020	0.020	0.40	25.0–27.5	0.50[f]	0.75–1.50	0.015	0.20[f]	—	—	0.05–0.20
25-4-4	S44635	0.025	1.00	0.040	0.030	0.75	24.5–26.0	3.5–4.5	3.5–4.5	0.035	—	—	Ti + Nb = [0.20 + 4(C + N)] − 0.80	—
29-4	S44700	0.010[g]	0.30	0.025	0.020	0.20	28.0–30.0	0.15	3.5–4.2	0.020[g]	0.15	—	—	—
29-4C	S44735	0.030	1.00	0.040	0.030	1.00	28.0–30.0	1.00	3.6–4.2	0.045	—	—	Ti + Nb = 0.20–1.00 Ti + Nb = 6 (C + N) min.	—
29-4-2	S44800	0.010[g]	0.30	0.025	0.020	0.20	28.0–30.0	2.00–2.50	3.5–4.2	0.020[g]	0.15	—	—	—

Source: Data from ASTM A176, A240, and A268.

[a]A single value is a maximum unless otherwise indicated.

[b]As given in ASTM A240/A240M-96a and prior versions.

[c]Beginning with ASTM A240/A240M-97a, the carbon in 409 was reduced and provision was made for three different stabilization approaches. Type 409 or UNS S40900 now includes S40910, S40920, and S40930.

[d]ASTM A240/A240M-00a and prior versions allowed up to 0.07% C in Type 439 under UNS S43035. ASTM A240/A240M-01 reduced the carbon of Type 439 to 0.030% maximum.

[e]Ti + Nb = 0.20 + 4(C + N) min., 0.80 max.

[f]Ni + Cu = 0.50% maximum.

[g](C + N) = 0.025% maximum.

TABLE 5.2 Composition of AWS Ferritic Stainless Steel Filler Metal Classifications[a]

Classification	UNS No.	Composition (wt%)										
		C	Mn	P	S	Si	Cr	Ni	Mo	Cu	Ti	Nb
E409Nb-XX	—	0.03	1.0	0.04	0.03	0.90	11.0–14.0	0.6	0.75	0.75	—	0.50–1.50
ER409	S40900	0.08	0.8	0.03	0.03	0.8	10.5–13.5	0.6	0.50	0.75	$10 \times C - 1.5$	—
ER409Cb	S40940	0.08	0.8	0.04	0.03	1.0	10.5–13.5	0.6	0.50	0.75	—	$10 \times C - 0.75$
E409TX-X	W41031	0.10	0.80	0.04	0.03	1.0	10.5–13.5	0.60	0.5	0.5	$10 \times C - 1.5$	—
E430-XX	W43010	0.10	1.0	0.04	0.03	0.9	15.0–18.0	0.6	0.75	0.75	—	—
E430Nb-XX	—	0.10	1.0	0.04	0.03	0.90	15.0–18.0	0.6	0.75	0.75	—	0.50–1.50
ER430	S43080	0.10	0.6	0.03	0.03	0.5	15.5–17.0	0.6	0.75	0.75	—	—
ER446LMo	S44687	0.015	0.4	0.02	0.02	0.4	25.0–27.5	[b]	0.75–1.50	[b]	—	—

Source: AWS A5.4, A5.9, A5.22.

[a] A single value is a maximum.

[b] Ni + Cu = 0.5% maximum.

consumables. This combination results in weld metals containing a mixture of austenite and ferrite. As described later in the chapter, this microstructure improves the toughness and ductility of the weld metal. However, in welding the third-generation ferritic stainless steels, selection of austenitic stainless steel filler metals should be done with caution, because a principal use of third-generation ferritic stainless steels is for resistance to chloride stress corrosion cracking (SCC). Many austenitic stainless steels and filler metals are susceptible to SCC in chloride-containing environments.

5.2 PHYSICAL AND MECHANICAL METALLURGY

The physical metallurgy of ferritic stainless steels has been studied extensively since the early 1940s. For a thorough treatment of this subject the reader is referred to comprehensive reviews by Thielsch [1] in 1951 and, more recently, by Demo [2] and Lacombe et al. [3]. The purpose of this section is to present the physical metallurgy background required for understanding the welding metallurgy of these steels, including a review of phase equilibria and microstructure, strengthening mechanisms, and embrittlement phenomena.

The Fe–Cr–C ternary system [4] can be used to describe the phase transformations that occur in the ferritic stainless steels. The pseudobinary diagram at 17%Cr is useful to illustrate and describe the physical metallurgy of these alloys. This diagram, shown in Figure 2.3, is presented again in Figure 5.1 with a nominal carbon

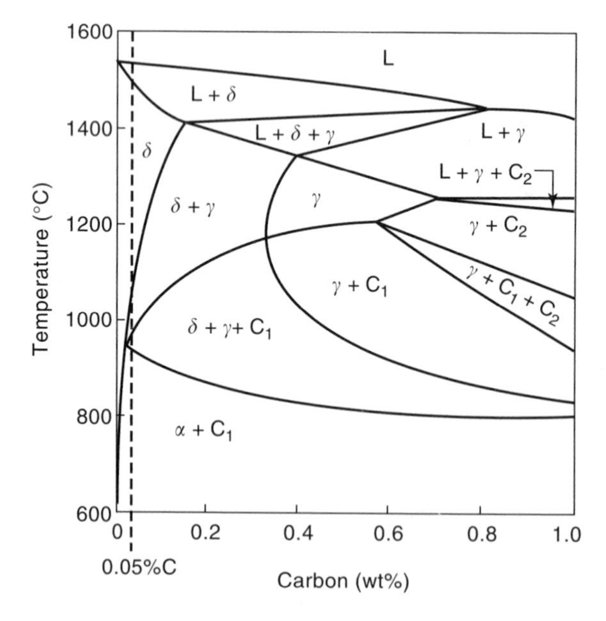

Figure 5.1 17% Cr pseudobinary phase diagram. (From Castro and Tricot [4].)

concentration of 0.05 wt% superimposed. This composition approximates that of the medium-chromium alloy, Type 430.

Note that primary solidification occurs as ferrite, and the diagram predicts that the structure will be fully ferritic at the end of solidification. It will remain ferritic in the solid state until it cools below approximately 1100°C (2010°F). At this temperature, some transformation to austenite will occur, and at slightly lower temperatures some $Cr_{23}C_6$ carbide will form. Under equilibrium cooling conditions, the austenite will transform to ferrite and carbide and the final structure will be a mixture of ferrite and $Cr_{23}C_6$ carbide.

Typical base metal microstructures of wrought ferritic stainless steels are shown in Figures 5.2 to 5.4. Figure 5.2 is representative of annealed Type 409 sheet, used widely for automotive exhaust applications. The microstructure is fully ferritic and contains scattered titanium nitride and/or carbide precipitates. Figure 5.3 shows the microstructure of Type 430 hot-rolled sheet that has been slowly cooled from a rolling temperature of approximately 850°C (1560°F). This microstructure consists of ferrite and carbides that form in the rolling direction of the sheet. Figure 5.4 is representative of the microstructure of a Type 430 grade that has been quenched from 1100°C (2012°F). This structure consists of martensite along the ferrite grain boundaries and intragranular carbide and nitride precipitates.

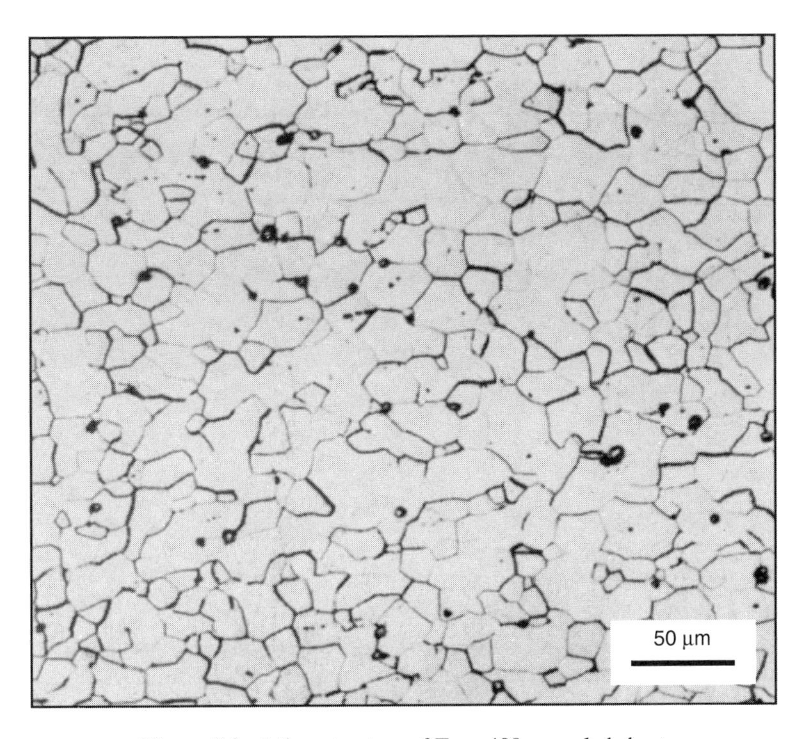

Figure 5.2 Microstructure of Type 409 annealed sheet.

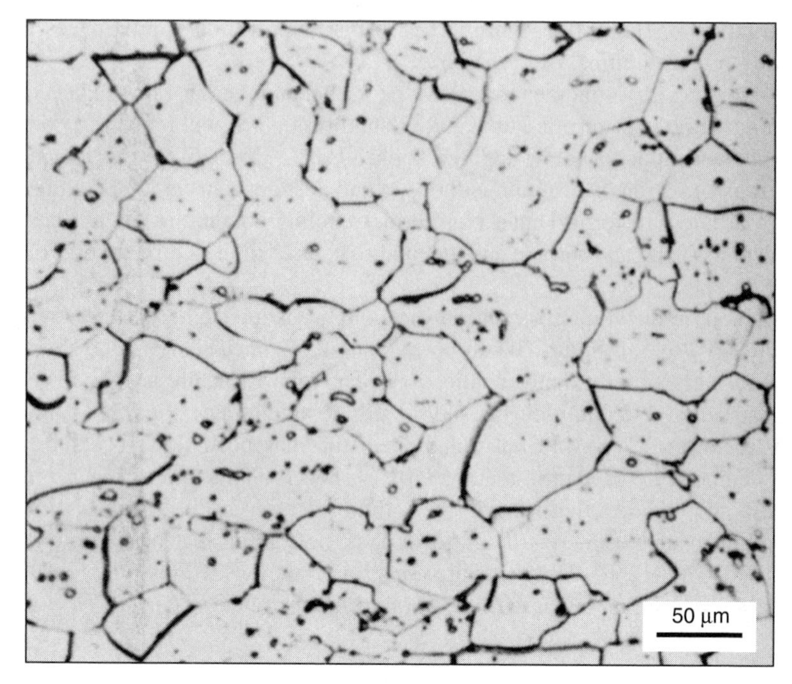

Figure 5.3 Microstructure of Type 430 hot-rolled sheet.

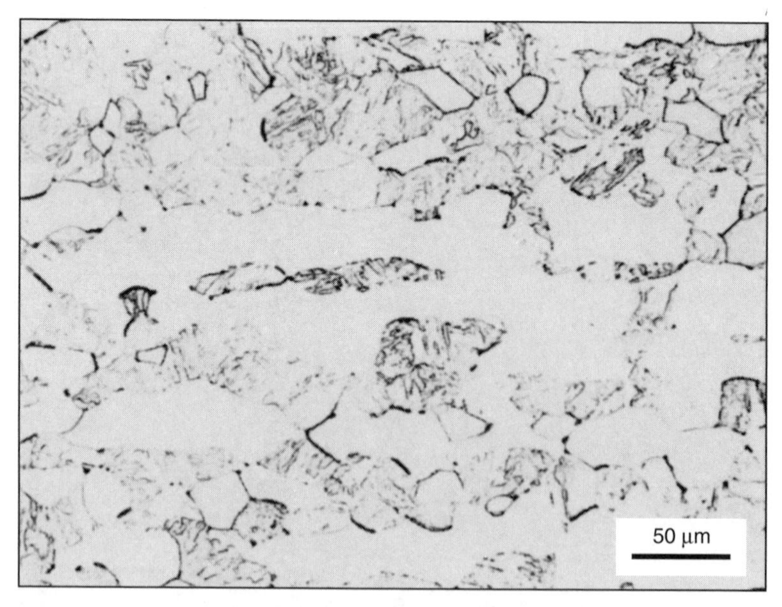

Figure 5.4 Microstructure of Type 430 annealed at 1100°C (2010°F) and quenched to room temperature.

5.2.1 Effect of Alloying Additions on Microstructure

The presence of other elements, either through intentional additions or as impurities, can significantly alter the shape and extent of the austenite regime or influence the microstructure of ferritic stainless steels. Nitrogen, normally present as an impurity rather than an intentional addition, has an effect similar to that of carbon in promoting austenite (i.e., expanding the austenite phase field). Baerlecken et al. [5] have demonstrated the effect of various levels of carbon + nitrogen on the expansion of the gamma loop in simple Fe–Cr alloys (Figure 2.2). Note, for example, additions of 0.04 wt% C and 0.03 wt% nitrogen result in a shift in the boundary of the austenite + ferrite phase field to over 20 wt% chromium. Thus, either reduction of carbon and nitrogen to extremely low levels (less than 100 ppm), or addition of alloying elements that promote ferrite formation, is necessary to maintain a primarily ferritic microstructure in low- and medium-chromium steels.

In addition to chromium, ferrite-promoting elements commonly added to ferritic stainless steels include silicon, titanium, niobium, molybdenum, and aluminum (see Table 5.1). Titanium and niobium are particularly useful in small concentrations due to their high affinity for both carbon and nitrogen, while aluminum is effective in combining with nitrogen. Aluminum is also added for improved oxidation resistance, particularly at elevated temperature. Silicon is normally added as a deoxidizer and to provide oxide-scaling resistance. Molybdenum is added to some alloys, especially the third generation of ferritic stainless steels, to improve corrosion resistance, particularly with respect to pitting.

Austenite-promoting elements (those that tend to expand the gamma loop) include manganese, nickel, and copper in addition to carbon and nitrogen. Manganese is traditionally added to control sulfur and therefore to improve castability and subsequent hot-working characteristics. Nickel and copper are not normally added to ferritic stainless steels, although small amounts of nickel may be effective in improving notch toughness [1,2]. A number of equivalency relationships have been developed to ensure that ferritic stainless steels are fully ferritic (i.e., free from martensite). These have been reviewed in Chapter 3 and the reader is referred to Section 3.5. Additional discussion regarding prediction of microstructure in ferritic stainless steel weldments is included later in this chapter.

5.2.2 Effect of Martensite

Under normal thermo-mechanical processing conditions, any austenite that forms at elevated temperature will generally transform to martensite upon cooling to room temperature (Figure 5.4). Only under very slow cooling, or isothermal holding at temperatures just below the austenite solvus [e.g., at about 900°C (1650°F) for the 0.05 wt% C–17 wt% Cr alloy of Figure 5.1], will the elevated temperature austenite transform to ferrite and carbides, as predicted by the equilibrium phase diagram. Martensite in ferritic stainless steels has been shown to have both beneficial and detrimental effects. The presence of martensite in appreciable quantities is purported to promote hydrogen-induced cracking, similar to that experienced in structural steels [6], although there are few data in the literature to support this. Martensite is also cited as a source of embrittlement due to its fracture and deformation characteristics relative to ferrite [1,7,8]. In sharp contrast to this, work by Hayden and Floreen [9] and Wright and Wood [10] has shown that duplex ferritic–martensitic microstructures based on the Fe–Cr–Ni and Fe–Cr–Mn systems, respectively, have superior impact toughness properties relative to either fully ferritic or fully martensitic steels of similar composition.

Depending on the carbon content of the alloy and volume fraction of martensite present, the martensite that forms in the ferritic stainless steels is typically of a low-carbon variety, with the hardness of the martensite itself at levels generally below 30 HRC. At elevated temperature, where the austenite is stable, carbon will partition from the ferrite to the austenite because of the higher solubility of carbon is austenite. For example, between 1000 and 1200°C (1832 and 2190°F) (Figure 5.1) the austenite carbon content ranges from 0.05 to over 0.3 wt% in a 17Cr–0.05C alloy. If the austenite carbon content reaches its equilibrium value at 1200°C (2190°F) and is then cooled rapidly to room temperature, the martensite hardness may approach 50 HRC. This would require complete dissolution of the original carbides and ample time for diffusion to occur.

In general, the elevated temperature microstructure does not reach equilibrium composition, and the martensite that forms does not reach these high levels. Thus, the loss of ductility and toughness usually associated with untempered martensite in structural steels with carbon contents exceeding 0.15 wt% is usually not an issue in these steels, and some martensite can be accommodated without a significant penalty with respect to mechanical properties. It will be seen later that martensite formation in the weld metal and HAZ of low- and medium-chromium ferritic stainless steels is quite common.

The presence of martensite in low-chromium ferritic stainless steels has been related to a loss of corrosion resistance [11]. Corrosion attack in simulated Type 409 HAZ microstructures in a boiling aqueous solution of 2% sulfuric acid containing 600 ppm Cu^+ ions was found to increase as a function of martensite content up to 20 vol%. There is also some evidence that the martensite–ferrite interface is a preferential site for intergranular stress corrosion cracking [12]. This is discussed in more detail in Section 5.6.

5.2.3 Embrittlement Phenomena

Based on the reviews of Thielsch [1] and Demo [2], there are three embrittlement phenomena that influence the mechanical properties of ferritic stainless steels: (1) 475°C (885°F) embrittlement, (2) sigma phase precipitation, and (3) high-temperature embrittlement. Detailed treatments of these phenomena can be found in the reviews referenced above. Although extensive literature is available regarding sigma phase precipitation and "475°C embrittlement," these are not normally a problem when welding ferritic stainless steels, since embrittlement due to these phenomena is normally associated with long exposure times at intermediate temperatures. As a consequence, *intermediate-temperature embrittlement* (ITE) of welded ferritic stainless steels is essentially insensitive to welding process/procedure selection and more an issue of engineering application. In practice, limiting service exposure temperatures to below 400°C (750°F) avoids the problem of ITE. Both of these embrittlement phenomena are accelerated as the chromium content of the alloy or filler metal increases, and special precautions may be required when considering postweld heat treatment of the high-chromium alloys. This is discussed in Section 5.7.

Reduction of mechanical and/or corrosion properties or catastrophic failure of welded ferritic stainless steels is, therefore, linked almost exclusively to *high-temperature embrittlement* (HTE), notch sensitivity, or a combination of the two. Due to the importance of HTE during welding of ferritic stainless steels and the influence of ITE on the service performance of these alloys, the metallurgical basis of these embrittlement phenomena and the influence of notch sensitivity are reviewed briefly in the following sections.

5.2.3.1 475°C Embrittlement Fe–Cr alloys containing from 15 to 70 wt% chromium can be severely embrittled when heated into the temperature range 425 to 550°C (800 to 1020°F). The metallurgical basis for embrittlement in this temperature range is still controversial. The predominant theory associates the onset of embrittlement with the formation of a coherent precipitate at temperatures below 550°C (1020°F) due to the presence of a miscibility gap in the Fe–Cr equilibrium phase diagram (Figure 2.1). Alloys aged below 550°C (1020°F) were found to form a chromium-rich ferrite (alpha-prime) and an iron-rich ferrite (alpha) [13–15]. The alpha-prime precipitate was found to be nonmagnetic, with a BCC crystal structure and contained 61 to 83% chromium.

The rate and degree of embrittlement is a function of chromium content, with high-chromium alloys embrittling in much shorter times and at slightly higher temperatures [16]. The lowest Cr ferritic stainless steels, such as Types 405 and 409, seem to be immune to 475°C embrittlement. In general, aging times of at least 100 hours are required to cause embrittlement in low- and medium-chromium alloys [17]. High-chromium alloys may exhibit loss of ductility and toughness after much shorter times. Alloying additions such as molybdenum, niobium, and titanium tend to accelerate the onset of 475°C embrittlement. The influence of these and other alloying and impurity elements are summarized in Table 5.3.

Cold work acts to promote the precipitation of alpha-prime and thus accelerates the onset of embrittlement. 475°C embrittlement also results in a severe reduction in corrosion resistance [18,19], probably due to selective attack of the iron-rich ferrite. The embrittlement can be eliminated and mechanical and corrosion properties restored to the unaged condition upon heating into the range 550 to 600°C (1020 to 1110°F) for a short time. Excessive time in this temperature range will produce sigma phase embrittlement, as described below.

5.2.3.2 Sigma and Chi Phase Embrittlement Sigma phase forms in Fe–Cr alloys containing 20 to 70 wt% chromium upon extended exposure in the temperature range 500 to 800°C (930 to 1470°F). As with 475°C embrittlement, alloys with higher

TABLE 5.3 Effect of Alloying Additions on 475°C Embrittlement of Iron–Chromium Alloys

Element	Effect
Aluminum	Intensifies
Carbon	No effect/intensifies
Chromium	Intensifies
Cobalt	Intensifies
Molybdenum	Intensifies
Nickel	Variable
Niobium	Intensifies
Nitrogen	Small effect/intensifies
Phosphorus	Intensifies
Silicon	Intensifies
Titanium	Intensifies

Source: Compilation by Demo [2].

Cr concentrations are more susceptible to sigma formation and the rate of formation is more rapid. In alloys containing less than 20 wt% chromium, sigma phase does not form readily, often requiring hundreds of hours of exposure in the critical temperature range. In high-Cr alloys, sigma formation is more rapid, requiring only a few hours of exposure in the sigma formation temperature range [16], as illustrated in Figure 5.5. Alloying additions such as molybdenum, nickel, silicon, and manganese shift the sigma formation range to higher temperatures, lower chromium concentrations, and shorter times. As with other precipitation phenomena, cold work accelerates the formation of sigma phase. The deleterious effects of sigma precipitation can be eliminated by heating for short times to temperatures above 800°C (1470°F). In the high-Cr high-Mo alloys (e.g., 29-4 and 29-4-2), Kiesheyer and Brandis [20] report that chi phase, variously described as $Fe_{36}Cr_{12}Mo_{10}$ or Fe_3CrMo, can form along with sigma, and these brittle intermetallic phases can be stable up to 900°C (1600°F) or higher.

5.2.3.3 High-Temperature Embrittlement High-temperature embrittlement (HTE) results from metallurgical changes that occur during exposure to temperatures above approximately $0.7T_m$ (melting temperature). Since this temperature range is well above the recommended service temperatures for ferritic stainless steels, HTE normally occurs during thermo-mechanical processing or welding. Exposure to these elevated temperatures can also result in a severe loss in corrosion resistance [2]. Susceptibility to HTE is influenced primarily by composition, particularly chromium and interstitial concentration, and grain size, as discussed in the following sections. Low-chromium and stabilized grades are relatively insensitive to HTE.

Composition Effects The level of interstitial elements, particularly carbon, nitrogen, and oxygen, has a strong influence on the HTE characteristics of ferritic stainless steels. At elevated temperatures the interstitial elements are present in a solid

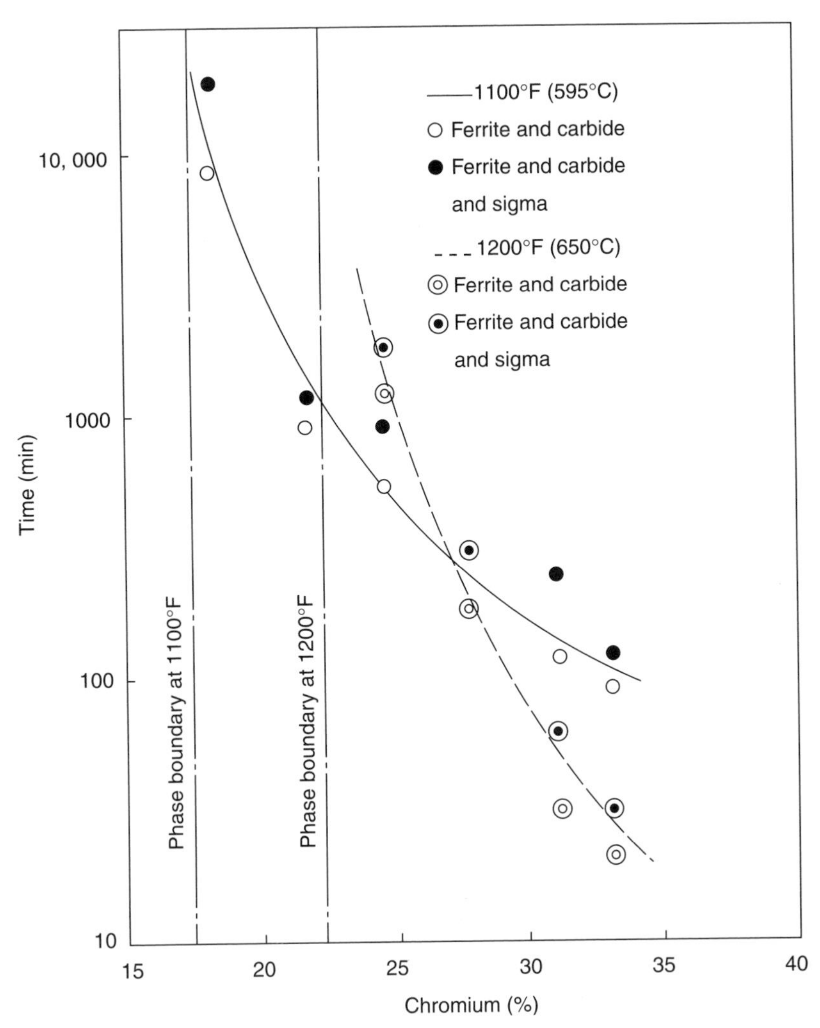

Figure 5.5 Effect of chromium concentration on the threshold time for sigma phase formation at 593 and 649°C (1100 and 1200°F). (From Kiesheyer and Brandis [20].)

solution of a ferrite or ferrite + austenite matrix. Upon cooling, these interstitials form precipitates, normally chromium-rich carbides, nitrides, or carbonitrides [2]. Precipitation may occur both inter- and intragranularly, with the former promoting intergranular corrosion and the latter a loss of tensile ductility and toughness. The effect of high-temperature exposure on the impact toughness of medium-chromium alloys containing various levels of carbon and nitrogen is shown in Figure 5.6. Note that nitrogen levels above 0.02 wt% result in a severe reduction in toughness. Increased nitrogen at a constant carbon level was found to have a similar effect, and thus it is the total interstitial content (C + N) that is critical. Similar behavior has

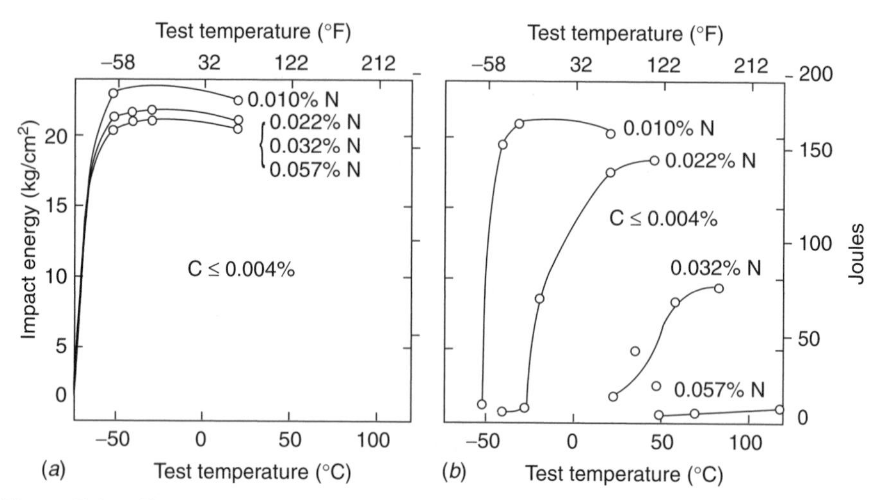

Figure 5.6 Effect of elevated temperature exposure on the impact toughness of 17Cr alloys with varying nitrogen levels heat treated at (*a*) 815°C (1500°F)/1hr/WQ, and (*b*) 1150°C (2100°F)/1hr/WQ. (From Semchysen et al. [21].)

been observed in high-chromium alloys. Plumtree and Gullberg [22] reported a shift in Charpy V-notch ductile-to-brittle fracture transition temperature of over 200°C (390°F) in 18Cr–2Mo and 25Cr alloys when C + N content increased from 0.02 wt% to 0.06 wt%.

There is general agreement that precipitation of chromium-rich carbides and nitrides during cooling from high temperatures ($>0.7T_m$) contributes significantly to HTE [2,5,21]. As a result, HTE is exacerbated by high levels of chromium, carbon, and nitrogen. The low-chromium alloys appear to be relatively insensitive to HTE. The rate of cooling from elevated temperature also influences HTE, but this effect is dependent on composition [23]. In alloys with low C + N contents, rapid cooling from above 1000°C (1830°F) tends to reduce embrittlement (lower DBTT) due to either the retention of carbon and nitrogen in solid solution or the tendency to form intragranular precipitates. At slower cooling rates, intergranular carbide and/or nitride precipitation is predominant and is accompanied by a loss in ductility and toughness [24].

In high-chromium alloys with high levels of C + N, on the order of 1000 ppm or more, rapid cooling appears to promote embrittlement and increase DBTT [25]. At these C + N levels, it is not possible to suppress precipitation via rapid cooling, particularly in high-chromium alloys, since carbon and nitrogen solubility decrease with increasing chromium concentration [5]. Alloying additions such as molybdenum, titanium, aluminum, and niobium also influence HTE, although their importance is secondary relative to that of chromium. Titanium and niobium have a high affinity for carbon, forming relatively stable carbides, and thus may reduce the deleterious effects resulting from precipitation of chromium-rich carbides and carbonitrides. The formation of aluminum-rich nitrides and oxides also reduces susceptibility to HTE. The presence of these precipitates in the microstructure also retards grain growth during elevated temperature exposure.

Grain-Size Effects Since HTE occurs upon exposure at elevated temperatures, grain growth is also a factor that influences subsequent mechanical properties, although grain size alone cannot account for the embrittling effects. In alloys that are fully ferritic (i.e., contain no austenite) at temperatures above approximately 1100°C (2010°F), grain growth can be quite dramatic, particularly in alloys that have been cold worked. For example, grain size on the order of ASTM 2-3 may be present in the HAZ of fusion welds.

The work of Plumtree and Gullberg [22] on 25Cr and 18Cr–2Mo alloys has demonstrated the combined effect of grain size and interstitial content on HTE. Impact toughness as a function of both grain size and C + N content is shown in Figure 5.7. Note that with a low interstitial content (350 ppm), small increases in grain size promote a large shift in the DBTT, on the order of 26°C per ASTM grain-size number. As interstitial content increases, the effect of grain size is less important (approximately 6°C per ASTM grain-size number), as embrittlement due to extensive precipitation dominates. Thus, high-purity alloys would be expected to show the greatest drop in toughness and ductility as a function of increasing grain size.

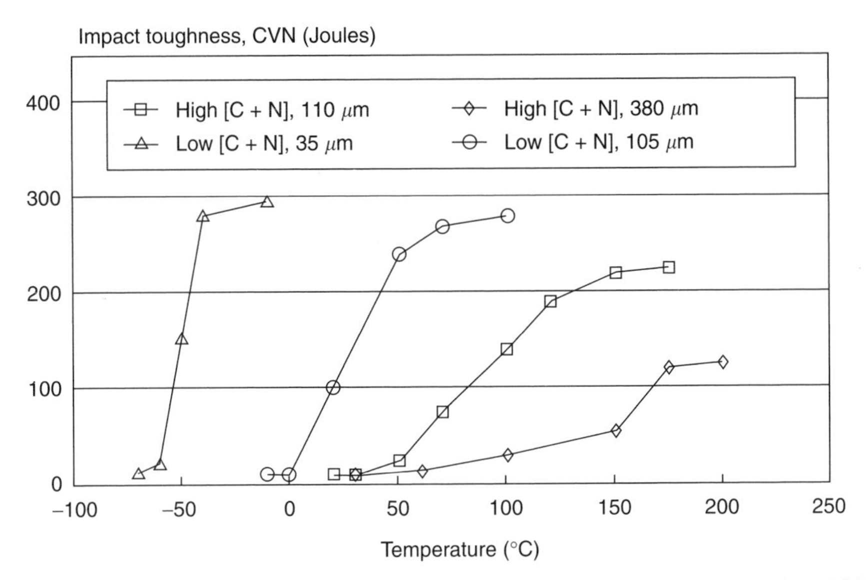

Figure 5.7 Effect of grain size and interstitial content on the impact toughness of Fe–25Cr alloys. (Replotted from Plumtree and Gullberg [22]. Courtesy of ASTM.)

TABLE 5.4 Effect of Composition and Microstructure on High-Temperature Embrittlement

Variable	Effect
Carbon + nitrogen	Intensifies severely
Chromium	Intensifies
Grain size	Small for high C + N, large for high Cr and low C + N
Oxygen	Intensifies slightly
Titanium, niobium	Reduces

Summary of High-Temperature Embrittlement Embrittlement due to elevated temperature exposure is influenced by a number of factors, both compositional and microstructural in nature, including (1) chromium and interstitial concentration, (2) grain size, and (3) the nature and distribution of precipitates. The effect of these various factors on susceptibility to HTE is summarized in Table 5.4. In general, high levels of interstitial elements (C, N, and O) are most damaging, and as a result, most commercial alloys (particularly the high-Cr alloys) contain extremely low levels of interstitials (<200 ppm). At these low levels, however, the effect of grain size becomes more important, and short-term elevated temperature exposure such as that experienced during welding may result in severe HTE.

The actual mechanism of HTE is the subject of considerable debate centering on the location of the precipitates in the microstructure. One theory [2,22] argues that intragranular precipitation is damaging, due to the restriction of dislocation motion. Another theory attributes HTE to intergranular precipitation and the localized embrittlement of the grain boundaries. Since fracture in materials susceptible to HTE is typically *transgranular*, it would seem that intragranular precipitation would most seriously influence HTE. Plumtree and Gullberg [22] have suggested, however, that intergranular precipitates can significantly influence crack initiation, thus reducing the energy needed for nucleation of transgranular cleavage cracks. Others have also associated the onset of embrittlement with intergranular precipitation [24–26]. The actual mechanism is probably a combination of the two, with intragranular precipitation becoming more dominant as the cooling rate increases. Both mechanisms can be used to explain the severe loss of corrosion resistance that occurs in ferritic stainless steels after high-temperature exposure.

Elimination of HTE may be possible in high-interstitial alloys by heating between 730 and 790°C (1350 and 1450°F) [1]. This heat treatment probably acts to overage the precipitates and thus reduces the deleterious effect on toughness and ductility. Some caution should be exercised when performing such a heat treatment, however, since sigma phase may precipitate during extended exposure in this temperature range. In low-interstitial alloys embrittled primarily due to grain growth at high temperature, thermal treatments will be of little help.

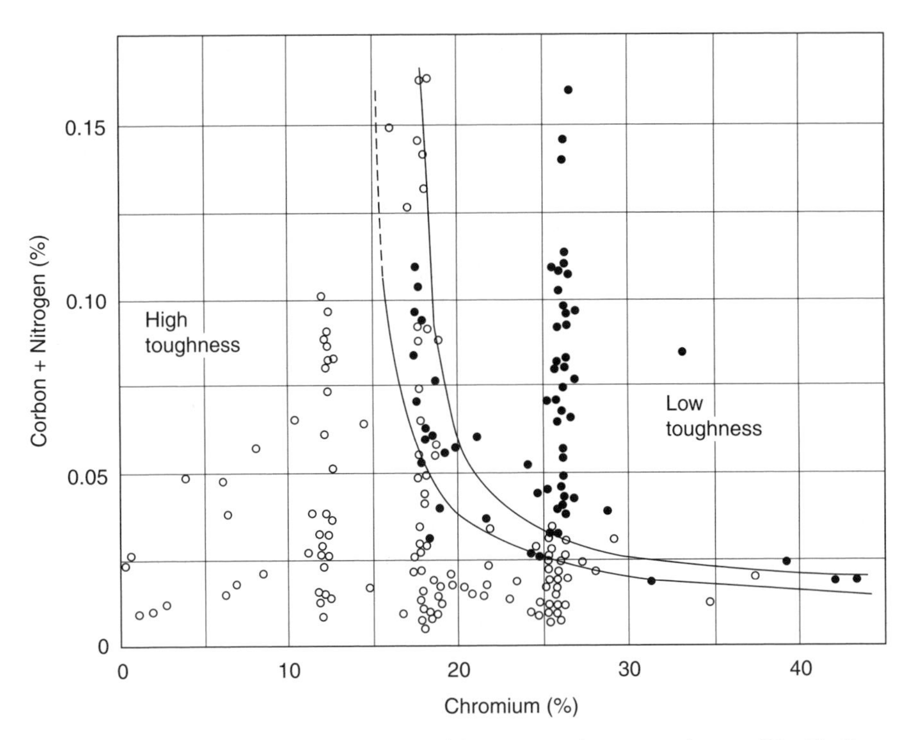

Figure 5.8 Effect of chromium and interstitial content on impact toughness of Fe–Cr alloys. Open circles, high impact toughness; solid circles, low impact toughness. (From Binder and Spendelow [28]. Courtesy of ASM International.)

5.2.3.4 Notch Sensitivity Even in the absence of ITE or HTE, ferritic stainless steels are highly notch sensitive. This behavior is influenced by chromium concentration as first described by Krivobok [27]. For low-chromium alloys in the annealed condition, impact toughness does not appear to be influenced by interstitial content. For medium- and high-chromium alloys, interstitial content plays a much more critical role. The combined effect of chromium and interstitial content was demonstrated by Binder and Spendelow [28] and is shown in Figure 5.8 for a wide range of chromium contents. Note that for medium-chromium alloys (17 to 19 wt% Cr), good impact toughness resistance is obtained only when the total C + N content is below approximately 0.05 wt% (500 ppm). In high-chromium alloys, the effect of C + N content is even more pronounced, and levels below 250 ppm must be maintained to ensure adequate toughness. This work also showed that carbon and nitrogen contribute equally to the notch sensitivity of ferritic stainless steels.

In the absence of metallurgical variables such as grain size and precipitation effects, composition influences notch sensitivity significantly. In medium- and high-chromium alloys, notch sensitivity can be reduced by maintaining extremely low interstitial levels, and thus most commercial alloys are formulated with extremely low C + N contents or contain "control" elements such as titanium, niobium, and aluminum to neutralize the interstitials. Extreme care must be taken in the fabrication of these alloys to avoid contributory embrittlement effects due to HTE.

5.2.4 Mechanical Properties

As described in previous sections, ferritic stainless steels are formulated such that the microstructure is essentially fully ferritic from the melting point to room temperature. As a result, strengthening via the austenite-to-martensite transformation is not normally possible or has essentially no effect due to the small amount of martensite that forms. Small increases in strength can be achieved via solid-solution hardening, particularly with carbon and nitrogen, although the effects are apparent in high-chromium alloys only where increased C + N does not also promote martensite formation. Various precipitation reactions can also be used to strengthen these steels, although these thermal treatments normally lead to embrittlement, loss of toughness, or both. In practice, the most widely used method for substantially strengthening ferritic stainless steels is cold work [1–3].

Most ferritic stainless steel wrought products are supplied in the annealed or hot-rolled conditions. The properties for plate or sheet forms of the ferritic stainless steels of Table 5.1 are provided in Table 5.5. These properties are representative of material that has been annealed in the temperature range indicated and then air cooled or water quenched to room temperature. The properties of weld metal from the ferritic stainless steel filler metals of Table 5.2 are provided in Table 5.6.

5.3 WELDING METALLURGY

5.3.1 Fusion Zone

5.3.1.1 Solidification and Transformation Sequence As shown in Figure 5.1, initial solidification of ferritic stainless steel welds always occurs as primary ferrite. The fusion zone microstructure may either be fully ferritic or consist of a mixture of ferrite and martensite, with the martensite located at the ferrite grain boundaries. Three solidification and transformation sequences are possible for the ferritic stainless steels. The first of these, and the simplest, is described as follows.

Transformation path 1: fully ferritic microstructure

$$L \rightarrow L + F \rightarrow F$$

This path is dominant when the ratio of ferrite-promoting elements to austenite-promoting elements is high and austenite formation at elevated temperature is

TABLE 5.5 Mechanical Property Requirements for Wrought Ferritic Stainless Steels

Type	UNS No.	Minimum Tensile Strength		Minimum Yield Strength		Minimum Elongation in 50 mm (2 in.) (%)	Maximum Hardness	
		MPa	ksi	MPa	ksi		Brinell	Rockwell B
405	S40500	415	60	170	25	20.0	179	88
409	S40900[a]	380	55	170	25	20	179	88
430	S43000	450	65	205	30	22.0	183	89
434	S43400	450	65	240	35	22.0	—	89
436	S43600	450	65	240	35	22.0	—	89
439	S43035	415	60	205	30	22.0	183	89
442	S44200	450	65	275	40	20.0	217	96
444	S44400	415	60	275	40	20.0	217	96
446	S44600	450	65	275	40	20.0	217	96
468	S46800	415	60	205	30	22.0	—	90
XM-27	S44627	450	65	275	40	22.0	187	90
25-4-4	S44635	620	90	515	75	20.0	269	28[b]
29-4	S44700	550	80	415	60	20.0	223	20[b]
29-4C	S44735	550	80	415	60	18.0	255	25[b]
29-4-2	S44800	550	80	415	60	20.0	223	20[b]

[a]Including S40910, S40920, and S40930.
[b]Rockwell C scale.

TABLE 5.6 Mechanical Property Requirements for Ferritic Stainless Steel Filler Metals

AWS Classification	UNS No.	Minimum Tensile Strength		Minimum Elongation in 50 mm (2 in.) (%)	Postweld Heat Treatment
		Mpa	ksi		
E409Nb-XX	—	450	65	20	[a]
ER409	S40900	[b]	[b]	[b]	[b]
ER409Cb	S40940	[b]	[b]	[b]	[b]
E409TX-X	W41031	450	65	15	None
E430-XX	W43010	450	65	20	[a]
E430Nb-XX	—	450	65	20	[a]
ER430	S43080	[b]	[b]	[b]	[b]
ER446LMo	S44687	[b]	[b]	[b]	[b]

[a]760°C (1400°F) to 790°C (1450°F) for 2 hours, furnace cooling at a rate not exceeding 55°C (100°F) per hour down to 595°C (1100°F), then air cooling to ambient.
[b]AWS A5.9 does not specify mechanical properties for wires and rods.

suppressed completely. This transformation path is common in the following alloys.

1. Low-chromium alloys, such as Types 405 and 409, when the carbon content is low
2. Medium-chromium alloys, such as Types 439, 444, and 468, where Ti and Nb are added as carbide stabilizers, effectively negating the austenite-promoting tendency of carbon
3. High-chromium alloys, such as Type XM-27 (popularly known as E-Brite or E-Brite 26-1), 25-4-4, 29-4, and 29-4-2, where the high-chromium content dominates

Because of the absence of elevated-temperature austenite, ferrite grain growth upon cooling from the solidification temperature range can be quite dramatic, particularly in high heat input welds. Examples of fully ferritic fusion-zone microstructures in gas–tungsten arc welds in Types 409 and 439 sheet alloys are shown in Figure 5.9. Note that the grain size is very large and there is no evidence of solidification subgrains (cells or dendrites). The absence of apparent solidification segregation (or partitioning) is a consequence of several factors: (1) a relatively narrow solidification temperature range, (2) little or no partitioning of chromium during solidification, and (3) rapid diffusion within the ferrite that eliminates any composition gradients that may have formed due to solidification. In particular, carbon diffusion in ferrite at elevated temperature is extremely rapid.

If martensite is present in the fusion zone, there are two transformation paths by which that can occur. In the sequence described below, solidification occurs

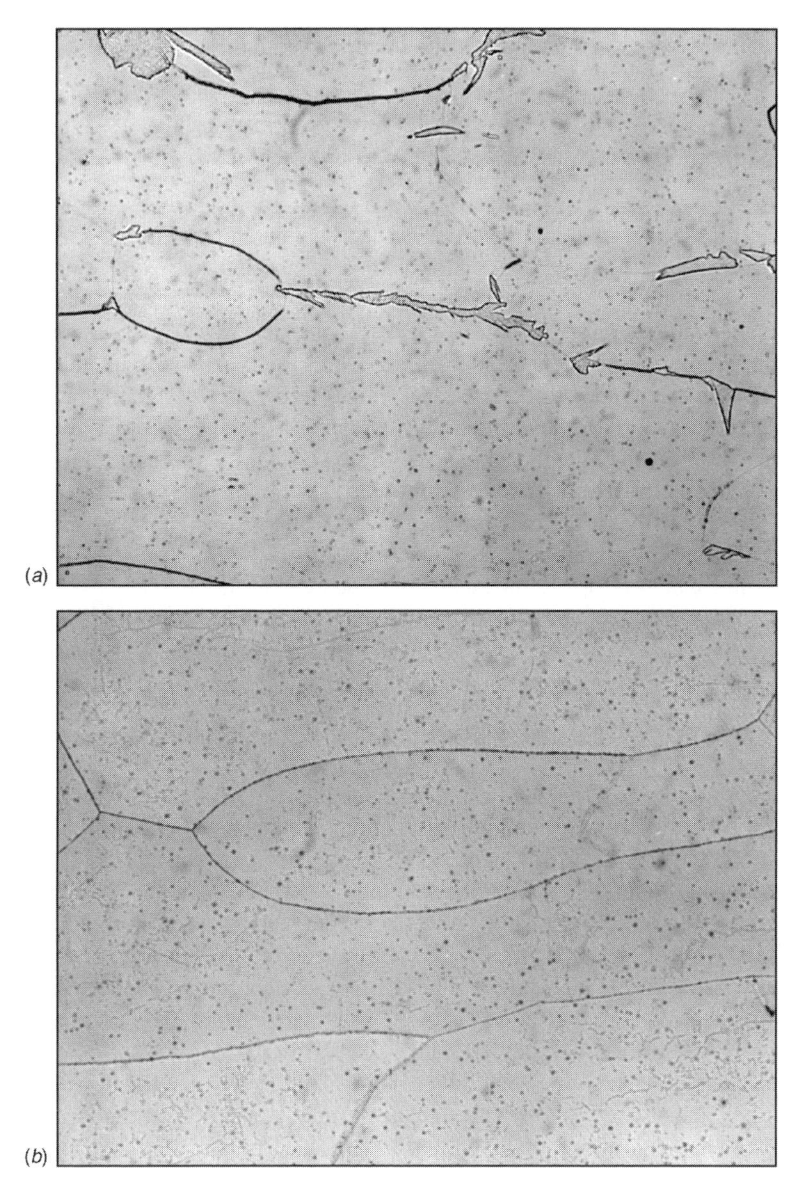

Figure 5.9 Fusion zone microstructure of fully ferritic stainless steels: (*a*) Type 409 (note some martensite along grain boundaries); (*b*) Type 439.

entirely as ferrite and that ferrite is stable in the solid state over some temperature range.

Transformation path 2: ferrite and martensite

$$L \rightarrow L + F \rightarrow F \rightarrow F + A \rightarrow F + M$$

Upon cooling, some austenite forms at elevated temperature along the ferrite grain boundaries. This austenite then transforms to martensite as the fusion zone cools to room temperature. This sequence is described in Figure 5.1 when considering carbon contents in the range 0.05 to 0.15 wt%. Note that as the carbon content (or austenite-promoting element content) increases, the ferrite solvus temperature increases and the temperature range over which ferrite alone is stable in the solid state decreases. This can have important implications with respect to ferrite grain growth, since growth of the ferrite grains will stop once austenite starts to form along the grain boundary. This transformation path is common in Types 430, 434, the lower-carbon portion of the range of 442 and 446 alloys, in higher-carbon 405, and in the older higher-carbon version of 409.

Austenite can also form at the end of solidification, as described by the following transformation path.

Transformation path 3: ferrite and martensite

$$L \rightarrow L + F \rightarrow L + F + A \rightarrow F + A \rightarrow F + M$$

Note that solidification begins as primary ferrite but some austenite forms at the end of solidification by way of a complicated peritectic–eutectic reaction. (This reaction is described in some detail in Chapter 6.) This transformation sequence can also be described using Figure 5.1 if a carbon level above 0.15 wt% is considered. Note that a three-phase region consisting of ferrite, austenite, and liquid exists below the primary ferrite plus liquid phase field. The alloy then cools in the solid state through the two-phase ferrite + austenite regime. Upon rapid cooling to room temperature, the austenite transforms to martensite. This transformation path is common to higher-carbon versions of Types 442 and 446.

Transformation path 3 will generally result in a higher-volume fraction of martensite in the fusion zone microstructure relative to path 2. However, the formation of austenite at the end of solidification and its presence in the solid state at elevated temperatures will restrict ferrite grain growth relative to path 2, where a fully ferritic structure exists over some temperature range. It is within this fully ferritic high-temperature range that ferrite grain growth occurs very rapidly. Later, it is discussed how large ferrite grain size in the weld metal and HAZ can result in reduced toughness and ductility.

A two-phase ferrite plus martensite fusion zone microstructure is shown in Figure 5.10. The martensite is present along the ferrite grain boundaries and is generally present as a continuous grain boundary phase. As the martensite content increases, it will also be present as Widmanstätten side plates that nucleate from the

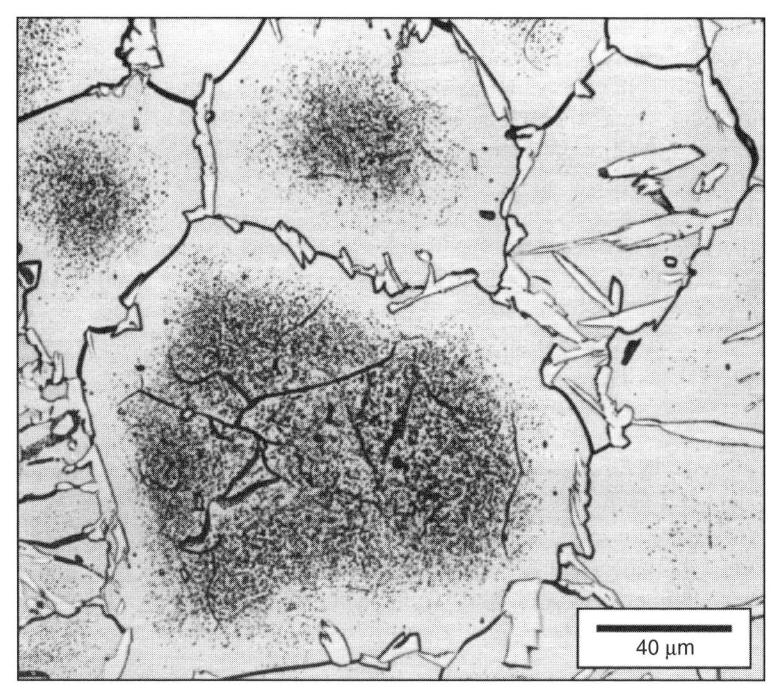

Figure 5.10 Two-phase ferrite and intergranular martensite microstructure in a Type 430 weld metal deposit. Note the precipitate-free zone surrounding the grain boundary martensite.

grain boundary and also form intragranularly. Remember that the martensite simply reflects the morphology of the austenite that formed at elevated temperature.

5.3.1.2 Precipitation Behavior Ferritic stainless steel weld metals often exhibit a fine dispersion of precipitates within the ferrite or at the ferrite–martensite boundary. In unstabilized alloys such as Type 430, these precipitates are primarily chromium-rich $M_{23}C_6$ and $M_{23}(C,N)_6$ carbides, or chromium-rich nitrides [29]. In stabilized grades (Types 444, 439, and 468), MC-type carbides are also possible. An example of this precipitation in the fusion zone of Type 439 can be seen in Figure 5.9. A similar precipitation behavior is observed in the HAZ. These precipitates form due to the supersaturation of carbon and nitrogen in the ferrite phase at elevated temperatures. Upon cooling, precipitates may form both inter- and intragranualarly, with the site dependent on cooling rate. At high cooling rates, significant intragranular precipitation is observed, while at slow cooling rates grain boundaries sites are preferred [30]. The nature and degree of precipitation has been found to influence both the mechanical and corrosion properties of welded ferritic stainless steels [2,29].

As shown in Figure 5.1 and previously in Figure 2.3, the solubility of carbon in ferrite drops dramatically as the fusion zone cools in the solid state. In 13 wt% Cr alloys, up to 0.1 wt% carbon is soluble in ferrite at 1400°C (2550°F), but the solubility drops to essentially zero at 1100°C (2010°F). The solubility of carbon in 17 wt% Cr alloys also drops dramatically upon cooling, decreasing from 0.15 wt% at 1400°C (2550°F) to approximately 0.03 wt% at 1000°C (1830°F). In commercial low- and medium-chromium ferritic stainless steels containing up to 0.05 wt% carbon, carbide precipitation in the fusion zone is inevitable in the absence of carbide-stabilizing elements (Ti and Nb) and/or elevated-temperature austenite. Many modern ferritic stainless steels contain 0.02 to 0.03 wt% carbon to avoid extensive carbide precipitation, but even in these steels (in the absence of stabilization) some carbide precipitation is expected.

Nitrogen behaves in a similar manner, as shown in the pseudobinary Fe–Cr–N phase diagram at 18 wt% Cr in Figure 5.11. According to this diagram, nitrogen solubility in ferrite drops from about 0.08 wt% at 1300°C (2370°F) to less than 0.02 wt% at 900°C (1650°F). Since most commercial first-generation ferritic stainless steels contain on the order of 0.05 wt% nitrogen, precipitation of nitrogen-rich precipitates within the ferrite is expected. Similar to carbon, the precipitation of nitrides or carbonitrides can be avoided by the addition of stabilizing elements such as Ti and Al, both of which are strong nitride formers.

If austenite is present at elevated temperature (e.g., along a ferrite grain boundary), precipitation will not occur in the austenite, due to its high solubility for carbon and nitrogen. As shown in Figures 5.1 and 5.11, the solubility of carbon and nitrogen in a 17 to 18 wt% Cr alloy at 1200°C (2210°F) will be approximately 0.32 and 0.41 wt%, respectively. Thus, austenite acts as a "sink" for these interstitial elements at elevated temperature. When austenite is present at elevated temperature (resulting in martensite in the room-temperature microstructure), there will often be a precipitate-free zone in the ferrite just adjacent to the martensite. This results from

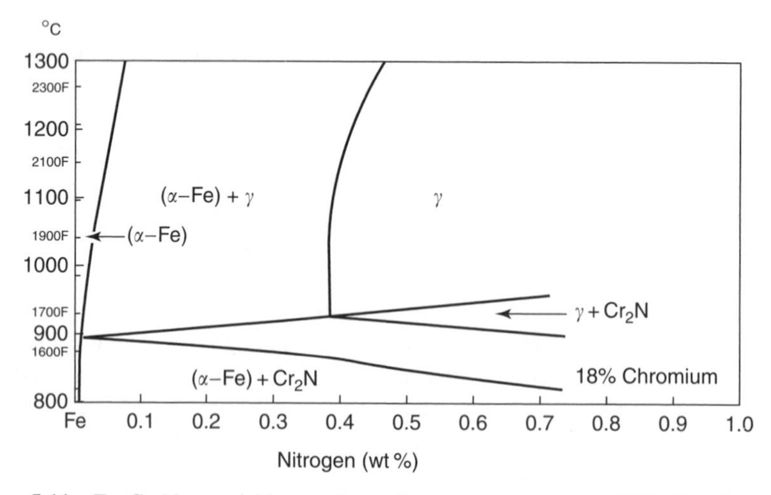

Figure 5.11 Fe–Cr–N pseudobinary phase diagram at a constant 18% Cr. (From *ASM Metals Handbook* [31]. Courtesy of ASM International.)

the high affinity of the parent austenite for carbon and nitrogen at elevated temperature. Short-range diffusion of carbon and nitrogen from the ferrite into the austenite reduces the local concentration in the ferrite, and upon cooling through the precipitation range there is little or no driving force for precipitation. An example of precipitate-free zones in a Type 430 weld metal deposit is shown in Figure 5.10.

Precipitation can also occur along the ferrite–ferrite or ferrite–martensite boundary. These precipitates are normally Cr-rich $M_{23}C_6$ and $M_{23}(C,N)_6$ [2]. This precipitation can lead to a local reduction in chromium adjacent to the boundary, making the boundary potentially "sensitive" to corrosion attack. This phenomenon, known as *sensitization*, is discussed in more detail in Section 5.6. Precipitation of carbides, carbonitrides, and nitrides can be effectively suppressed by reducing carbon and nitrogen to extremely low levels (less than 0.01 wt%), by the addition of stabilizing elements, or both. Many of the third-generation high-Cr ferritic grades, such as E-Brite 26-1, restrict carbon to less than 0.01 wt%, nitrogen to below 0.02 wt%, and contain additions of Ti and Nb. As discussed later in this chapter, avoiding precipitation in these alloys is critical to maintaining their mechanical and corrosion properties.

5.3.1.3 Microstructure Prediction As described in Chapter 3, a number of equivalency relationships and diagrams have evolved for the predication of microstructure in ferritic stainless steel welds. The Schaeffler diagram (Figure 3.4) covers the composition range for the ferritic stainless steels but has been proven to be relatively inaccurate for predicting whether martensite will be present in ferritic alloys. The Kaltenhauser ferrite factor can be used to predict whether the microstructure is fully ferritic but does not provide any information regarding the amount of martensite that will be present [32]. The K-factor, shown below, cannot be applied equally to all alloys, and low- and medium-chromium alloys must be considered separately.

$$\text{K-factor} = Cr + 6Si + 8Ti + 4Mo + 2Al - 40(C + N) - 2Mn - 4Ni \quad (5.1)$$

For Types 405 and 409 stainless steels, Kaltenhauser determined that this factor must exceed 13.5 to prevent martensite formation in welds. In medium-chromium alloys, such as Types 430 and 439, this factor must exceed 17.0. In practice, many commercial low-chromium alloys are now formulated such that the K-factor approaches or exceeds 13.5, and thus welds in these alloys are normally fully ferritic or contain only small amounts of martensite. This K-factor level, and the resulting microstructural effect, is normally achieved by reducing carbon, adding titanium, or a combination of the two.

The Balmforth diagram [33] was developed to predict the microstructure of both martensitic and ferritic stainless weld metals, as discussed in Chapter 3. It is reproduced here (Figure 5.12) to demonstrate its use with the ferritic stainless steels. On this diagram the composition ranges of Types 409 (UNS S40910), 430, and 439 have been superimposed. Note that both 409 and 430 overlap the martensite plus ferrite region of the diagram, while the Ti-stabilized 439 alloy is completely within the ferritic region. The composition of Type 409 with more restrictive carbon content

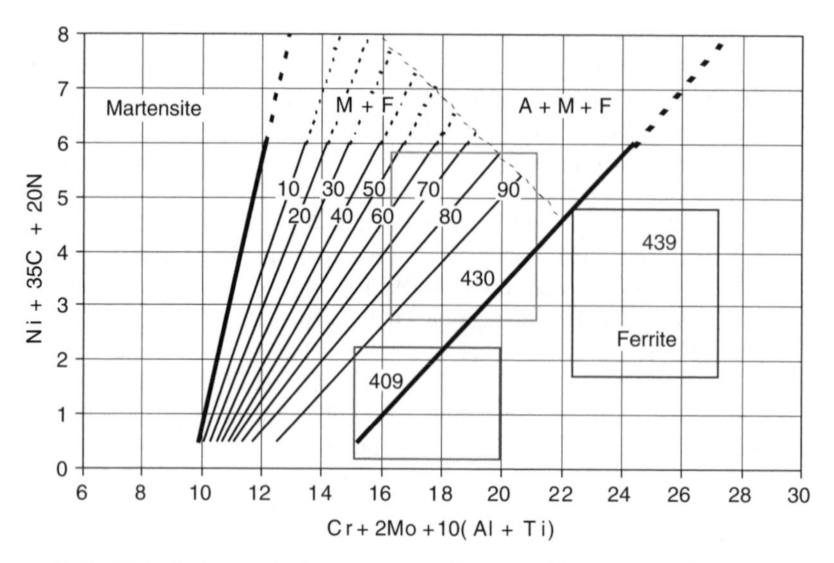

Figure 5.12 Balmforth constitution diagram with composition ranges of Types 409, 430, and 439 superimposed. (From Balmforth and Lippold [33]. Courtesy of the American Welding Society.)

(0.03 wt% max., Table 5.1, UNS S40910) lies predominantly within the ferrite region of the diagram. If the higher-carbon version of this alloy (UNS S40900, 0.08% C max.) were to be plotted on the Balmforth diagram, the composition space would be shifted much higher into the two-phase ferrite + martensite region of the diagram.

Currently, this diagram is the most accurate available for predicting the microstructure in ferritic stainless steel weld metals. It should be noted that the diagram was developed using alloys within the composition range shown in Figure 3.22. The diagram may be inaccurate when considering alloys whose compositions lie outside this range, particularly those with very low carbon content (less than 0.03 wt%) or levels of Al + Ti exceeding 1.0 wt%.

5.3.2 Heat-Affected Zone

The microstructure of most wrought ferritic stainless steels consists of a mixture of ferrite and carbides (or carbonitrides). When this structure is heated to elevated temperature in the HAZ surrounding the fusion zone, a number of metallurgical reactions can occur. During the heating cycle, any carbides and other precipitates will tend to dissolve. If any martensite is present (such as around all but the last pass in a multipass weld), it will also tend to dissolve either by reverting to ferrite (accompanied by carbide formation) or transforming back to austenite during reheating due to deposition of subsequent passes.

Depending on the composition of the alloy, the elevated-temperature HAZ can either be fully ferritic or a mixture of ferrite and austenite. In the absence of precipitates and

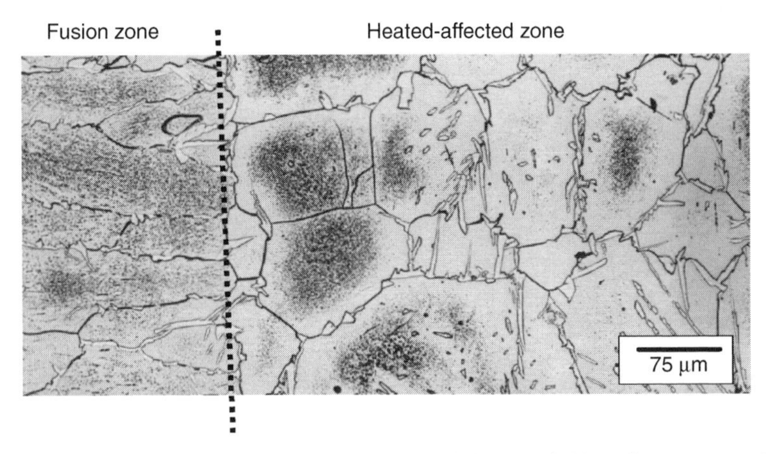

Figure 5.13 HAZ microstructure in a Type 430 stainless steel. Note the presence of grain boundary martensite and heavy intragranular precipitation.

grain boundary austenite, the ferrite grains will grow quite rapidly and the HAZ of most ferritic stainless steels exhibits relatively large ferrite grains. If austenite is stable in the microstructure at elevated temperature, it can act to inhibit ferrite grain growth by pinning the grain boundaries. Upon cooling, some reprecipitation of carbides and nitrides may occur. Since carbon and nitrogen have relatively low solubility in the ferrite at low temperatures, there can be a strong driving force for precipitation. Any austenite that may have formed at elevated temperature will transform to martensite during the cooling cycle. This martensite will normally be distributed along the ferrite grain boundaries. It is often identified mistakenly as austenite.

Figure 5.13 shows the HAZ in a Type 430 stainless steel. The ferrite grains are quite large and the grain boundaries have a continuous layer of martensite that has formed from the elevated temperature austenite. The "peppery" structure at the grain interiors represents the carbides, carbonitrides, or nitrides that have formed during cooling. A precipitate-free zone similar to that formed in the fusion zone (Figure 5.10) is also evident in the HAZ. Figure 5.14 shows the HAZ in a gas–tungsten arc weld in Type 409. In this alloy, virtually no elevated-temperature austenite forms and the HAZ is fully ferritic. The HAZ in this alloy is also free of apparent precipitation. This is due to the low chromium content of this alloy and the addition of titanium, which ties up carbon and nitrogen, thereby preventing the precipitation of $M_{23}C_6$ and $M_{23}(C,N)_6$ carbides and carbonitrides.

5.3.3 Solid-State Welds

Where applicable, the use of solid-state processes that result in grain refinement can produce a substantial improvement in properties relative to fusion welds. In tube welding applications, processes using high-frequency (HF) induction or HF resistance have become more popular than arc welding, due to both mechanical property

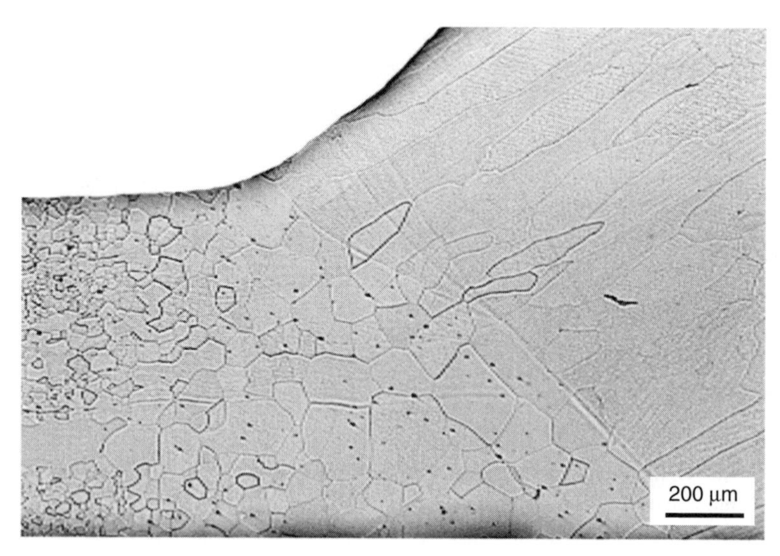

Figure 5.14 HAZ microstructure of a Type 409 stainless steel that is essentially fully ferritic.

and economic/productivity issues. Under optimum conditions the ductility and toughness of HF-welded tube is superior to that produced via arc welding because of the grain refinement that occurs in the weld region. The microstructure of an HF-induction weld in Type 409 is shown in Figure 5.15. Note the extreme grain refinement relative to the fusion zone and HAZ in a GTA weld in the same material (Figures 5.9 and 5.14). This refinement is the product of the combined effects of temperature and mechanical forging that result in dynamic recrystallization within the weld region. Control of both temperature and the degree of forging is critical in producing optimum welds. At excessive temperatures, melting occurs, usually resulting in localized defects. At low temperatures the forging loads become excessive and insufficient deformation at the weld interface results in a "cold" weld (one that may contain unbonded regions and little grain refinement).

5.4 MECHANICAL PROPERTIES OF WELDMENTS

The mechanical properties of welded ferritic stainless steel are very dependent on the alloy type and chromium concentration. For convenience, this section is subdivided into low-, medium-, and high-chromium alloys.

5.4.1 Low-Chromium Alloys

In the absence of martensite, the most significant microstructural effect on mechanical properties of welded low-chromium alloys is ferrite grain growth. Thomas and Apps [11] evaluated the effect of welding process on the impact toughness of the

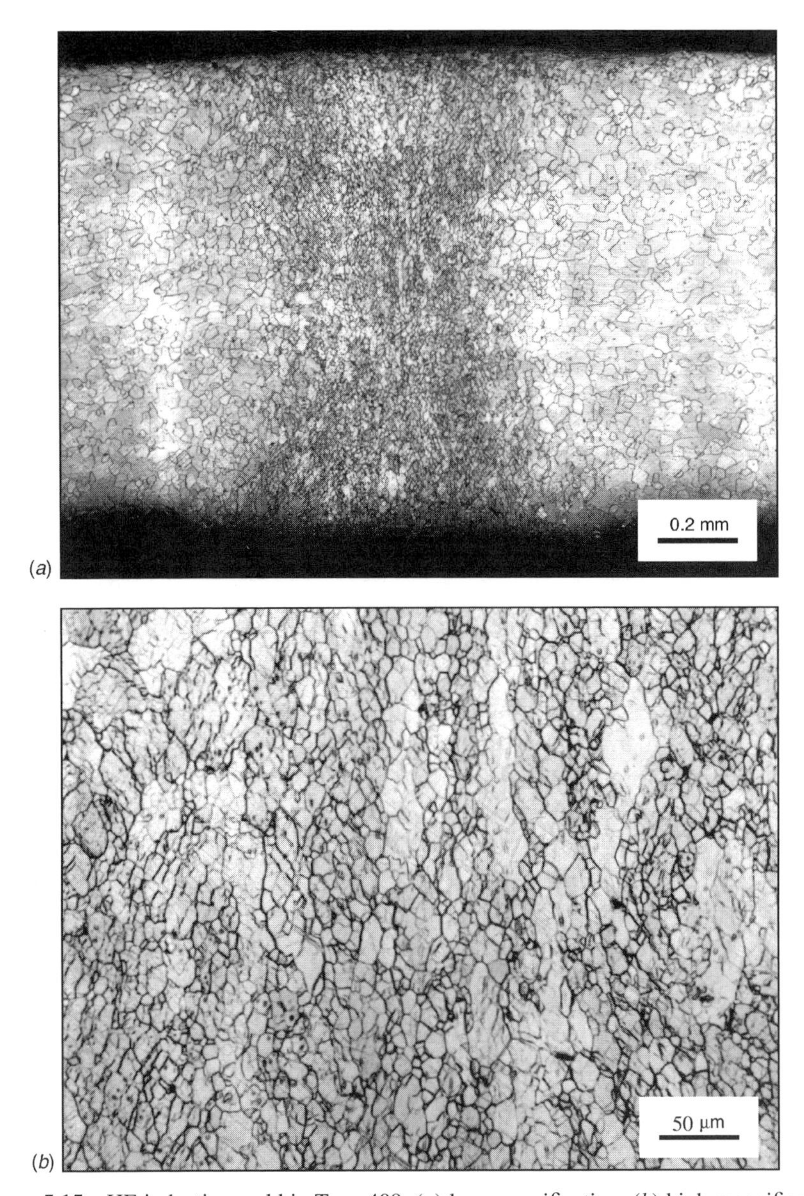

Figure 5.15 HF-induction weld in Type 409: (*a*) low magnification; (*b*) high magnification showing recrystallized region in center of weld.

HAZ in Type 409 by testing both actual welds and simulated HAZ samples. Their results (Figure 5.16) show that the DBTT of either actual or simulated HAZs in Type 409 increases from 30 to 70°C (86 to 158°F) relative to the base metal. The correlation between grain size and toughness shown in Figure 5.16 may not be solely related to ferrite grain growth, however, since the welds and simulated HAZs also contained a small amount of martensite. (Note that their work was published in

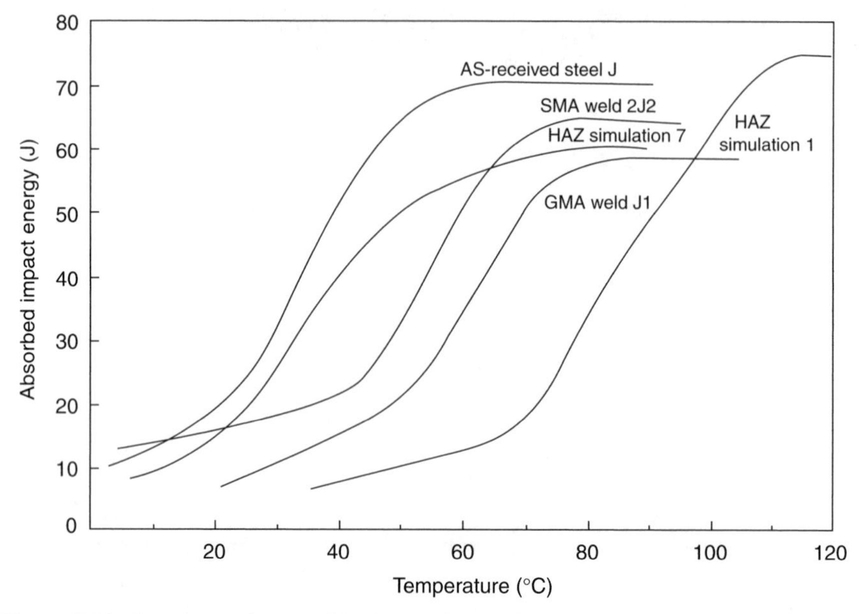

Figure 5.16 Impact toughness of both actual and simulated HAZs in Type 409. (From Thomas and Apps [11]. Courtesy of ASTM.)

1980, many years prior to the redefinition of 409 by ASTM.) Nevertheless, it was their opinion that the HAZ properties of low-chromium stainless steels can be improved by reducing the as-welded grain size.

5.4.2 Medium-Chromium Alloys

Relative to the low-chromium grades of ferritic stainless steels, considerable work has been undertaken to determine the effect of welding on the mechanical properties of medium-chromium alloys, including Types 430, 434, 436, 439, and 444. As with the low-chromium grades, mechanical behavior is dependent on microstructure and a differentiation must be made between fully ferritic and duplex ferritic/martensitic structures. In the former case, grain growth and precipitation significantly influence the properties, while in the latter case, the amount and nature of the martensite also contribute to the mechanical response. Unfortunately, many investigators do not adequately describe the weld microstructure or comment on the relative contribution of the metallurgical factors.

Sufficient data exist to demonstrate the negative effect of ferrite grain size on both toughness and ductility. The results of Charpy impact tests on simulated HAZ specimens from two stabilized alloys are summarized in Table 5.7. Increased grain size could be correlated with a dramatic increase in DBTT and a drop in the upper shelf energy in both alloys [6,34]. In simulated Type 436 HAZ samples, the increase in DBTT was found to be approximately 14°C (25°F) per ASTM grain-size number over

TABLE 5.7 Effect of Grain Size on the Toughness of Simulated HAZs in 17% Cr-Stabilized Alloys

	Grain Size		Toughness (J) at Temperature (°C)			
Alloy	Average Grain Diameter (μm)	ASTM No.	0	20	60	100
430Nb[a]	65	5	—	15	28	56
	350	0	—	3	8	18
	470	00	—	—	4	9
436[b]	22	8	5	11	13	13
	45	6	2	3	8	11
	75	4.5	2	4	9	9
	105	3.5	2	2	2	4

[a]From [6], 5-mm-thick samples.
[b]From [34], 3-mm-thick CVN samples per ASTM E23, Type A.

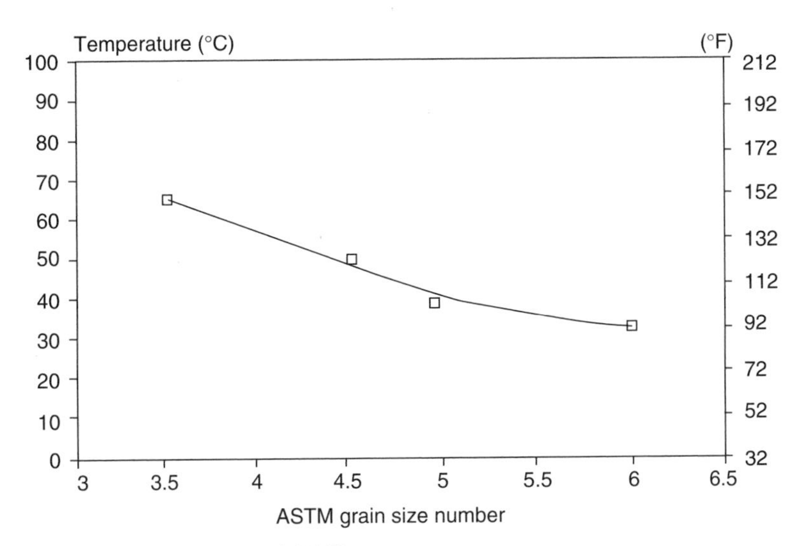

Figure 5.17 DBTT for simulated HAZ samples in Type 436. (From Lippold and Shademan [34].)

the range from ASTM 6 to 3.5, as shown in Figure 5.17. The bend ductility of simulated HAZs in Type 436 was also found to decrease as grain size increased, as illustrated in Figure 5.18. As with the data presented in Figure 5.17, the microstructures were fully ferritic and failures were characterized by transgranular cleavage. Nishio et al. [6] also showed the effects of high-temperature exposure and subsequent grain growth on mechanical properties in simulated HAZ samples of Type 430Nb

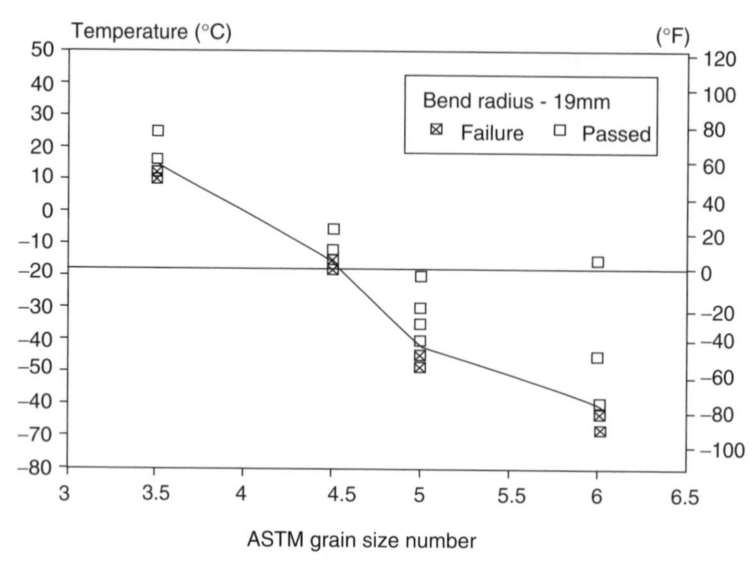

Figure 5.18 Effect of grain size on the bend ductility of simulated HAZ samples in Type 436. (From Lippold and Shademan [34].)

(Figure 5.19). A significant drop in tensile ductility was observed in specimens heated to 1350°C (2460°F), although this effect was complicated by the presence of both grain boundary martensite and retained austenite.

It is apparent that the degradation of toughness and ductility in medium-chromium stainless steel welds is not related strictly to grain size. Thomas and Robinson [30] suggested that intragranular precipitation due to rapid cooling also contributed. Rapidly cooled samples were found to exhibit inferior toughness relative to samples of equivalent grain size that were cooled slowly. Hunter and Eagar [35] related loss in ductility of welds in Type 444 to both increased grain size and the precipitation of extremely fine particles in the weld metal and HAZ. These precipitates were most likely Ti- or Cr-rich nitrides or carbonitrides, with the Cr-rich nitrides (Cr_2N) presumed to be more kinetically favorable during rapid cooling.

The combined influence of grain size and precipitation behavior on weld metal and HAZ toughness and ductility is analogous to the high-temperature embrittlement (HTE) phenomenon that is characteristic in medium- and high-chromium alloys heated above $0.7T_m$ (see Section 5.2.3). Unlike ferritic base metals, the control of HTE during welding is more difficult due to the high temperatures that promote grain growth and precipitate dissolution on heating and more uniform intragranular precipitation on cooling. In addition, stabilizing elements, such as Ti and Nb, are less effective during welding since the rapid cooling cycles tend to favor precipitation of Cr-rich precipitates. Thus, in fully ferritic microstructures, the nature of precipitation appears to have a strong influence on toughness and ductility, particularly when coupled with coarse grain size.

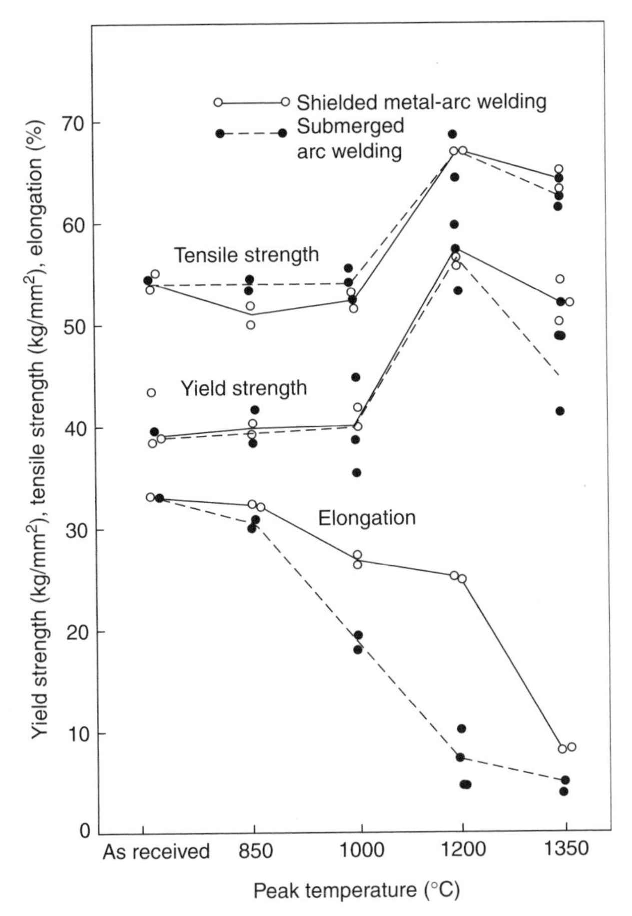

Figure 5.19 Effect of peak temperature on the mechanical properties of simulated HAZ samples in Type 430Nb. (From Nishio et al. [6]. Courtesy of the American Welding Society.)

Postweld thermal treatments may act to mitigate this effect by promoting coarsening of the precipitates, thereby reducing their negative influence. In alloys where austenite forms at elevated temperatures, the precipitation effect may be reduced since carbon and nitrogen will diffuse to the austenite, thus reducing the amount available for precipitation in the ferrite. This must be balanced against the potentially damaging contribution of the martensite that forms on cooling, as described by Nishio et al. [6]

5.4.3 High-Chromium Alloys

The toughness and ductility of as-welded high-carbon, high-chromium alloys such as Type 446 is extremely poor, due to the coupled effect of large grain size and high interstitial content. Unlike the low- and medium-chromium steels, weld properties are more sensitive to composition than microstructure and these alloys are extremely susceptible to HTE in the weld metal and HAZ. For example, the work of Pollard [36] and Wright [37] has shown that with 26% Cr alloys the maximum allowable interstitial content to ensure weld ductility is on the order of 150 ppm. Above this level, both ductility and toughness deteriorate significantly. The use of carbon stabilizers such as Ti and Nb is effective in minimizing the deleterious effects of HTE, but excess stabilizer content has been shown to be detrimental [37]. In addition, these stabilizers are only marginally successful in controlling grain growth and the embrittlement that results from this phenomenon. Ti is generally considered to be more effective than Nb for grain-size control. Demo [38] has also suggested the addition of "ductilizing" agents such as Al, Cu, V, Pt, Pd, or Ag as a means of improving weld ductility, although there are few data to support their effectiveness. In general, weld properties are controlled primarily by alloy selection and can be maximized by using materials with low interstitial contents and/or appropriate levels of stabilizing elements.

Most of the published information on high-chromium ferritic stainless steels addresses the use of arc welding processes, since these materials were originally to be used in moderate section thickness as replacements for austenitic stainless steels. As a result, GTAW, GMAW, and SMAW have been studied as methods to join these alloys. Krysiak [39,40] has reported CVN impact toughness for a number of high-chromium alloys welded using both the GTAW and SMAW processes. The toughness of GTAW welds in both a standard high-purity 26Cr–1Mo alloy and a Ti-stabilized grade of the same alloy is shown in Figure 5.20. Note that some improvement in toughness can be achieved by adjusting the welding conditions for the standard grade, but that the stabilized grade shows poor toughness despite these adjustments. (The author does not specify the welding conditions.) The use of SMAW with various austenitic filler metals (Figure 5.20*b*) avoids the sharp DBTT exhibited by GTA welds, but the upper shelf weld toughness is generally low. It is also likely that the HAZ toughness and ductility of these welds would be poor. The use of PWHT to improve weld properties was not investigated.

Pollard [36] found that weld heat input had little effect on the ductility of 26 wt% Cr alloys, probably since grain growth was significant at all heat input levels. Despite this observation, it is recommended as "good practice" to minimize heat input with these materials. In addition, due to the tendency for extreme grain growth, preheat and interpass temperature are to be avoided and minimized, respectively [39]. As with the lower-chromium alloys, PWHT has been shown to be beneficial in improving as-welded properties, presumably by altering the nature of the precipitates that induce HTE. The effect of PWHT at 850°C (1560°F) on the ductility of a 26 wt% Cr alloy is shown in Figure 5.21. Rapidly cooling from the PWHT temperature results in an improvement in ductility over a range of tensile test temperatures, while slow cooling promotes severe degradation in ductility [36]. PWHT followed by rapid cooling acts to dissolve precipitates formed during welding and retard their

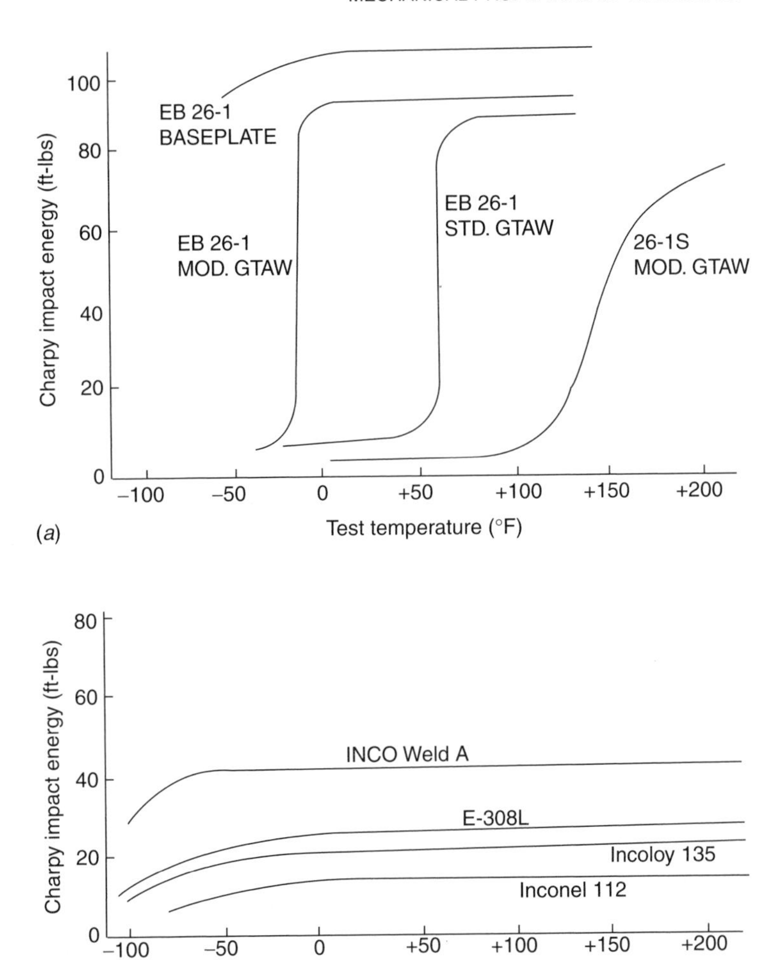

Figure 5.20 Weld metal toughness of 26Cr–1Mo alloys, as measured using subsize Charpy V-notch samples: (*a*) GTAW with matching filler (the modified technique provided for better weld shielding than the standard technique); (*b*) SMAW with various austenitic filler materials. (From Krysiac [39]. Courtesy of ASTM.)

reformation, while slow cooling after PWHT allows precipitates to re-form along grain boundaries. It is expected that PWHT would have the same effect with respect to weld toughness.

Although the use of austenitic stainless steel filler materials improves the weld metal properties of high-chromium alloys (Figure 5.20*b*) by promoting duplex, austenitic–ferritic microstructures, these filler alloys must be used with caution since

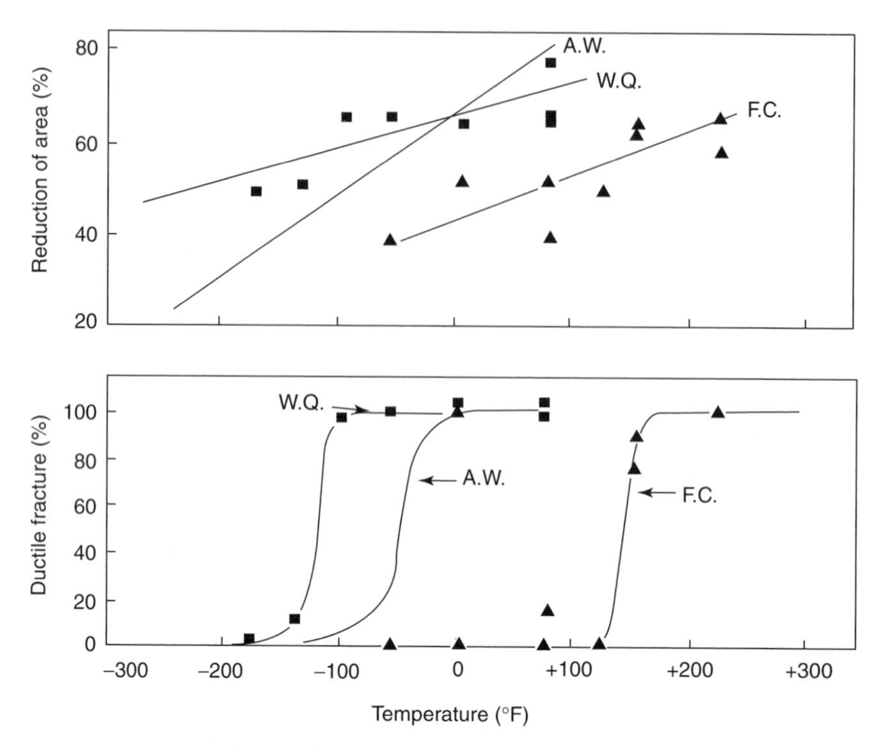

Figure 5.21 Effect of cooling rate from the PWHT temperature of 850°C (1560°F) on the ductility of a 26Cr alloy. AW, as welded; WQ, water quench; FC, furnace cool. (From Pollard [36]. Courtesy of the American Welding Society.)

they may compromise the corrosion properties of the welded structure. Nickel-base filler materials are also a possibility, although the compositional mismatch with the base material may also reduce corrosion resistance in certain environments. The use of nickel-base filler metals may also promote the formation of intermetallic phases at the fusion boundary [41]. As a consequence, matching filler metals are generally preferred for joining the high-chromium alloys and often contain stabilizing elements (particularly Ti) to help promote weld metal grain refinement and minimize the effect of HTE.

Although little information is available on the use of high-energy-density (HED) processes (laser and electron-beam welding) to join these materials, HED processes show promise due to the low weld heat input and rapid weld cooling rates inherent to both processes. Solid-state welding processes are also appropriate, based on their potential to refine weld grain size and minimize the effects of HTE. Unfortunately, there is little information available on the welding characteristics or properties of high-chromium alloys welded using either HED or solid-state processes.

5.5 WELDABILITY

Most information regarding the weldability of ferritic stainless steels is associated with either hydrogen-induced cracking (HIC) or weld solidification cracking. The research on these topics is generally limited to the medium-chromium alloys. The low-chromium alloys tend to have low susceptibility to weld-related cracking, while the high-chromium alloys are generally welded under very controlled conditions to avoid cracking, or used in applications where welding is not required. The medium-chromium alloys have been reported to be susceptible to both HIC and weld solidification cracking [6,42–44].

5.5.1 Weld Solidification Cracking

Weld solidification cracking occurs during the final stages of freezing due to the combined effects of impurity and alloying element segregation, liquid film formation at grain boundaries, and imposed thermo-mechanical restraint. Cracking is most often associated with solidification grain boundaries, where elemental segregation is the highest resulting in the lowest solidification temperature. When the primary solidification phase is ferrite, weld solidification cracking susceptibility is generally low. All of the ferritic stainless steels solidify as primary ferrite, and thus cracking in these steels is relatively rare. Addition of alloying elements, such as Ti and Nb, and high impurity levels will increase susceptibility to solidification cracking since segregation during solidification can still lead to the formation of low-melting liquid films along solidification grain boundaries. These alloys also exhibit a relatively narrow solidification temperature range which limits the level of restraint that develops due to shrinkage during solidification. A more detailed discussion of the effect of solidification behavior on cracking susceptibility is included in Chapter 6.

The relative solidification cracking susceptibility of Type 430, a 26Cr–1Mo alloy (E-Brite®), and austenitic alloy Type 304 as measured in the Varestraint test is shown in Figure 5.22. Solidification cracking was related to the concentration of C, S, P, and N in Type 430, and additionally to Nb and Ti in the stabilized 26Cr–1Mo alloy. Significant reduction in cracking was observed when C + N was less than 0.04% and titanium was restricted below 0.65%. Although not reported, it is likely that HAZ liquation cracking is also possible in these alloys.

Nishio et al. [6] and Sawhill et al. [44] also reported solidification cracking in Nb-bearing weld metals that were fully ferritic. In the case of Type 430Nb, the addition of nitrogen appeared to reduce the cracking susceptibility by promoting a finer solidification structure via the inoculation effect of Nb-rich nitrides. The addition of nitrogen also resulted in the formation of martensite in the weld metal. These weld metals also contained Ti, which probably contributed to the cracking susceptibility. Alloys containing stabilizing elements (Ti and Nb) have been shown to undergo a columnar-to-equiaxed grain transition at the weld centerline [45,46]. This transition

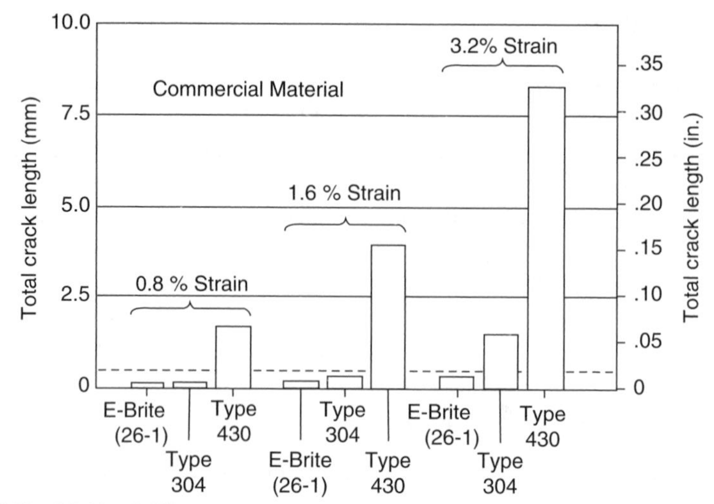

Figure 5.22 Weld solidification cracking susceptibility of several commercial stainless steels obtained using the Varestraint test. (From Kah and Dickinson [42]. Courtesy of the American Welding Society.)

results in a grain refinement along the centerline that can be effective in increasing the resistance to weld solidification cracking.

5.5.2 High-Temperature Embrittlement

High-temperature embrittlement is one of the most serious problems associated with the welding of the ferritic stainless steels. As described in Section 5.2.3, this form of embrittlement is a function of both composition and microstructure and is most damaging in the high-Cr alloys. High levels of interstitial elements, particularly carbon and nitrogen, have an additive effect. Large grain size also contributes to embrittlement, particularly in the HAZ. The combined effect of grain size and interstitial content was shown in Figure 5.7. In the weld metal and HAZ, HTE can result in a dramatic loss in toughness and ductility relative to the base metal. An example of this is shown in Figure 5.23, where failure has occurred in the HAZ of Type 436 during cold deformation. This failure occurred in the coarse-grained HAZ and the fracture path is transgranular. The fracture morphology is transgranular cleavage (Figure 5.24), which is characteristic of brittle fracture in ferritic steels.

A subsequent study of the effect of HAZ grain size on toughness was conducted by producing a range of HAZ microstructures in Type 436 using a Gleeble™ thermo-mechanical simulator. Over the range of ASTM grain size from 6 (fine grain) to 3.5, the ductile–brittle transition temperature (DBTT) increases and the upper shelf toughness drops precipitously, as shown in Figure 5.25. Note

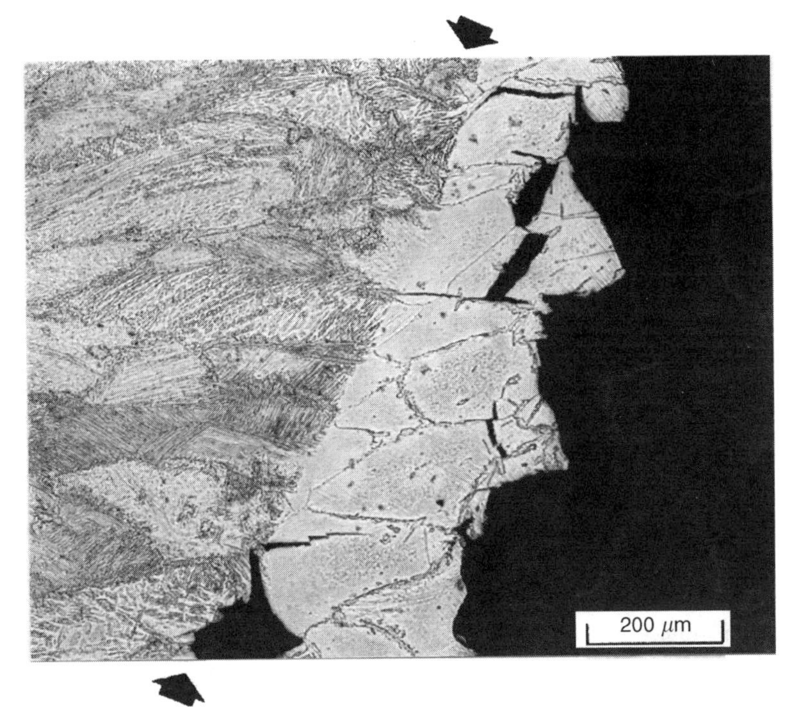

Figure 5.23 HAZ fracture during cold deformation of a GTAW weld in Type 436 made with Type 308L filler metal. Note that fracture is through the coarse-grained HAZ.

Figure 5.24 Fracture surface of failure in Figure 5.23, showing a transgranular cleavage fracture mode.

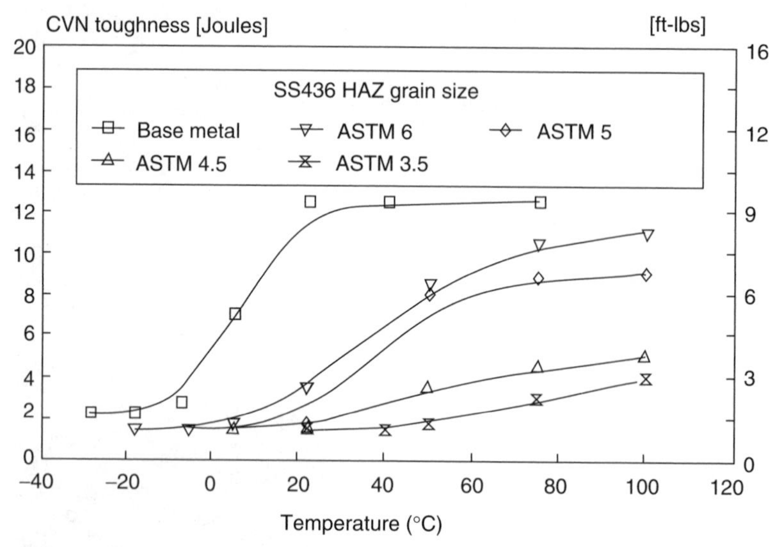

Figure 5.25 Effect of grain size on the toughness of simulated HAZ microstructures in Type 436.

that even moderate grain growth in this alloy increases the DBTT to above room temperature.

5.5.3 Hydrogen-Induced Cracking

Nishio et al. [6] also studied the HIC susceptibility of Type 430Nb and found that both the weld metal and HAZ were crack susceptible when martensite was present along ferrite grain boundaries. Cracking was often of the "delayed type," requiring an incubation time for hydrogen diffusion. Proper preheat and PWHT conditions were found to eliminate this problem (see Section 5.7).

5.6 CORROSION RESISTANCE

The corrosion resistance of ferritic stainless steels can be reduced severely by welding. These steels may be susceptible to a variety of forms of corrosive attack, including intergranular corrosion (IGC), crevice corrosion and pitting. Due to the absence of nickel, these materials are generally resistant to stress-corrosion cracking and thus are promising replacements for austenitic stainless steels in chloride-containing environments.

Crevice and pitting corrosion are generally avoided by proper alloy selection, while IGC is extremely sensitive to welding procedure and postweld conditioning. The subject of IGC in ferritic stainless steel welds has been reviewed at length by Demo [2,38] and others [47,48]. Consensus opinion is that IGC results from a sensitization mechanism similar to that proposed for austenitic stainless steels. Via this mechanism, grain boundary precipitation of Cr-rich carbides and/or nitrides results in a "Cr-depleted" zone that is susceptible to corrosive attack. As a result, resistance to IGC is a strong function of interstitial content, as shown in Figure 5.26. The mechanism of IGC, or intergranular attack (IGA), is discussed in detail in Chapter 6.

During welding, carbon and nitrogen are completely dissolved in the fusion zone and regions of the HAZ heated above approximately 1000°C (1830°F) (this temperature varies depending on heating rate). Upon cooling, Cr-rich $M_{23}C_6$ and Cr_2N may precipitate at either inter- or intragranular sites, the location depending on cooling rate and interstitial content. At high cooling rates, intragranular sites will be preferred while slow cooling favors intergranular sites. In as-welded microstructures of medium- and high-chromium alloys, precipitates are normally observed at both locations.

For high interstitial C + N levels (>1000 ppm), intergranular precipitation cannot be suppressed even by rapid quenching, and these alloys are inherently susceptible to IGC after welding. In alloys with low to moderate interstitial levels

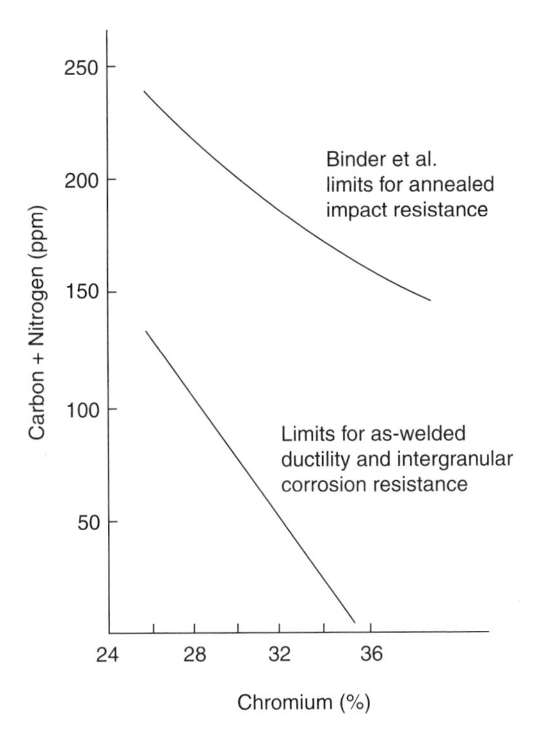

Figure 5.26 Effect of interstitial C + N level and chromium content on as-welded ductility and IGC resistance. (From Demo [2]. Courtesy of McGraw-Hill.)

TABLE 5.8 Carbon and Nitrogen Limits for As-Welded Ductility and Intergranular Corrosion Resistance as a Function of Chromium Content

Chromium (wt%)	Limit of Carbon + Nitrogen (ppm)	
	IGC Resistance[a]	Ductility[b]
19	60–80	<700
26	100–130	200–500
30	130–200	80–100
35	250	<20

Source: Demo [2].

[a]Tested in boiling ferric sulfate–50% sulfuric acid solution.

[b]Determined by bending around 5.08-mm mandrel (2.54-mm-thick material).

(approximately 200 to 500 ppm) rapid cooling is effective in suppressing grain boundary precipitation, and thus a critical cooling rate range exists over which sensitization will occur. This range usually overlaps the typical cooling rates experienced during arc welding. Alloys with extremely low interstitial levels are the most resistant to IGC, but the critical interstitial level in high-chromium alloys must be extremely low to assure weld ductility (Figure 5.26). The interplay between IGC resistance and weld ductility as a function of chromium and interstitial content is shown in Table 5.8. The extremely low interstitial contents required in these high-chromium alloys illustrate their inherent susceptibility to both HTE and IGC during welding.

PWHT can significantly improve the IGC resistance of ferritic stainless steels. Heating into the range 700 to 950°C (1290 to 1740°F) is effective in "healing" the Cr-depleted regions around grain boundary carbides via bulk chromium diffusion. This treatment has been shown earlier to be effective in improving weld ductility and toughness. Rapid cooling after PWHT is normally recommended in high-chromium alloys to avoid potential embrittlement due to extensive grain boundary precipitation (see Figure 5.21).

As in the case of HTE, IGC resistance can be improved in the ferritic stainless steels through the addition of stabilizing elements such as titanium and niobium [2,48]. These elements form stable MC-type carbides that resist dissolution during high-temperature exposure. The amount of stabilizing element required varies from alloy to alloy, but generally a level of $6 \times (C + N)$ or $0.20 + 4 \times (C + N)$ is considered a safe minimum in high-chromium alloys [48]. The effect of the Ti/(C + N) ratio on IGC in thermally treated samples of a 26Cr–1Mo alloy is shown in Figure 5.27. At ratios approaching 10, high-temperature exposure similar to that experienced during welding has little effect on corrosion resistance. In Type 409, a stabilization requirement of $Ti + Nb \geq 0.08 + 8 \times (C + N)$ has been proposed for resistance to IGC [49].

The effect of stabilizers in the fusion zone is not apparent, since dissolution of stabilized carbides/nitrides is likely during melting. During rapid cooling, it is

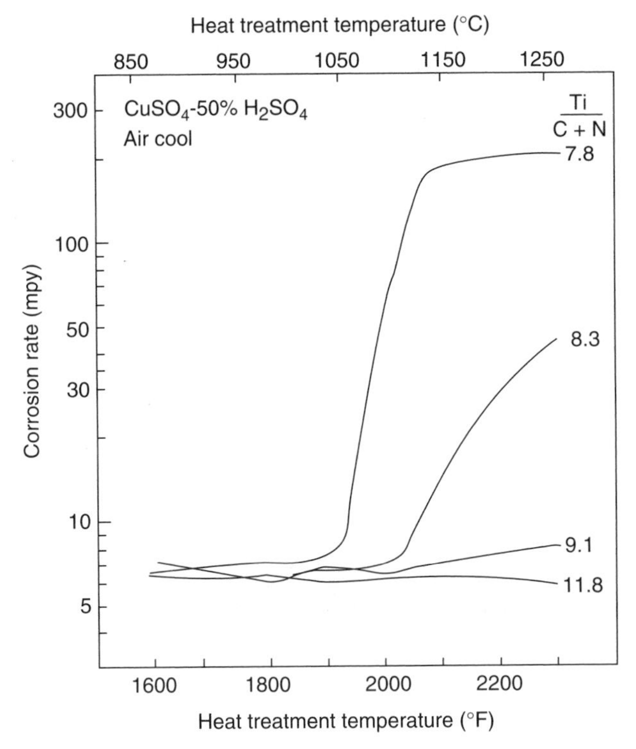

Figure 5.27 Effect of high-temperature exposure and Ti/(C + N) ratio on IGC of a 26Cr–1Mo alloy. (From Nichol and Davis [48]. Courtesy of ASTM.)

possible that Cr-rich precipitates may form, thereby sensitizing the weld metal. An example of IGC in the HAZ of Type 430 stainless steel is shown in Figure 5.28. This failure occurred in sheet material that was exposed to a marine environment. The combination of applied and residual stress and the presence of chloride ions led to the corrosion failure. Upon closer examination it can be seen that there is martensite along the ferrite grain boundaries (similar to Figure 5.13) and that the failure path is along the ferrite–martensite interface. This occurs due to the sensitization mechanism described previously, where the formation of Cr-rich carbides at the interface results in the depletion of chromium adjacent to the interface.

In summary, the IGC resistance of welds in ferritic stainless steels is controlled by a number of factors, including chromium and interstitial content and weld cooling rate. For welded applications, low interstitial alloys should be used in conjunction with processes/conditions that maximize the weld cooling rate. The use of PWHT, where applicable, is highly recommended, since this treatment has beneficial effects with respect to both IGC and HTE.

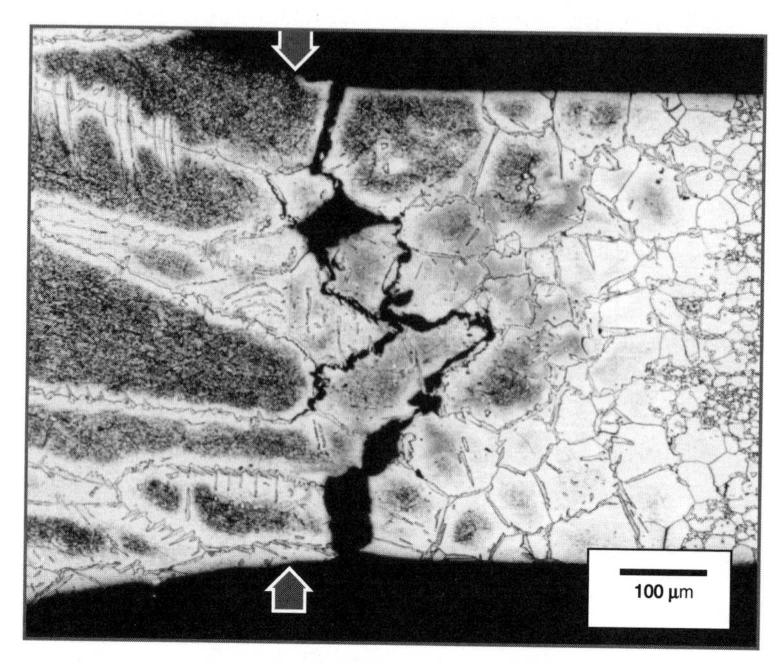

Figure 5.28 IGC in Type 430 stainless steel showing failure in the HAZ. The fracture path is along the ferrite–martensite boundary. Arrows indicate the location of the fusion boundary.

5.7 POSTWELD HEAT TREATMENT

As indicated in previous sections of this chapter, PWHT of ferritic stainless steels can be helpful in restoring toughness, ductility, and corrosion resistance of welded structures. In some cases, preheat and interpass temperatures are also controlled to optimize properties. Recommendations regarding preheat, interpass temperature and PWHT are extremely microstructure dependent [7,8]. In welds that are fully ferritic, preheat is not necessary and, in fact, may exacerbate grain growth and precipitation, since the cooling rate will be reduced. PWHT in the range from 750 to 800°C (1382 to 1472°F) may be used to reduce weld residual stresses, although this thermal treatment has little effect on microstructure.

However, such a PWHT can be dangerous for very high-Cr, high-Mo alloys such as 25-4-4, 29-4, or 29-4-2, since sigma and/or chi phase can form rapidly at these temperatures. Kiesheyer and Brandis [20] report sigma and chi formation in a matter of minutes in this temperature range for similar alloys. In welds that contain martensite, preheat and interpass temperatures in the range 200 to 300°C (392 to 572°F) are recommended, although these treatments must be balanced by the potential for ferrite grain growth and precipitation induced by slower weld cooling rates. A PWHT in the range 750 to 800°C is again used to reduce weld residual stresses and temper any martensite that is present. This temperature range is below that in which austenite will re-form.

Hooper [50] has also shown the beneficial effects of PWHT on the HAZ toughness of Types 405 and 430, as shown in Figure 5.29. It is presumed that the dramatic shift in the DBTT results from the tempering of the martensite. This thermal treatment may also improve the cold deformation characteristics of the weld region, although there are few data to support this premise. The improvement in toughness with PWHT is also shown in Figure 5.30 for weld metals of Types 430 and 430Nb

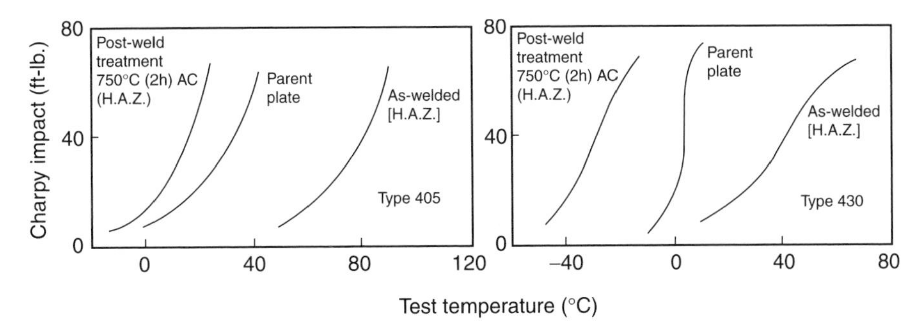

Figure 5.29 Effect of PWHT on the toughness of the HAZ in Types 405 and 430. (From Hooper [50].)

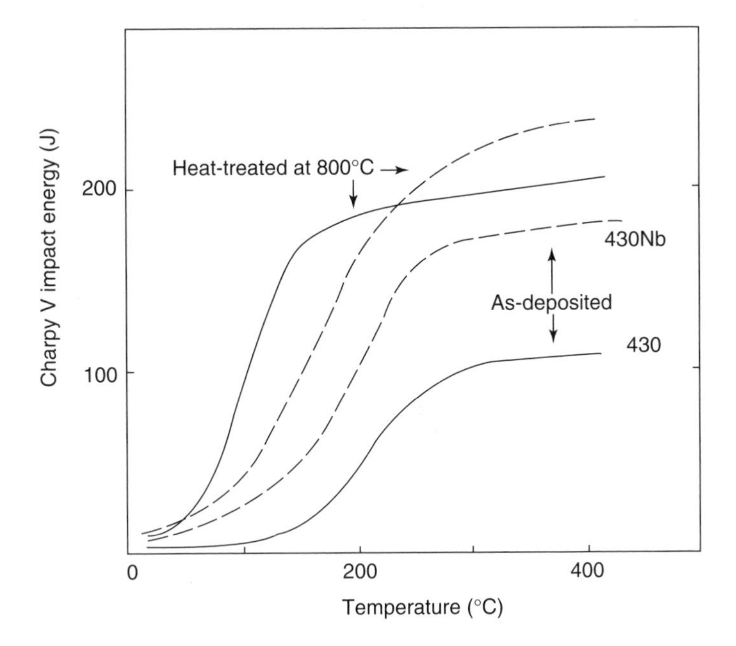

Figure 5.30 Variation in weld metal toughness (CVN) versus temperature for Type 430 (solid line) and Type 430Nb (dashed line). (From Castro and de Cadenet [7]. Courtesy of Cambridge University Press.)

in both the as-welded and PWHT conditions [7]. The inherent improvement in as-welded toughness of the stabilized grade, relative to Type 430, is probably due to the absence of martensite in the as-welded microstructure. The presence of martensite in Type 430 also accounts for the more dramatic improvement in toughness upon PWHT. As shown in Figure 5.30, PWHT results in a shift in DBTT of approximately 100°C (180°F) for Type 430, relative to a shift of 50°C (90°F) for Type 430Nb. The shift in toughness exhibited by Type 430Nb is related primarily to the reduction in HTE, resulting from the overaging of Cr- and Nb-rich precipitates [30,35].

5.8 FILLER METAL SELECTION

A variety of filler metals can be used with ferritic stainless steels, depending on the properties and service performance required. Matching, or near-matching, compositions are most often used since these will be most compatible with the base metal. In cases where matching filler metal is not commercially available, particularly with the very high alloy third-generation ferritic stainless steels, fabricators have resorted to shearing thin sheets of base metal into narrow strips to use as filler metal for GTAW. Austenitic stainless steels are also employed and will result in a two-phase, austenite + ferrite microstructure that has superior ductility and toughness relative to a ferritic stainless steel deposit. Austenitic stainless steel filler metals are often selected for use with the first- and second-generation ferritic stainless steel base metals. As noted previously, there is a risk in applying austenitic stainless filler metal to third-generation ferritic stainless steels because they are likely to be used in environments where chloride SCC resistance is essential. Austenitic stainless steels are generally inferior to ferritic stainless steels in chloride SCC resistance.

Ni-base filler metals are also compatible with ferritic stainless steels and will generally produce a fully austenitic weld deposit. These filler metals are normally used with the high-Cr grades in order to match the base metal corrosion resistance while providing good weld metal mechanical properties. In particular, properly chosen Ni-base filler metal, such as AWS A5.14 ERNiCrMo-3, can match the resistance of high-Cr ferritic stainless steel to chloride SCC. The austenitic and Ni-base filler metals can also be used in dissimilar welds between ferritic alloys and other materials, such as structural steels or other types of stainless steel. In such cases, the filler metal selection need only match the corrosion resistance of whichever of the two base metals is the less resistant in the design environment.

5.9 CASE STUDY: HAZ CRACKING IN TYPE 436 DURING COLD DEFORMATION

Many ferritic stainless steels are rolled to thin sheet stock for applications in the automotive and other industries. As described in Chapter 1, a Sendzimir mill is often used for producing this product. Hot-rolled bands of steel are used as the starting stock. To improve efficiency, hot-rolled bands are welded together to produce a longer starting coil of steel. In the case described here, two bands of hot-rolled Type 436, nominally 6.4 mm (0.25 in.) in thickness, are welded in a single pass using the hot-wire GTAW process and an austenitic Type 308 filler metal. This double band is then recoiled and stored prior to cold rolling.

Upon uncoiling this band and passing it through the Sendzimir mill, cracking sometimes occurred in the HAZ of the GTAW weld. When cracking occurred, it was usually on the first or second pass through the mill and was associated with the band stretching and bending as it passed over a "belly" roll prior to being cold reduced. In most cases, the band fractured completely, requiring the mill to be shut

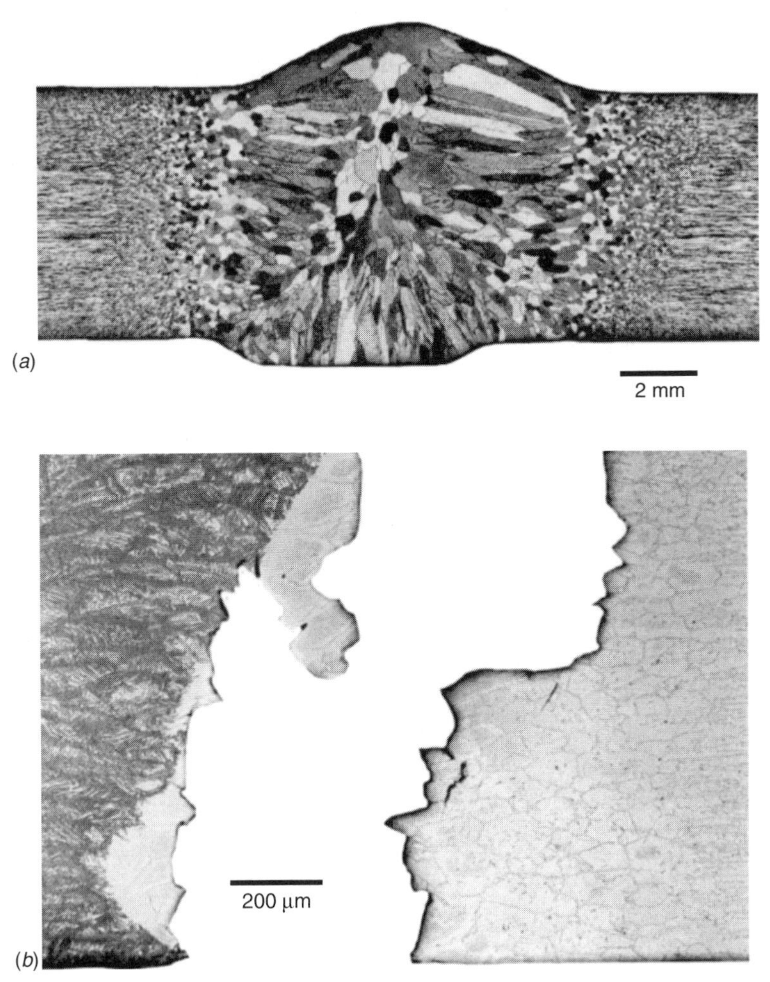

Figure 5.31 GTAW weld between two Type 436 hot-rolled steels made with Type 308 filler metal: (*a*) as-welded condition; (*b*) fracture after one pass through the cold-rolling mill.

down and the band removed. A cross section of the original weld and a representative through-thickness crack that has occurred after one cold reduction pass are shown in Figure 5.31. Note that cracking occurs in the coarse-grained region of the HAZ. A micrograph and fractograph representative of this failure was shown in Figures 5.23 and 5.24, respectively. Failure resulted from high-temperature embrittlement (HTE) in the HAZ. The effect of grain growth on the toughness of the simulated HAZ in Type 436 was shown in Figure 5.25. Increasing grain size

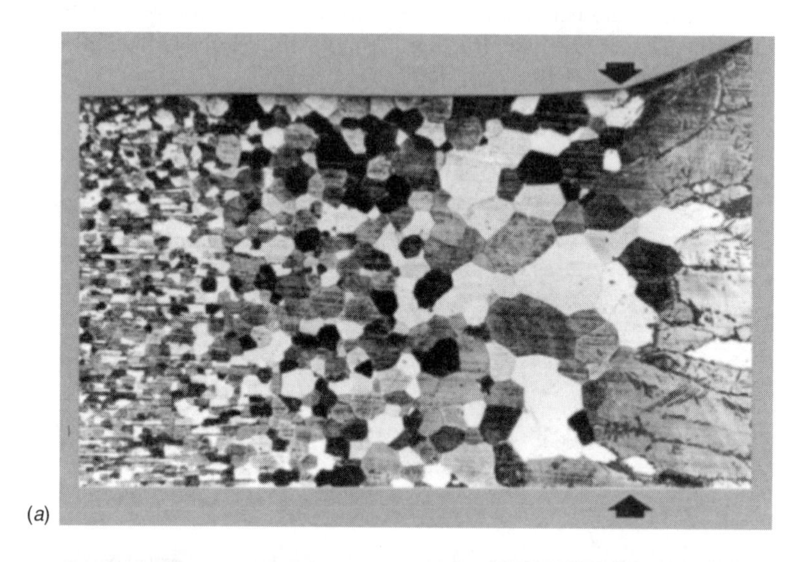
(*a*)

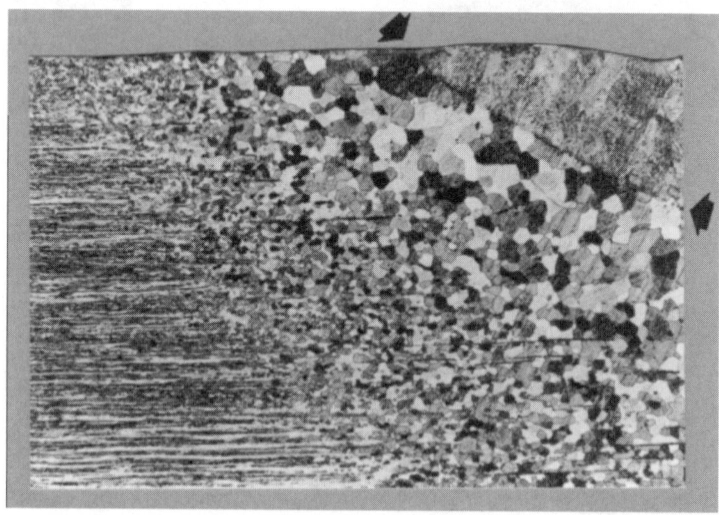
(*b*)

Figure 5.32 Reduction in grain size achieved by adopting a two-pass welding technique: (*a*) single-pass weld; (*b*) two-pass weld.

shifted the DBTT above room temperature and reduced the upper shelf toughness significantly.

Several approaches were taken to prevent cracking during cold reduction of Type 436. A two-pass GTAW coil joining process was developed that reduced the grain size in the HAZ, as shown in Figure 5.32. A short-time postweld stress relief heat treatment was also adopted, consisting of heating for 10 minutes to 800°C (1470°F). Following this heat treatment, the weld overbead was ground flush and the hot bands recoiled. Prior to cold reduction, the coiled hot bands were moved to a portion of the plant where their temperature would reach close to 20°C (70°F). Since most coil breakage occurred in the winter months with coils that where stored in an unheated portion of the plant, this measure was adopted to increase the toughness of the material based on the data shown in Figure 5.25. The combination of grain size control, PWHT, and coil temperature control prior to cold rolling was very successful, and incidents of coil breakage due to HTE in the weld HAZ were eliminated.

5.10 CASE STUDY: INTERGRANULAR STRESS CORROSION CRACKING IN THE HAZ OF TYPE 430

Special facilities are required to house critical electronic communication equipment for protection against electromagnetic radiation that can destroy or impair the equipment. Such protection can be achieved by constructing welded metal structures that are essentially hermetic to electromagnetic radiation. These are usually constructed of carbon steel and subsequently painted, but in an effort to reduce maintenance costs, a facility was constructed using Type 430 stainless steel. The Type 430 was supplied as 18-gauge (0.048-in.-thick) sheet metal and corrugated roof panels. Welding was performed using both GMAW and GTAW with a matching filler metal.

Almost immediately after construction of the facility was completed, cracking was detected in the welds joining the roof panels. Attempts were made to repair these cracks using the manual GTAW process. Within weeks, the cracks reappeared in the repair welds. A cross section through one of the GMAW fillet welds is shown in Figure 5.33. Note the extremely large grain size in the weld metal and HAZ. A typical failure that occurred in these welds was shown in Figure 5.28. A higher-magnification micrograph of the fracture path is shown in Figure 5.34. Note that the fracture path is along the ferrite grain boundary and at the interface between the ferrite and martensite at the boundary. A fractograph of this failure is shown in Figure 5.35.

In determining the nature of the failure, it was important to understand the environment. The facility was located on a tropical island that experienced almost daily afternoon rainstorms. The tropical sun would heat the roof to temperatures approaching 150°C (300°F). Because of the design of the roof, the thermal expansion of 10-m-long panels created a tensile bending stress in the vicinity of the weld connecting the roof panels. The combination of tensile stress due to thermal

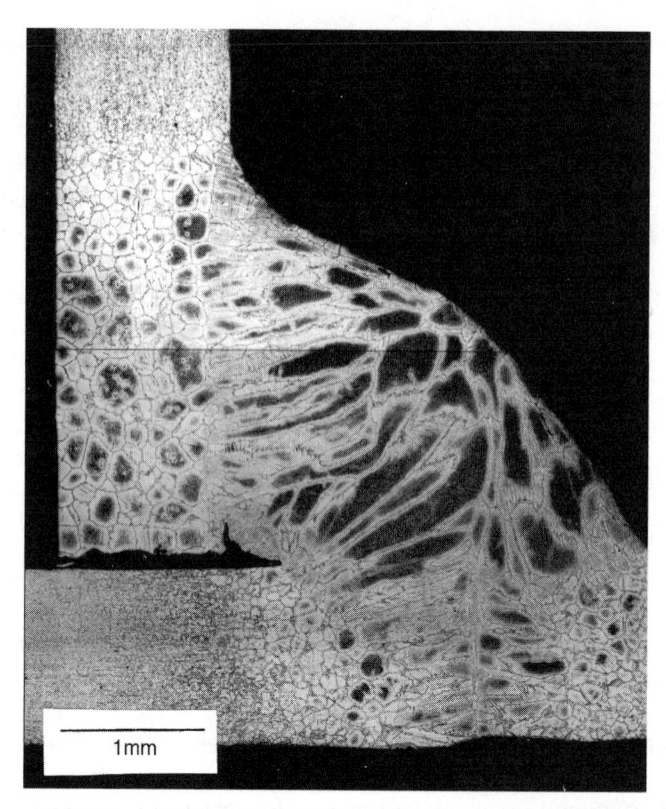

Figure 5.33 Cross section of GMA fillet weld between Type 430 sheet alloys.

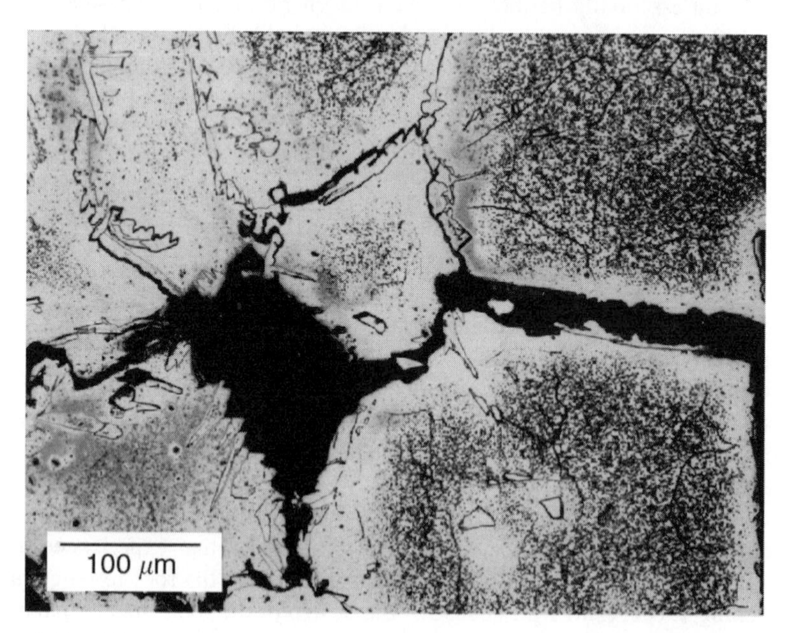

Figure 5.34 IGSCC along the ferrite grain boundaries in the HAZ of Type 430. Note that cracking propagates along the ferrite–martensite interface.

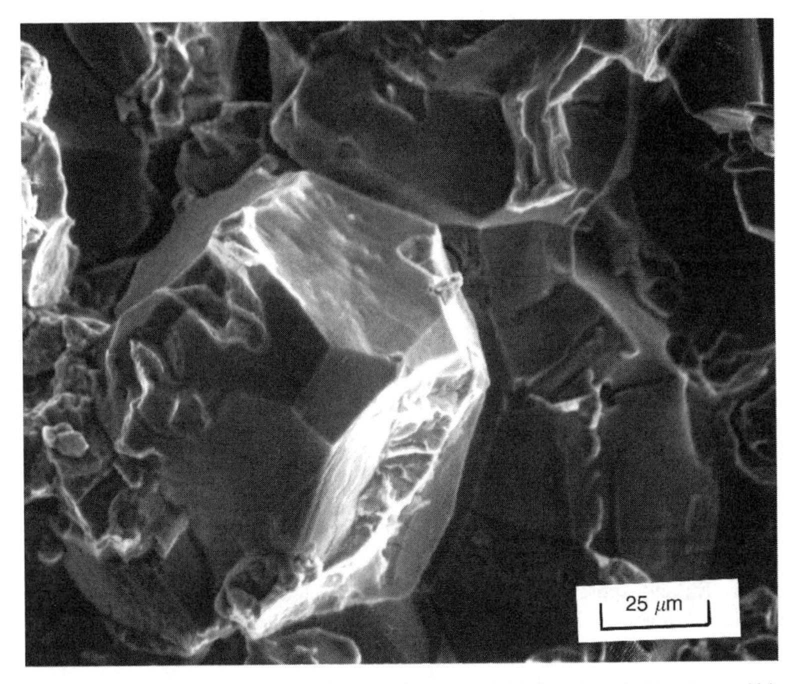

Figure 5.35 Fracture surface characteristic of IGSCC in the HAZ of Type 430.

expansion of the roof and chloride ions resulting from the tropical rain water promoted intergranular stress corrosion cracking (IGSCC). Repair welding actually accelerated the IGSCC process since the repair welds further overheated the structure, creating even larger grains in the HAZ and promoted additional sensitization. The repair welds also increased the degree of stress concentration at the point of tensile bending. This problem could have been prevented if the proper base metal was selected. Rather than Type 430, a stabilized grade such as Type 439 or 468 would have been immune to IGSCC in this environment. These alloys prevent the sensitization of the HAZ resulting from $M_{23}C_6$ precipitation at the ferrite grain boundaries, or ferrite–martensite interface, by the formation of more stable MC carbides enriched in Ti and Nb.

REFERENCES

1. Thielsch, H. 1951. Physical and welding metallurgy of chromium stainless steels, *Welding Journal*, 30(5):209s–250s.
2. Demo, J. J. 1977. Structure and constitution of wrought ferritic stainless steels, in *Handbook of Stainless Steels*, D. Peckner and I. M. Bernstein, eds., McGraw-Hill, New York.
3. Lacombe, P., Baroux, B., and Beranger, G., eds. 1993. *Stainless Steels*, Les Éditions de Physique, Les Ulis, France.

4. Castro, R., and Tricot, R. 1962. Études des transformations isothermes dans les aciers inoxydables semi-ferritiques á 17% de chrome, *Memoires Scientifiques de la Revue de Metallurgie*, Part 1, 59:571–586; Part 2, 59:587–596.
5. Baerlecken, E., Fischer, W., and Lorenz, K. 1961. Investigations concerning the transformation behavior, the notched impact toughness and the susceptibility to intergranular corrosion of iron–chromium alloys with chromium contents to 30%, *Stahl und Eisen*, 81(12):768.
6. Nishio, Y., Ohmae,T., Yoshida, Y., and Miura, A. 1971. Weld cracking and mechanical properties of 17% chromium steel weldment, *Welding Journal*, 50(1):9s–18s.
7. Castro, R. J., and de Cadenet, J. J. 1974. *Welding Metallurgy of Stainless and Heat-Resisting Steels*, Cambridge University Press, Cambridge.
8. Folkhard, E., *Welding Metallurgy of Stainless Steels*, Springer-Verlag, Berlin (in German; 1998, English translation).
9. Hayden, H. W., and Floreen, S. 1950. The influence of martensite and ferrite on the properties of two-phase stainless steels having microduplex microstructures, *Metallurgical Transactions*, Vol. 1, pp. 1955–1959.
10. Wright, R. N., and Wood, J. R. 1977. Fe–Cr–Mn micro-duplex ferritic martensitic stainless steels, *Metallurgical Transactions*, 8A:1977–2007.
11. Thomas, R., and Apps, R. L. 1980. Weld heat-affected zone properties of AISI 409 ferritic stainless steel, in *Toughness of Ferritic Stainless Steels*, ASTM STP 706, R. A. Lula, ed., American Society for Testing and Materials, West Conshohocken, PA, pp. 161–183.
12. Lippold, J. C. 1996. Unpublished research.
13. Williams, R. O. 1958. Further studies of the iron-chromium system, *Trans. AIME,* 212:497.
14. Williams, R. O., and Paxton, H. W. 1957. The nature of aging of binary iron-chromium alloys around 500°C, *JISI,* 185:358–374.
15. Marcincowski, M. J., Fisher, R. M., and Szirmae, A. 1964. Effect of 500°C aging on the deformation behavior of an iron-chromium alloy, *Trans. AIME,* 230:676–689.
16. Shortsleeve, F. J., and Nicholson, M. E. 1951. Transformations in ferritic chromium steels between 1100 and 1500°F (595 and 815°C), *Trans.ASM,* 43:142–156.
17. Grobner, P. J. 1973. The 885°F (475°C) embrittlement of ferritic stainless steels, *Met. Trans,* 4:251–260.
18. Zappfe, C. A., and Worden, C. O. 1951. A notch-bend test, *Welding Journal,* 30(1):47s–54s.
19. Bandel, G., and Tofaute, W. 1941. *Arch. Eisenhut.,*15(7):307.
20. Kiesheyer, H., and Brandis, H. 1977. Ausscheidungs- und Versprödungsverhalten nickelhaltiger Superferite (Precipitation and embrittlement of nickel containing Superferites), Zeitßchrift für Werkstoffech. 8(3):69–77.
21. Semchysen, M., Bond, A. P., and Dundas, H. J. 1971. Effects of composition on ductility and toughness of ferritic stainless steels, in *Proceedings of the Symposium Toward Improved Ductility and Toughness*, Kyoto, Japan, p. 239.
22. Plumtree, A., and Gullberg, R. 1980. The influence of interstitial and some substitutional alloying elements, in *Toughness of Ferritic Stainless Steels*, ASTM STP 706, R. A. Lula, ed., American Society for Testing and Materials, West Conshohocken, PA, pp. 34–55.
23. Grubb, J. F., and Wright, R. N. 1979. The role of C and N in the brittle fracture of Fe–26Cr, *Metallurgical Transactions,* 10A:1247–1255.

24. Wright, R. N. 1980. Toughness of ferritic stainless steels, in *Toughness of Ferritic Stainless Steels*, ASTM STP 706, R. A. Lula, ed., American Society for Testing and Materials, West Conshohocken, PA, pp. 2–33.
25. Pollard, B. 1974. Effect of titanium on the ductility of 26% chromium, low interstitial ferritic stainless steel, *Metals Technology*, 1:31.
26. Richter, J., and Finke, P. 1976. *Freiberger Forschungschefte Metallurgie*, B172:55.
27. Krivobok, V. N. 1935. *Transactions of the American Society for Metals*, pp. 1–56.
28. Binder, W. O., and Spendelow, H. R. 1951. The influence of chromium on the mechanical properties of plain chromium steels, *Transactions of the American Society for Metals*, 43:759–772.
29. Castro, R., and Tricot, R. 1964. Study of the isothermal transformations in 17% Cr stainless steels, 2: influence of carbon and nitrogen, *Metal Treatment and Drop Forging*, 31(231):469.
30. Thomas, C. R., and Robinson, F. P. A. 1978. Kinetics and mechanism of grain growth during welding in niobium-stabilized 17% chromium stainless steels, *Metals Technology*, 5(4):133.
31. ASM. 1973. *ASM Metals Handbook*, 8th ed., Vol. 8, ASM International, Materials Park, OH. p. 424.
32. Kaltenhauser, R. H. 1971. Improving the engineering properties of ferritic stainless steels, *Metals Engineering Quarterly*, 11(2):41–47.
33. Balmforth, M. C., and Lippold, J. C. 2000. A new ferritic–martensitic stainless steel constitution diagram, *Welding Journal*, 79(12):339s–345s.
34. Lippold, J. C., and Shademan, S. 1992. Unpublished research.
35. Hunter, G. B., and Eagar, T. W. 1980. Ductility of stabilized ferritic stainless steel welds, *Metallurgical Transactions*, 11A:213–218.
36. Pollard, B. 1972. Ductility of ferritic stainless weld metal, *Welding Journal*, 51(4):222s–230s.
37. Wright, R. N. 1971. Mechanical behavior and weldability of a high chromium ferritic stainless steel as a function of purity, *Welding Journal*, 50(10):434s–440s.
38. Demo, J. J. 1971. Mechanism of high temperature embrittlement and loss of corrosion resistance in AISI Type 446 stainless steel, *Corrosion*, 27(12):531.
39. Krysiak, K. F. 1980. Weldability of the new generation of ferritic stainless steels: update, in *Toughness of Ferritic Stainless Steels*, ASTM STP 706, R. A. Lula, ed., American Society for Testing and Materials, West Conshohocken, PA, pp. 221–240.
40. Krysiak, K. F. 1986. Welding behavior of ferritic stainless steels: an overview, *Welding Journal*, 65(4):37–41.
41. Grubb, J. F. Private communication, Allegheny-Ludlum.
42. Kah, D. H., and Dickinson, D. W. 1981. Weldability of ferritic stainless steels, *Welding Journal*, 60(8):135s–142s.
43. DeRosa, S., Jacobs, M. H., Jones, D. G., and Sherhod, C. 1979. Studies of TIG weld pool solidification and weld bead microstructure in stainless steel tubes, in *Solidification and Casting of Metals*, Metals Society, London, p. 416.
44. Sawhill, J. M., Jr., and Bond, A. P. 1976. Ductility and toughness of stainless steel welds, *Welding Journal*, 55(2):33s–41s.

45. Villafuerte, J. C., and Kerr, H. W. 1990. The effect of alloy composition and welding conditions on columnar-equiaxed transitions in ferritic stainless steel gas-tungsten arc welds, *Metallurgical Transactions A*, Vol. 21A(7):2009–2019.
46. Washko, S. D., and Grubb, J. F. 1991. The effect of niobium and titanium dual stabilization on the weldability of 11% chromium ferritic stainless steels, in *Proceedings of the International Conference on Stainless Steels*, Chiba, Japan, published by Iron and Steel Institute of Japan.
47. Bond, A. P. 1969. Mechanisms of intergranular corrosion in ferritic stainless steels, *Transactions of AIME*, 245(8):2127–2134.
48. Nichol, T. J., and Davis, J. A. 1978. Intergranular corrosion testing and sensitization of two high-chromium ferritic stainless steels, in *Intergranular Corrosion of Stainless Alloys*, ASTM STP 656, R. F. Steigerwald, ed., American Society for Testing and Materials, West Conshohocken, PA, pp. 179–196.
49. Fritz, J. D., and Franson, I. A. 1997. Sensitization and stabilization of Type 409 ferritic stainless steel, *Materials Performance*, August, pp. 57–61.
50. Hooper, R. A. E. 1972. Ferritic stainless steels, *Sheet Metal Industries*, 49(1):26.

CHAPTER 6

AUSTENITIC STAINLESS STEELS

Austenitic stainless steels represent the largest of the general groups of stainless steels and are produced in higher tonnages than any other group. They have good corrosion resistance in most environments. The austenitic stainless steels have strengths equivalent to those of mild steels, approximately 210 MPa (30 ksi) minimum yield strength at room temperature, and are not transformation hardenable. Low-temperature impact properties are good for these alloys, making them useful in cryogenic applications. Service temperatures can be up to 760°C (1400°F) or even higher, but the strength and oxidation resistance of most of these steels are limited at such high temperatures. Austenitic stainless steels can be strengthened significantly by cold working. They are often used in applications requiring good atmospheric or elevated temperature corrosion resistance. They are generally considered to be weldable, if proper precautions are followed.

Elements that promote the formation of austenite, most notably nickel, are added to these steels in large quantities (generally over 8 wt%). Other austenite-promoting elements are C, N, and Cu. Carbon and nitrogen are strong austenite promoters, as can be seen from the various values in nickel equivalency formulas (see Chapter 3). Carbon is added to improve strength (creep resistance) at high temperatures. Nitrogen is added to some alloys to improve strength, mainly at ambient and cryogenic temperatures, sometimes more than doubling it. Nitrogen-strengthened alloys are usually designated with a suffix N added to their AISI 300 series designation (e.g., 304LN). The AISI 200 series (e.g., 201) are also nitrogen strengthened and are commonly referred to under various trade names, such as Nitronic®.

Welding Metallurgy and Weldability of Stainless Steels, by John C. Lippold and Damian J. Kotecki
ISBN 0-471-47379-0

Composition Range of Standard Austenitic Stainless Steels

- 16 to 25 wt% chromium
- 8 to 20 wt% nickel
- 1 to 2 wt% manganese
- 0.5 to 3 wt% silicon
- 0.02 to 0.08 wt% carbon (<0.04 wt% designated *L grades*)
- 0 to 2 wt% molybdenum
- 0 to 0.15 wt% nitrogen
- 0 to 0.2 wt% titanium and niobium

Austenitic stainless steels generally have good ductility and toughness and exhibit significant elongation during tensile loading. They are more expensive than the martensitic and low to medium Cr ferritic grades, due to the higher alloy content of these alloys. Despite the steel cost, they offer distinct engineering advantages, particularly with respect to formability and weldability, that often reduce the overall cost compared to other groups of stainless steels.

Although there are a wide variety of austenitic stainless steels, the 300 series alloys are the oldest and most commonly used. Most of these alloys are based on the 18Cr–8Ni system, with additional alloying elements or modifications to provide unique or enhanced properties. Type 304 is the foundation of this alloy series, and along with 304L, represents the most commonly selected austenitic grade. Type 316 substitutes approximately 2%Mo for a nearly equal amount of Cr to improve pitting corrosion resistance.

The *stabilized grades*, 321 and 347, contain small additions of Ti and Nb, respectively, to combine with carbon and reduce the tendency for intergranular corrosion due to Cr-carbide precipitation. The *L grades* became popular in the 1960s and 1970s with the advent of AOD (argon–oxygen decarburization) melting practice that reduced the cost differential between standard (not low carbon) and L grades. These low-carbon grades (304L, 316L) have been widely used in applications where intergranular attack and intergranular stress corrosion cracking are a concern. These forms of corrosion attack are discussed later in the chapter.

Austenitic stainless steels are used in a wide range of applications, including structural support and containment, architectural uses, kitchen equipment, and medical products. They are widely used not only because of their corrosion resistance but because they are readily formable, fabricable, and durable. Some highly alloyed grades are used for very high temperature service [above 1000°C (1830°F)] for applications such as heat-treating baskets. In addition to higher chromium levels, these alloys normally contain higher levels of silicon (and sometimes aluminum) and carbon, to maintain oxidation and/or carburization resistance and strength, respectively.

It should be pointed out that the common austenitic stainless steels are not an appropriate choice in some common environments such as seawater or other chloride-containing media, or in highly caustic environments. This is due to their susceptibility to stress corrosion cracking, a phenomenon that afflicts the base metal, HAZ, and weld metal in these alloys. Care should be taken when selecting stainless steels that will be under significant stress in these environments. The aspects and consequences of stress corrosion cracking will be described later.

6.1 STANDARD ALLOYS AND CONSUMABLES

Austenitic stainless steels include both the 200 series and 300 series alloys, as designated by the American Iron and Steel Institute (AISI). The 200 series alloys contain high levels of carbon, manganese, and nitrogen and are used in specialty applications, such as where galling resistance is required. These alloys also have lower nickel content than the 300 series alloys to balance the high carbon and nitrogen levels. The 300 series alloys are by far the most widely used of the austenitic grades. A list of the most common 300 series alloys is provided in Table 6.1. A more comprehensive list of alloys is provided in Appendix 1.

The most widely used alloys, Types 304, 316, 321, and 347 and their variants, are of the "18–8" type with nominal values of 18Cr and 8–10Ni. L grades represent low-carbon variants with a nominal carbon level of 0.03 wt%. These alloys have improved resistance to intergranular attack in corrosive environments. H grades have carbon levels approaching 0.1 wt%. These alloys are used at elevated temperature since they have higher elevated temperature strength than standard or L grades. The N grades have nitrogen added as an intentional addition to levels as high as 0.20 wt % in the 300 series alloys (304N, 316N) and to even higher levels in alloys that contain high manganese contents. (Manganese increases the solubility of nitrogen in the austenite phase.) The higher nitrogen improves the strength, galling resistance, and pitting corrosion resistance of austenitic stainless steels.

Alloys containing titanium and niobium, such as Types 321 and 347, are known as *stabilized grades* since the addition of these two elements stabilizes the alloy against the formation of $M_{23}C_6$ chromium carbides. Since niobium and titanium both form stable MC-type carbides at elevated temperature, the formation of chromium-rich carbides is restricted. The addition of these elements at levels up to 1.0 wt% effectively reduces the matrix carbon content and makes chromium carbide precipitation more difficult. This reduces the possibility for sensitization that can lead to intergranular corrosion in austenitic alloys. This phenomenon is described in more detail in Section 6.6.

Many other specialty alloys exist, including the superaustenitic grades. These alloys are described separately in a later section because of their unique characteristics and weldability concerns.

Austenitic stainless steel filler metals are listed in Table 6.2. Note that this table is divided into three parts, reflecting three AWS specifications for consumables based on type: (1) AWSA5.4 for covered electrodes (SMAW), (2) AWS A5.9 for bare wire and tubular metal-cored electrodes (GTAW and GMAW), and (3) AWS A5.22 for gas-shielded flux-cored electrodes (FCAW). Many of the same consumables

TABLE 6.1 Composition of Standard Wrought Austenitic Stainless Steels

Type	UNS No.	Composition (wt%)[a]									
		C	Mn	P	S	Si	Cr	Ni	Mo	N	Other
201	S20100	0.15	5.5–7.5	0.06	0.03	1.0	16.0–18.0	3.5–5.5	—	0.25	—
302	S30200	0.15	2.0	0.045	0.03	1.0	17.0–19.0	8.0–10.0	—	—	—
304	S30400	0.08	2.0	0.045	0.03	1.0	18.0–20.0	8.0–10.5	—	—	—
304L	S30403	0.03	2.0	0.045	0.03	1.0	18.0–20.0	8.0–12.0	—	—	—
304H	S30409	0.04–0.1	2.0	0.045	0.03	1.0	18.0–20.0	8.0–10.5	—	—	—
308	S30800	0.08	2.0	0.045	0.03	1.0	19.0–21.0	10.0–12.0	—	—	—
309	S30900	0.20	2.0	0.045	0.03	1.0	22.0–24.0	12.0–15.0	—	—	—
310	S31000	0.25	2.0	0.045	0.03	1.0	24.0–26.0	19.0–22.0	—	—	—
316	S31600	0.08	2.0	0.045	0.03	1.0	16.0–18.0	10.0–14.0	2.0–3.0	—	—
316L	S31603	0.03	2.0	0.045	0.03	1.0	16.0–18.0	10.0–14.0	2.0–3.0	—	—
317	S31700	0.08	2.0	0.045	0.03	1.0	18.0–20.0	11.0–15.0	3.0–4.0	—	—
321	S32100	0.08	2.0	0.045	0.03	1.0	17.0–19.0	9.0–12.0	—	—	Ti: 5 × C-0.70
330	S33000	0.10	2.0	0.045	0.03	0.75–1.5	17.0–20.0	34.0–37.0	—	—	—
347	S34700	0.08	2.0	0.045	0.03	1.0	17.0–19.0	9.0–13.0	—	—	Nb: 10 × C-1.00

[a]A single value is a maximum.

TABLE 6.2 Austenitic Stainless Steel AWS Filler Metal Classifications

Type	UNS No.	Composition (wt%)[a]									
		C	Mn	P	S	Si	Cr	Ni	Mo	N	Other
Part 1: Covered Electrodes from AWS A5.4											
219	W32310	0.06	8.0–10.0	0.04	0.03	1.00	19.0–21.5	5.5–7.0	0.75	0.10–0.30	—
308	W30810	0.08	0.5–2.5	0.04	0.03	1.0	18.0–21.0	9.0–11.0	0.75	—	—
308H	W30810	0.04–0.08	0.5–2.5	0.04	0.03	1.0	18.0–21.0	9.0–11.0	0.75	—	—
308L	W30813	0.04	0.5–2.5	0.04	0.03	1.0	18.0–21.0	9.0–11.0	0.75	—	—
309	W30910	0.15	0.5–2.5	0.04	0.03	1.0	22.0–25.0	12.0–14.0	0.75	—	—
309L	W30917	0.04	0.5–2.5	0.04	0.03	1.0	22.0–25.0	12.0–14.0	0.75	—	—
310	W31010	0.08–0.20	1.0–2.5	0.03	0.03	0.75	25.0–28.0	20.0–22.5	0.75	—	—
316	W31610	0.08	0.5–2.5	0.04	0.03	1.0	17.0–20.0	11.0–14.0	2.0–3.0	—	—
316H	W31610	0.04–0.08	0.5–2.5	0.04	0.03	1.0	17.0–20.0	11.0–14.0	2.0–3.0	—	—
316L	W31613	0.04	0.5–2.5	0.04	0.03	1.0	17.0–20.0	11.0–14.0	2.0–3.0	—	—
317	W31710	0.08	0.5–2.5	0.04	0.03	1.0	18.0–21.0	12.0–14.0	3.0–4.0	—	—
317L	W31713	0.04	0.5–2.5	0.04	0.03	1.0	18.0–21.0	12.0–14.0	3.0–4.0	—	—
330	W88331	0.18–0.25	1.0–2.5	0.04	0.03	0.90	14.0–17.0	33.0–37.0	0.75	—	—
347	W34710	0.08	0.5–2.5	0.04	0.03	1.0	18.0–21.0	9.0–11.0	0.75	—	Nb: 8 × C-1.00
Part 2: Bare Electrodes, Bare Rods, Tubular Metal-Cored Electrodes, and Strips from AWS A5.9											
219	S21980	0.05	8.0–10.0	0.03	0.03	1.00	19.0–21.5	5.5–7.0	0.75	0.10–0.30	—
308	S30880	0.08	1.0–2.5	0.03	0.03	0.30–0.65	19.5–22.0	9.0–11.0	0.75	—	—
308H	S30880	0.04–0.08	1.0–2.5	0.03	0.03	0.30–0.65	19.5–22.0	9.0–11.0	0.75	—	—
308L	S30883	0.03	1.0–2.5	0.03	0.03	0.30–0.65	19.5–22.0	9.0–11.0	0.75	—	—
308Si	S30881	0.08	1.0–2.5	0.03	0.03	0.65–1.00	19.5–22.0	9.0–11.0	0.75	—	—
308LSi	S30888	0.03	1.0–2.5	0.03	0.03	0.65–1.00	19.5–22.0	9.0–11.0	0.75	—	—
309	S30980	0.12	1.0–2.5	0.03	0.03	0.30–0.65	23.0–25.0	12.0–14.0	0.75	—	—
309L	S30983	0.03	1.0–2.5	0.03	0.03	0.30–0.65	23.0–25.0	12.0–14.0	0.75	—	—

(continued)

TABLE 6.2 *(Continued)*

Type	UNS No.	Composition (wt%)[a]									
		C	Mn	P	S	Si	Cr	Ni	Mo	N	Other
309Si	S30981	0.12	1.0–2.5	0.03	0.03	0.65–1.00	23.0–25.0	12.0–14.0	0.75	—	—
309LSi	S30988	0.03	1.0–2.5	0.03	0.03	0.65–1.00	23.0–25.0	12.0–14.0	0.75	—	—
310	S31080	0.08–0.15	1.0–2.5	0.03	0.03	0.30–0.65	25.0–28.0	20.0–22.5	0.75	—	—
316	S31680	0.08	1.0–2.5	0.03	0.03	0.30–0.65	18.0–20.0	11.0–14.0	2.0–3.0	—	—
316H	S31680	0.04–0.08	1.0–2.5	0.03	0.03	0.30–0.65	18.0–20.0	11.0–14.0	2.0–3.0	—	—
316L	S31683	0.03	1.0–2.5	0.03	0.03	0.30–0.65	18.0–20.0	11.0–14.0	2.0–3.0	—	—
316Si	S31681	0.08	1.0–2.5	0.03	0.03	0.65–1.00	18.0–20.0	11.0–14.0	2.0–3.0	—	—
316LSi	S31688	0.03	1.0–2.5	0.03	0.03	0.65–1.00	18.0–20.0	11.0–14.0	2.0–3.0	—	—
317	S31780	0.08	1.0–2.5	0.03	0.03	0.30–0.65	18.5–20.5	13.0–15.0	3.0–4.0	—	—
317L	S31783	0.03	1.0–2.5	0.03	0.03	0.30–0.65	18.5–20.5	13.0–15.0	3.0–4.0	—	—
330	N08331	0.18–0.25	1.0–2.5	0.03	0.03	0.30–0.65	15.0–17.0	34.0–37.0	0.75	—	—
347	S34780	0.08	1.0–2.5	0.03	0.03	0.30–0.65	19.0–21.5	9.0–11.0	0.75	—	Nb: 10 × C-1.0
347Si	S34788	0.08	1.0–2.5	0.03	0.03	0.65–1.00	19.0–21.5	9.0–11.0	0.75	—	Nb: 10 × C-1.0
					Part 3: Gas-Shielded Flux-Cored Electrodes from AWS A5.22[b]						
308	W30831	0.08	0.5–2.5	0.04	0.03	1.0	18.0–21.0	9.0–11.0	0.5	—	—
308L	W30835	0.04	0.5–2.5	0.04	0.03	1.0	18.0–21.0	9.0–11.0	0.5	—	—
308H	W30831	0.04–0.08	0.5–2.5	0.04	0.03	1.0	18.0–21.0	9.0–11.0	0.5	—	—
309	W30931	0.10	0.5–2.5	0.04	0.03	1.0	22.0–25.0	12.0–14.0	0.5	—	—
309L	W30935	0.04	0.5–2.5	0.04	0.03	1.0	22.0–25.0	12.0–14.0	0.5	—	—
310	W31031	0.20	1.0–2.5	0.03	0.03	1.0	25.0–28.0	20.0–22.5	0.5	—	—
316	W31631	0.08	0.5–2.5	0.04	0.03	1.0	17.0–20.0	11.0–14.0	2.0–3.0		
316L	W31635	0.04	0.5–2.5	0.04	0.03	1.0	17.0–20.0	11.0–14.0	2.0–3.0	—	—
317L	W31735	0.04	0.5–2.5	0.04	0.03	1.0	18.0–21.0	12.0–14.0	3.0–4.0	—	—
347	W34731	0.08	0.5–2.5	0.04	0.03	1.0	18.0–21.0	9.0–11.0	0.5	—	Nb: 8 × C-1.0

[a] A single value is a maximum.

[b] Self-shielded flux-cored stainless steel electrodes are also included in the AWS A5.22 standard. They have virtually identical composition limits to those of the gas-shielded electrodes except that the Cr limits tend to be slightly higher in the ferrite-containing grades, to compensate for expected higher nitrogen in the self-shielded deposit.

appear in all three specifications, but the composition limits may be slightly different based on consumable type.

6.2 PHYSICAL AND MECHANICAL METALLURGY

Austenitic stainless steels are formulated and thermo-mechanically processed such that the microstructure is primarily austenite. Depending on the balance of ferrite-promoting to austenite-promoting elements, the wrought or cast microstructure will be either fully austenitic or a mixture of austenite and ferrite. Two examples of wrought austenitic stainless steel microstructures are shown in Figure 6.1. In one case, the microstructure consists of equiaxed austenite grains. In the other case, some residual high-temperature ferrite (delta ferrite) is aligned along the rolling direction. This ferrite results from the segregation of ferrite-promoting elements (primarily chromium) during solidification and thermo-mechanical processing. It is usually present in relatively low volume fraction (less than 2 to 3%). Although not considered deleterious in most applications, the presence of ferrite in the wrought microstructure can reduce the ductility and, potentially, the toughness of austenitic stainless steels. It can also be a preferential site for the precipitation of $M_{23}C_6$ carbides and sigma phase, of which the latter is an embrittling agent in stainless steels.

The transformation behavior of austenitic stainless steels can be described using the Fe–Cr–Ni pseudobinary diagram at 70% constant iron [1], as described in Chapter 2. This diagram is reproduced here as Figure 6.2. Note that primary solidification of austenitic stainless steels can occur as either austenite or ferrite. The demarcation between these two primary phases of solidification is at approximately 18Cr–12Ni in the ternary system. At higher chromium/nickel ratios, primary solidification occurs as delta ferrite and at lower ratios as austenite. Note that there is a small triangular region within the solidification temperature range where austenite, ferrite, and liquid coexist. Alloys that solidify as austenite to the left of this triangular region are stable as austenite upon cooling to room temperature. However, when alloys solidify as ferrite, they may be either fully ferritic or consist of a mixture of ferrite and austenite at the end of solidification. Because of the slope of the ferrite and austenite solvus lines, most or all of the ferrite transforms to austenite under equilibrium cooling conditions, as can be seen for a nominal 20Cr–10Ni alloy where the structure becomes fully austenitic upon cooling to 1000°C (1830°F). For the rapid cooling conditions experienced during welding, this transformation is suppressed and some ferrite will remain in the microstructure. The implications of these are described in more detail in Section 6.3.1.

The effect of other alloying elements on the phase equilibrium in austenitic stainless steels can be determined using phase diagrams generated using computational thermodynamics, such as ThermCalc™ [2]. As an example, a Fe–18Cr–10Ni–1.5Mn–0.5Si–0.04N system with variable carbon and a Fe–10Ni–1.5Mn–0.5Si 0.04C–0.04N system with variable chromium are shown in Figure 6.3. These diagrams are very similar to Figure 2.5*a*, but additionally, show the carbide and nitride precipitation regimes.

A variety of precipitates may be present in austenitic stainless steels, depending on composition and heat treatment. A list of those precipitates, their structure, and

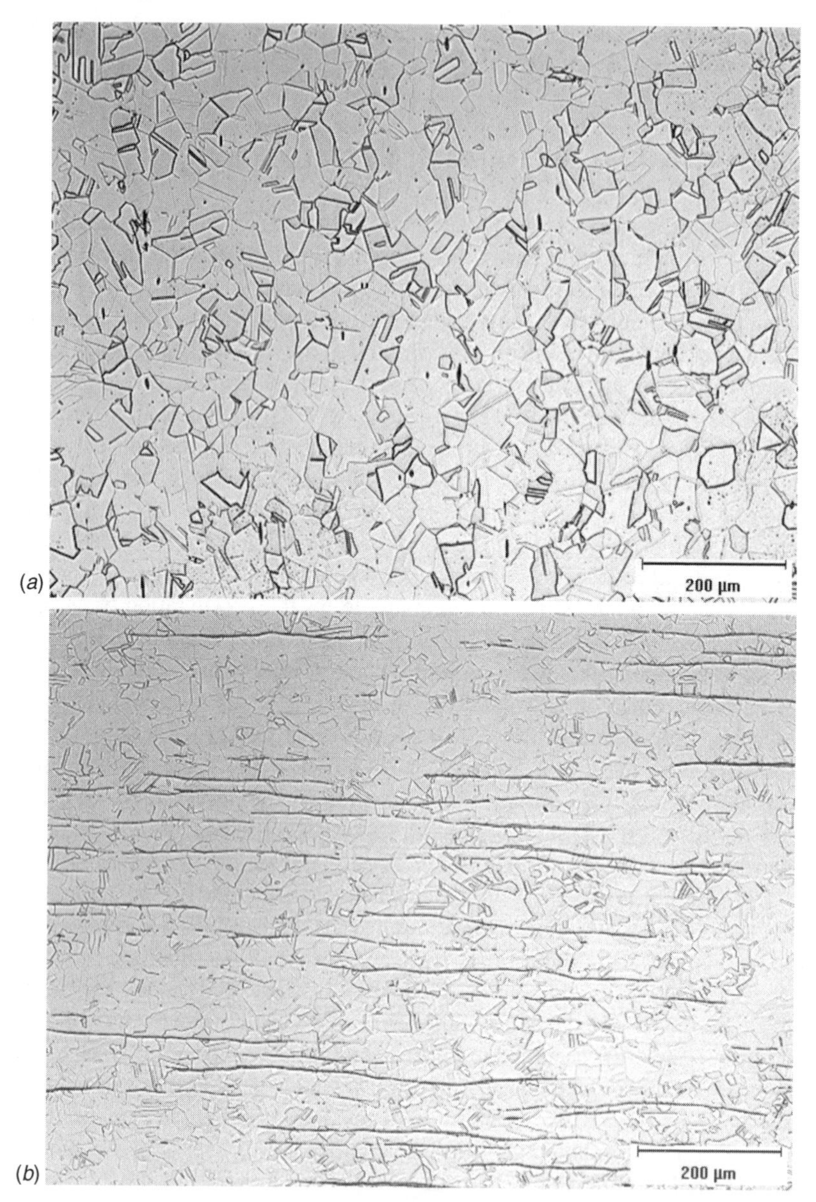

Figure 6.1 Microstructure of Type 304 plate: (*a*) fully austenitic; (*b*) austenite with ferrite stringers.

stoichiometry is provided in Table 6.3. Carbides are present in virtually every austenitic stainless steel, since chromium is a strong carbide former. Additions of other carbide formers, including Mo, Nb, and Ti, also promote carbide formation. The nature of carbide formation, including the effect of composition and the temperature range of formation, is quite complex. The reader is referred to other, more comprehensive sources for more details on carbide formation [3,4].

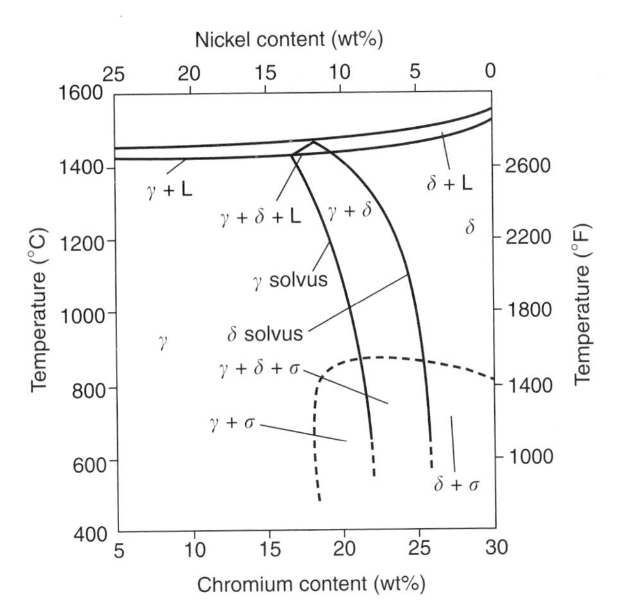

Figure 6.2 Pseudobinary section of the Fe–Cr–Ni system at 70% iron. (From Lippold and Savage [1]. Courtesy of the American Welding Society.)

Precipitation of $M_{23}C_6$ carbides has received considerable attention because of its effect on corrosion resistance. As shown in Figure 6.4, these carbides precipitate very rapidly along grain boundaries in the temperature range 700 to 900°C (1290 to 1650°F). At slightly longer times, the presence of these grain boundary carbides can lead to intergranular corrosion when exposed to certain environments [5]. This precipitation reaction is accelerated in alloys that are strengthened by cold work.

Sigma, chi, eta, G, and Laves phases (Table 6.3) may also form in austenitic stainless steels, particularly in those that contain additions of Mo, Nb, and Ti. Typically, these phases form after longtime exposure at elevated temperature and result in embrittlement of the steel. An example of the effect of sigma phase on the impact toughness of Fe–Cr–Ni alloys is shown in Figure 6.5. Note that less than 5% sigma is effective in reducing the impact toughness by over 50%.

6.2.1 Mechanical Properties

Minimum room-temperature mechanical properties of a number of austenitic stainless steels are provided in Table 6.4. These properties reflect the hot-finished and annealed condition. Considerable strength can be imparted to these alloys by cold working. In general, the austenitic stainless steels cannot be substantially strengthened by either precipitation or transformation. Some specialty alloys containing higher nickel contents and titanium additions can be strengthened by precipitation of the gamma-prime precipitate, $Ni_3(Al,Ti)$, that is a common strengthening agent in nickel-base superalloys. These alloys are discussed in Chapter 8. It is also possible to form martensite in some austenitic grades, but this occurs under only very special

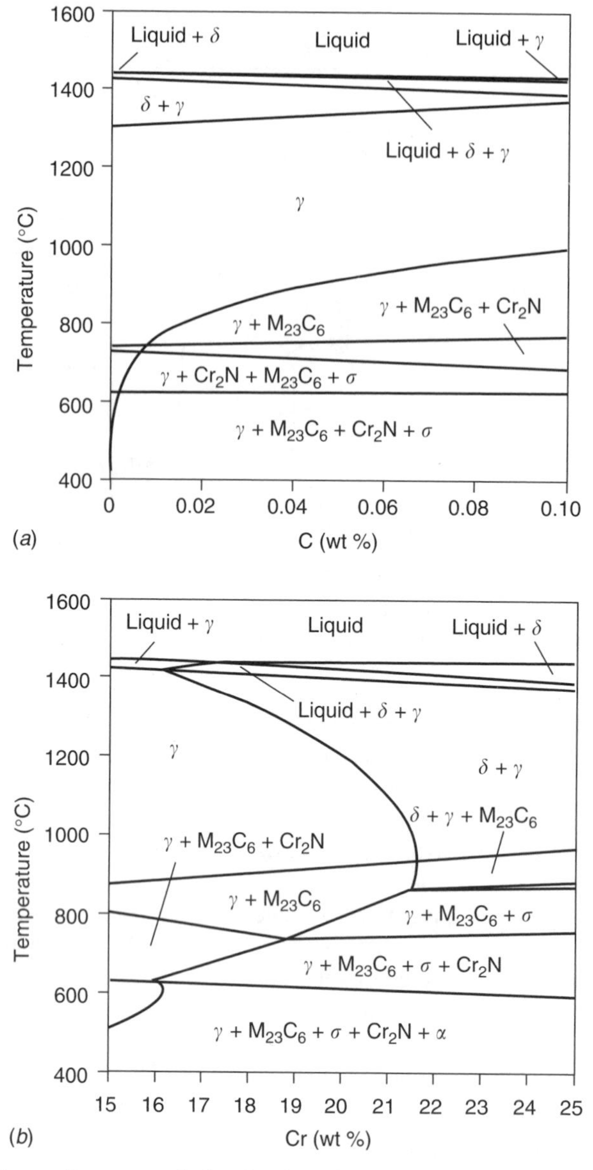

Figure 6.3 Phase diagrams calculated using ThermoCalc: (*a*) Fe–18Cr–10Ni–1.5Mn–0.5Si–0.04N, variable C; (*b*) Fe–10Ni–1.5Mn–0.5Si–0.04C–0.04N, variable Cr. (Courtesy of Antonio Ramirez.)

TABLE 6.3 Precipitates in Austenitic Stainless Steels

Precipitate	Crystal Structure	Lattice Parameters (nm)	Stoichiometry
MC	FCC	$a = 0.424–0.447$	TiC, NbC
M_6C	Diamond cubic	$a = 1.062–1.128$	$(FeCr)_3Mo_3C$, Fe_3Nb_3C, Mo_5SiC
$M_{23}C_6$	FCC	$a = 1.057–1.068$	$(Cr,Fe)_{23}C_6$, $(Cr,Fe,Mo)_{23}C_6$
NbN	FCC	$a = 0.440$	NbN
Z phase	Tetragonal	$a = 0.307$, $c = 0.7391$	CrNbN
Sigma phase	Tetragonal	$a = 0.880$, $c = 0.454$	Fe–Ni–Cr–Mo
Laves phase (η)	Hexagonal	$a = 0.473$, $c = 0.772$	Fe_2Mo, Fe_2Nb
Chi phase (χ)	BCC	$a = 0.8807–0.8878$	$Fe_{36}Cr_{12}Mo_{10}$
G phase	FCC	$a = 1.12$	$Ni_{16}Nb_6Si_7$, $Ni_{16}Ti_6Si_7$
R	Hexagonal	$a = 1.0903$, $c = 1.9342$	Mo–Co–Cr
	Rhombohedral	$a = 0.9011$ $\alpha = 74°27.5'$	Mo–Co–Cr
ε Nitride (Cr_2N)	Hexagonal	$a = 0.480$, $c = 0.447$	Cr_2N
Ni_3Ti	Hexagonal	$a = 0.9654$, $c = 1.5683$	Ni_3Ti
$Ni_3(Al,Ti)$	FCC	$a = 0.681$	Ni_3Al

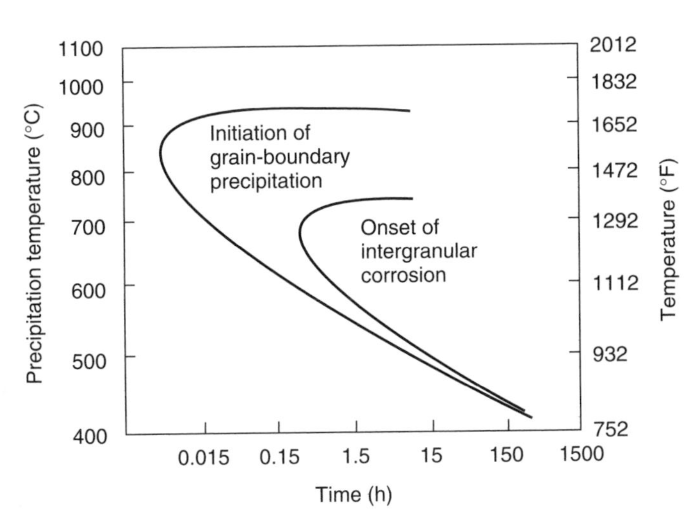

Figure 6.4 Precipitation of $M_{23}C_6$ carbides in Type 304 stainless steel with 0.05 wt% carbon. (From Cihal [5].)

conditions. Martensite has been observed in very heavily cold-worked alloys and/or when materials are cooled to cryogenic temperatures [3,7].

6.3 WELDING METALLURGY

The room-temperature microstructure of the fusion zone of austenitic stainless steels is dependent both on the solidification behavior and subsequent solid-state

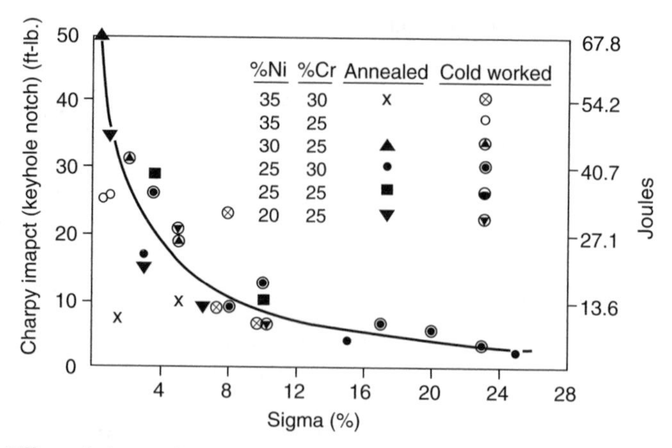

Figure 6.5 Effect of sigma phase on the room-temperature impact strength of Fe–Cr–Ni alloys. (From Talbot and Furman [6]. Courtesy of ASM International.)

Table 6.4 Minimum Room-Temperature Mechanical Properties of Wrought Austenitic Stainless Steels

Alloy	Tensile Strength		Yield Strength		Elongation (%)	Reduction in Area (%)
	MPa	ksi	MPa	ksi		
302	515	75	205	30	40	50
304	515	75	205	30	40	50
304L	480	70	170	25	40	50
308	515	75	205	30	40	50
309	515	75	205	30	40	50
310	515	75	205	30	40	50
316	515	75	205	30	40	50
316L	480	70	170	25	40	50
317	515	75	205	30	40	50
321	515	75	205	30	40	50
330	480	70	205	30	30	—
347	515	75	205	30	40	50

Source: ASM Handbook, Vol. 6, p. 468.

transformations. All stainless steels solidify with *either* ferrite or austenite as the primary phase. The austenitic stainless steels may solidify as primary ferrite or primary austenite, depending on the specific composition. Small changes in composition within a given alloy system may promote a shift from primary ferrite to primary austenite. The composition range of many austenitic stainless steels is broad enough that both solidification modes are possible. Following solidification, additional transformations can occur in the solid state upon cooling to room temperature. These transformations are most important in alloys undergoing primary ferrite solidification, since most of the ferrite will transform to austenite.

6.3.1 Fusion Zone Microstructure Evolution

There are four solidification and solid-state transformation possibilities for austenitic stainless steel weld metals. These reactions are listed in Table 6.5 and related to the Fe–Cr–Ni phase diagram in Figure 6.6. Note that A and AF solidification modes are associated with primary austenite solidification, whereby austenite is the first phase to form upon solidification. Types FA and F solidification have delta ferrite as the primary phase. Following solidification, additional microstructural modification

TABLE 6.5 Solidification Types, Reactions, and Resultant Microstructures

Solidification Type	Reaction	Microstructure
A	$L \rightarrow L + A \rightarrow A$	Fully austenitic, well-defined solidification structure
AF	$L \rightarrow L + A \rightarrow L + A + (A + F)_{eut} \rightarrow A + F_{eut}$	Ferrite at cell and dendrite boundaries
FA	$L \rightarrow L + F \rightarrow L + F + (F + A)_{per/eut} \rightarrow F + A$	Skeletal and/or lathy ferrite resulting from ferrite-to-austenite transformation
F	$L \rightarrow L + F \rightarrow F \rightarrow F + A$	Acicular ferrite or ferrite matrix with grain boundary austenite and Widmanstätten side plates

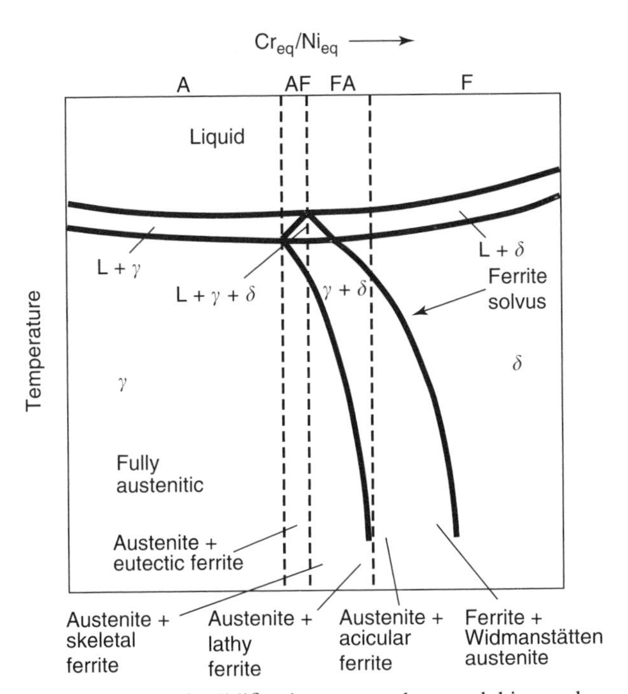

Figure 6.6 Relationship of solidification type to the pseudobinary phase diagram.

occurs in the solid state for the FA and F types, due to the instability of the ferrite at lower temperatures. The various microstructures that are possible in austenitic stainless steel weld metals and their evolution are described in the following sections.

6.3.1.1 Type A: Fully Austenitic Solidification When solidification occurs as primary austenite, two weld metal microstructures are possible. If the microstructure is fully austenitic at the end of solidification, it will remain austenitic upon cooling to room temperature and exhibit a distinct solidification structure when viewed metallographically. This is defined as Type A solidification and is shown schematically in Figure 6.7. An example of Type A solidification is shown metallographically in Figure 6.8. Note that the solidification substructure (cells and dendrites) is readily apparent in this microstructure. This is characteristic of solidification as primary austenite due to the segregation of alloying and impurity elements that occurs during solidification and the relatively low diffusivity of these elements at elevated temperature, which preserves the segregation profile that develops during solidification. When alloys such as Types 304 and 316 solidify as Type A, Cr and Mo have been shown to partition to the cell and dendrite boundaries.

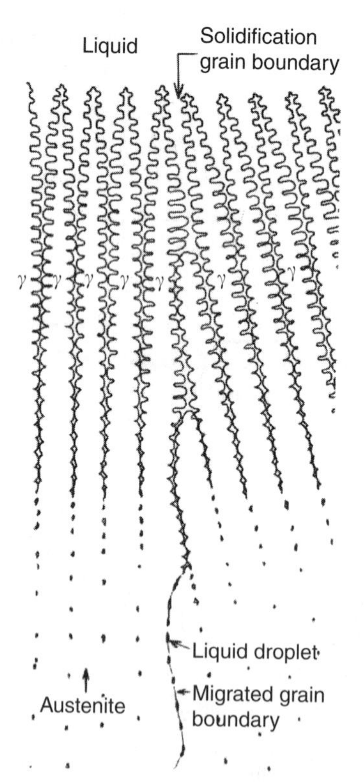

Figure 6.7 Type A solidification, fully austenitic. (From Katayama et al. [14]. Courtesy of the Japanese Welding Research Institute.)

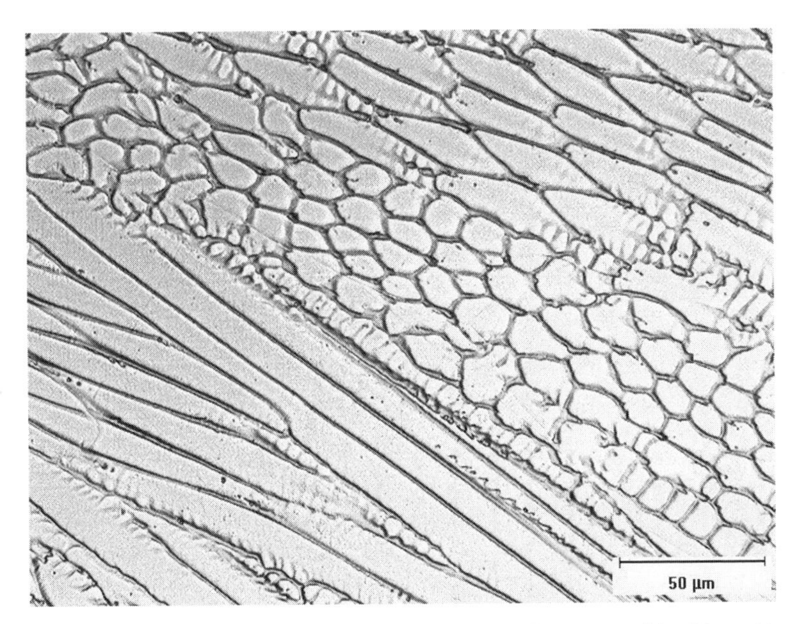

Figure 6.8 Fusion zone microstructure resulting from fully austenitic (Type A) solidification.

6.3.1.2 Type AF Solidification If some ferrite forms at the end of the primary austenite solidification process via a eutectic reaction, solidification is termed Type AF. This occurs if sufficient ferrite-promoting elements (primarily Cr and Mo) partition to the solidification subgrain boundaries during solidification to promote the formation of ferrite as a terminal solidification product. This is thought to occur by a eutectic reaction and is represented by the three-phase triangular region of the phase diagram in Figures 6.2 and 6.6. The ferrite that forms along the boundary is relatively stable and resists transformation to austenite during weld cooling since it is already enriched in ferrite-promoting elements. A schematic of AF solidification is shown in Figure 6.9. An example of a microstructure that exhibits ferrite along solidification subgrain boundaries is shown in Figure 6.10. Note that because this is primary austenite solidification, the solidification substructure is readily apparent.

6.3.1.3 Type FA Solidification When solidification occurs as primary ferrite, there are also two possibilities. If some austenite forms at the end of solidification, it is termed Type FA. This austenite forms via a peritectic–eutectic reaction and exists at the ferrite solidification boundaries at the end of solidification. This reaction has been studied extensively by David et al. [8,9], Lippold and Savage [1,10,11], Brooks et al. [12], Arata et al. [13], Katayama et al. [14], Leone and Kerr [15], and others [16,17].

Based on these investigations, the following solidification and transformation sequence occurs to give rise to the ferrite morphologies resulting from FA solidification (Figures 6.11 and 6.12).

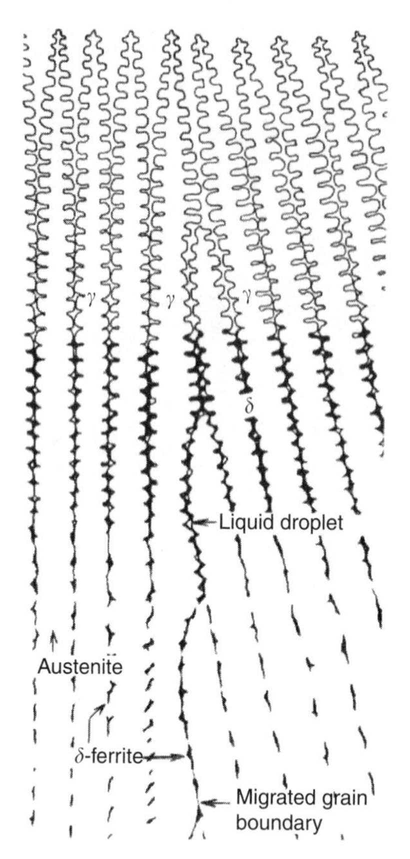

Figure 6.9 Fusion zone microstructure resulting from Type AF solidification. (From Katayama et al. [14]. Courtesy of the Japanese Welding Research Institute.)

1. At the end of primary ferrite solidification, a peritectic–eutectic reaction results in the formation of austenite along the ferrite cell and dendrite boundaries. This reaction occurs within and along the triangular three-phase region shown in Figures 6.2 and 6.6. It is called a peritectic–eutectic reaction because it is composition dependent and results from a transition from a peritectic reaction in the Fe–Ni system to a eutectic reaction in the Fe–Cr–Ni system (see Figure 2.4).
2. When solidification is complete, the microstructure consists of primary ferrite dendrites with an interdendritic layer of austenite. The amount of austenite that is present depends on the solidification conditions and the Cr_{eq}/Ni_{eq} value. As Cr_{eq}/Ni_{eq} increases, the amount of austenite decreases until solidification is entirely ferritic. At this point the solidification type shifts from FA to F.
3. As the weld metal cools through the two-phase delta ferrite + austenite field, the ferrite becomes increasingly unstable and the austenite begins to consume the ferrite via a diffusion-controlled reaction. There was some early debate about the nature of this reaction [1,8,9,11,15], but it is now generally agreed

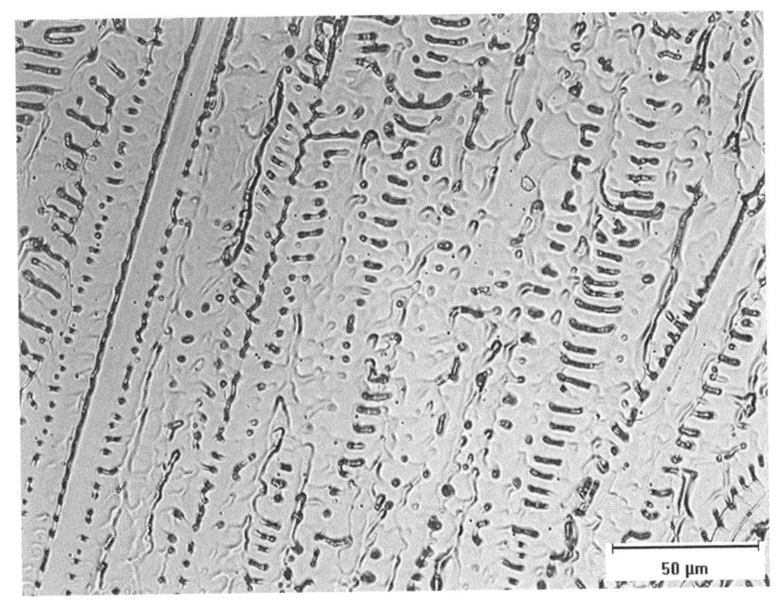

Figure 6.10 Fusion zone microstructure resulting from Type AF solidification.

that diffusion across the austenite–ferrite interface controls the rate and nature of the transformation.

4. When weld cooling rates are moderate and/or when the Cr_{eq}/Ni_{eq} is low but still within the FA range (Figure 6.6), a vermicular, or skeletal, ferrite morphology results. This is a consequence of the advance of the austenite consuming the ferrite until the ferrite is sufficiently enriched in ferrite-promoting elements (chromium and molybdenum) and depleted in austenite-promoting elements (nickel, carbon, and nitrogen) that it is stable at lower temperatures where diffusion is limited. This skeletal microstructure is shown schematically in Figure 6.11*a* and in the micrograph in Figure 6.12*a*.
5. When cooling rates are high and/or when the Cr_{eq}/Ni_{eq} increases within the FA range in Figure 6.6, a lathy ferrite morphology results. The lathy morphology forms in place of the skeletal morphology due to restricted diffusion during the ferrite–austenite transformation. When diffusion distances are reduced it is more efficient for the transformation to proceed as more tightly spaced laths, resulting in a residual ferrite pattern that cuts across the original dendrite or cell growth direction. This is shown schematically in Figure 6.11*b* and in the micrograph in Figure 6.12*b*.
6. When solidification and cooling rates are extremely high, such as during laser or electron beam welding, a complete transformation from ferrite to austenite may be possible due to a diffusionless, "massive" transformation. A shift in primary solidification mode from ferrite to austenite may also occur at high solidification rates. This behavior is described in more detail in Section 6.5.1.5.

Figure 6.11 Type FA solidification: (*a*) skeletal ferrite; (*b*) lathy morphology. (From Katayama et al. [14]. Courtesy of the Japanese Welding Research Institute.)

Considerable research into the ferrite–austenite transformation sequence was conducted in the late 1970s and 1980s. For more details regarding this transformation, the reader is referred to the review by Brooks and Thompson [18].

6.3.1.4 Type F Solidification If solidification occurs completely as ferrite, it is termed Type F. In this case the microstructure is fully ferritic at the end of solidification, as shown in Figure 6.6. When the weld metal cools below the ferrite solvus, austenite forms within the microstructure, usually first at the ferrite grain boundaries. Because the structure was fully ferritic in the solid state between the solidus and ferrite solvus, diffusion eliminates most or all composition gradients resulting from solidification, and thus, when the transformation starts, the microstructure consists

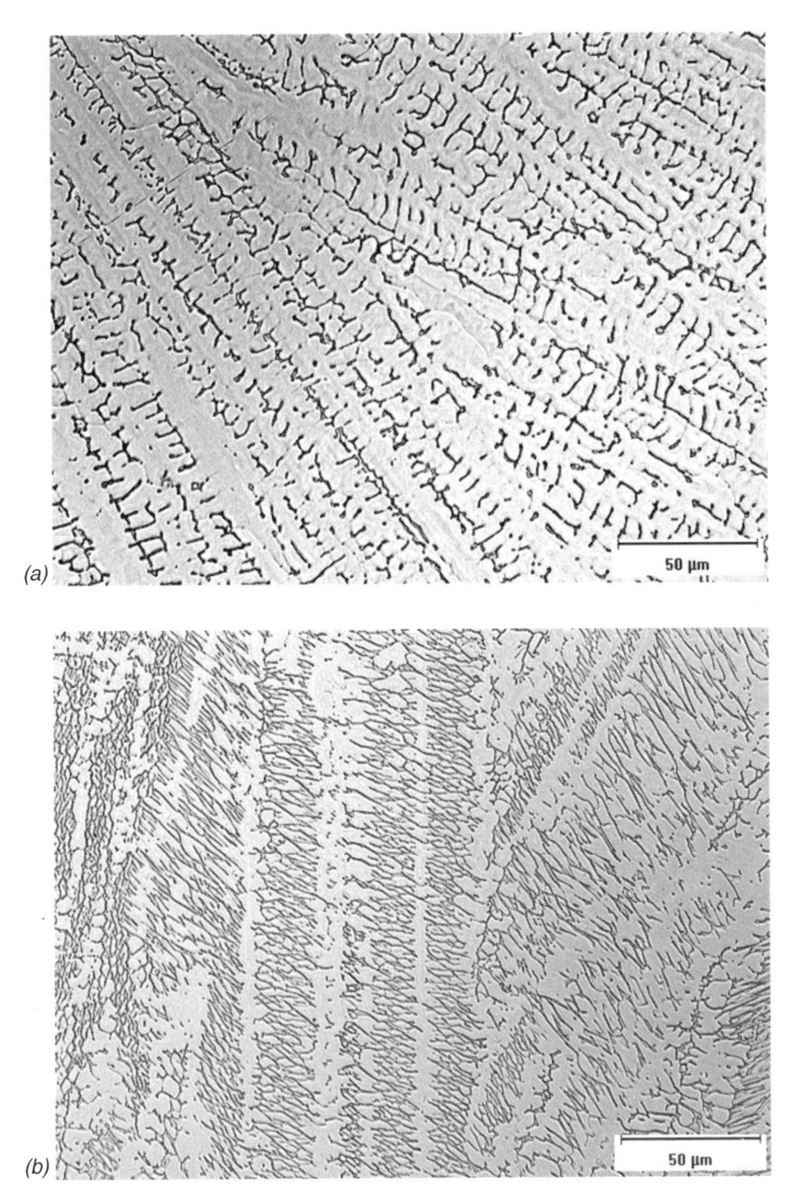

Figure 6.12 Fusion zone microstructure resulting from FA solidification: (*a*) skeletal ferrite morphology; (*b*) lathy ferrite morphology.

of large relatively homogeneous ferrite grains. The degree of transformation to austenite again depends on Cr_{eq}/Ni_{eq} and cooling rate. At low Cr_{eq}/Ni_{eq} values within the F range (Figure 6.6), the transformation begins at a higher temperature, and at low to moderate weld cooling rates, much of the ferrite is consumed. With higher cooling rates, diffusion is suppressed and the austenite will not consume as much of the ferrite. Similarly, if the Cr_{eq}/Ni_{eq} value is increased within the F range, the ferrite solvus is depressed and transformation will occur at lower temperatures. In both cases, weld metals with high ferrite contents will result.

The microstructure that forms as a result of Type F solidification in austenitic stainless steels again is a function of composition and cooling rate. At low Cr_{eq}/Ni_{eq} values within the F range (Figure 6.6) an acicular ferrite structure will form within the ferrite grains. This structure is shown in the schematic in Figure 6.13*a*. Note that continuous austenite networks are present at the prior ferrite grain boundaries and that the acicular ferrite is no longer contained within the bounds of the original ferrite dendrites, as during FA solidification with lathy ferrite formation (Figure 6.11*b*). This occurs because of the absence of austenite within the ferrite grains during Type F solidification. The structure is completely ferritic in the solid state before the transformation to austenite begins. When this structure cools below the ferrite solvus, austenite first forms at the ferrite grain boundary, but the transformation front breaks down and parallel needles of austenite form within the ferrite. As in the case of FA

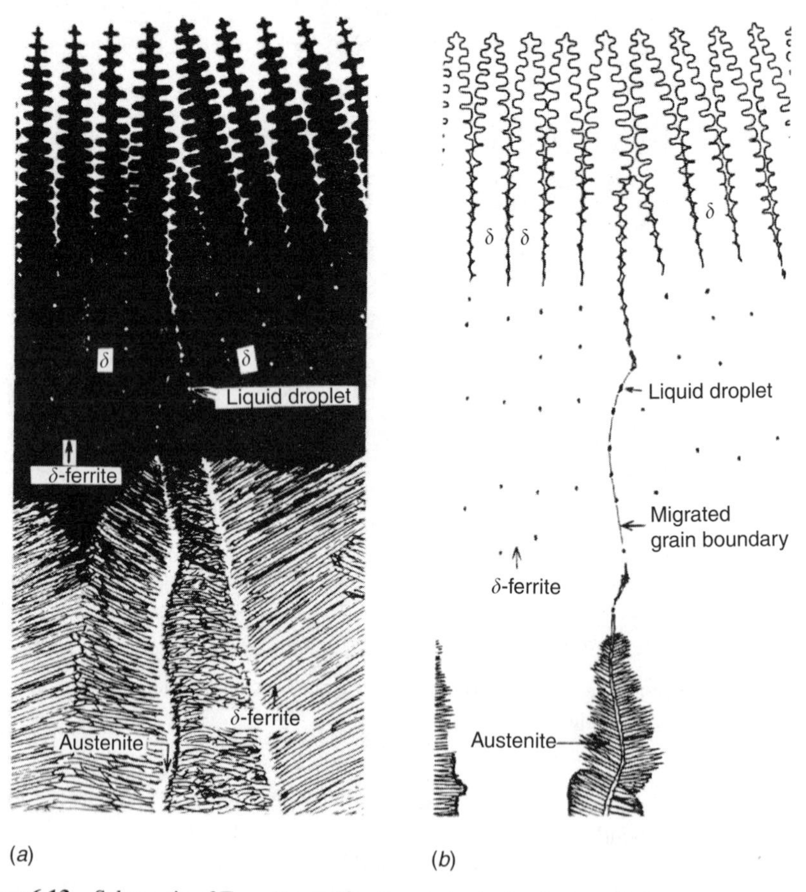

Figure 6.13 Schematic of Type F solidification: (*a*) acicular ferrite; (*b*) ferrite and Widmanstätten austenite. (From Katayama et al. [14]. Courtesy of the Japanese Welding Research Institute.)

solidification with lathy ferrite, the restriction of long-range diffusion at the lower transformation temperature forces the transformation to occur over shorter distances. This produces the acicular structure shown in Figure 6.13*a*.

At higher Cr_{eq}/Ni_{eq} values (given the same cooling rate) the microstructure will consist of a ferrite matrix with grain boundary austenite and Widmanstätten austenite plates that nucleate at the grain boundary austenite or within the ferrite grains. This microstructure is shown schematically in Figure 6.13*b* and in a micrograph in Figure 6.14. In this case, transformation does not occur entirely across the ferrite grain. Initial austenite again forms at the ferrite grain boundary, but transformation across the entire grain is suppressed by lower diffusion rates and lower driving force (the equilibrium microstructure contains more ferrite). This again can be understood from the pseudobinary diagram in Figure 6.6. As Cr_{eq}/Ni_{eq} increases, the ferrite solvus decreases and the equilibrium ferrite content increases, thus reducing the driving force for the ferrite-to-austenite transformation and the temperature at which the transformation begins.

In practice, Type F solidification is very unusual in austenitic stainless steel weld metals. Most filler metals are formulated such that solidification occurs in the FA mode, with weld metal ferrite contents ranging from 5 to 20 FN (Ferrite Number). Only highly alloyed filler metals such as Type 309LMo and Type 312* (30Cr–10Ni)

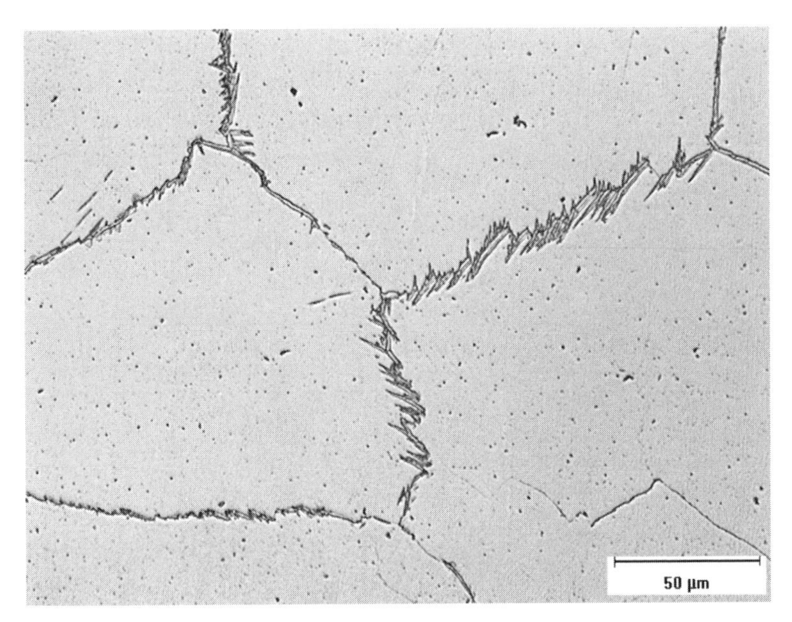

Figure 6.14 Fusion zone microstructure resulting from F solidification: Widmanstätten austenite nucleates from austenite along ferrite grain boundaries. This microstructure is very unusual in austenitic stainless steels.

*Type 312 can be considered as an austenitic or as a duplex filler metal. The microstructure of Type 312 deposits is more representative of duplex stainless steel weld metals (30 to 80 FN) than of austenitic stainless steel weld metals.

would be expected to exhibit microstructures with higher levels of ferrite. Type F solidification (such as that shown in Figure 6.14) is more characteristic of the duplex stainless steels, as described in Chapter 7.

6.3.2 Interfaces in Single-Phase Austenitic Weld Metal

It is important to understand the nature of the various boundaries, or interfaces, that are present in austenitic stainless steel weld metal, since many of the defects associated with the fusion zone both during fabrication and in service are associated with these boundaries. Boundaries are especially evident in weld metals that solidify in the A or AF mode, since the solidification structure is clearly apparent after polishing and etching. At least three different boundary types can be observed metallographically [19]. These are shown schematically in Figure 6.15 and described in the following sections.

6.3.2.1 Solidification Subgrain Boundaries The solidification subgrains represent the finest structure that can be resolved in the optical microscope. These subgrains are normally present as cells or dendrites and the boundary separating adjacent subgrains is known as a *solidification subgrain boundary* (SSGB). These boundaries are evident in the microstructure because their composition is different

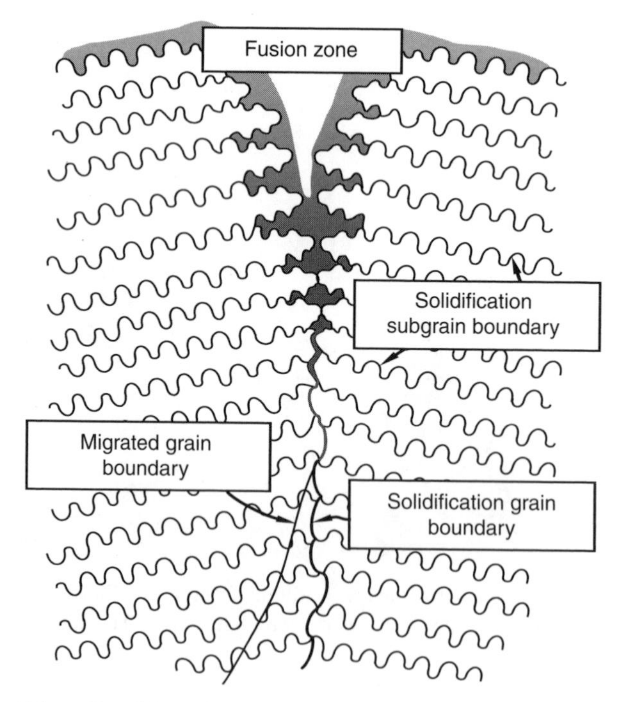

Figure 6.15 Schematic of boundaries observed in weld metals that solidify as primary austenite (A and AF mode).

from that of the bulk microstructure. Solute redistribution that creates this compositional gradient at the SSGB is dictated by Case 2 (microscopic) solute redistribution, also referred to as *Scheil partitioning*.

There is little crystallographic misorientation across the SSGB, and these boundaries are characterized crystallographically as *low-angle boundaries*. The low angular misorientation (approaching zero) results from the fact that subgrain growth during solidification occurs along preferred crystallographic directions (or easy growth directions). In FCC and BCC metals, these are ⟨100⟩ directions. Because of this, the dislocation density along the SSGB is generally low since there is not a large structural misorientation to accommodate.

6.3.2.2 Solidification Grain Boundaries The solidification grain boundary (SGB) results from the intersection of packets, or groups, of subgrains. Thus, SGBs are the direct result of competitive growth that occurs during solidification along the trailing edge of the weld pool. Because each of these packets of subgrains has a different growth direction and orientation, their intersection results in a boundary with high angular misorientation. These are often called *high-angle grain boundaries*. This misorientation results in the development of a dislocation network along the SGB.

The SGB also exhibits a compositional component resulting from solute redistribution during solidification. This redistribution can be modeled using Case 3 (macroscopic) solidification boundary conditions and often results in high concentrations of solute and impurity elements at the SGBs. These compositions may lead to the formation of low-melting liquid films along the SGBs at the conclusion of solidification that can promote weld solidification cracking. When weld solidification cracking occurs in stainless steels, it is almost always along SGBs.

6.3.2.3 Migrated Grain Boundaries The SGB that forms at the end of solidification has both a compositional and a crystallographic component. In some situations, it is possible for the crystallographic component of the SGB to migrate away from the compositional component. This new boundary that carries with it the high-angle misorientation of the "parent" SGB is called a *migrated grain boundary* (MGB).

The driving force for migration is the same as for simple grain growth in base metals, a lowering of boundary energy. The original SGB is quite tortuous since it forms from the intersection of opposing cells and dendrites. The crystallographic boundary can lower its energy by straightening and, in the process, it pulls away from the original SGB. Further migration of the boundary is possible during reheating, such as during multipass welding. Because it carries the crystallographic misorientation of the SGB with it, the MGB represents a high-angle boundary, normally with misorientations greater than 30°. The composition of the boundary varies locally, depending on the composition of the microstructure where it has migrated. It is also possible that some segregation can occur along MGBs, possibly by a "sweeping" mechanism.

MGBs are most prevalent in fully austenitic weld metals. When the weld metal undergoes AF solidification, ferrite forms at the end of solidification along the SSGBs and SGBs. This ferrite is effective in "pinning" the crystallographic component of the SGB, thus preventing it from migrating away from the parent SGB. In this case, an

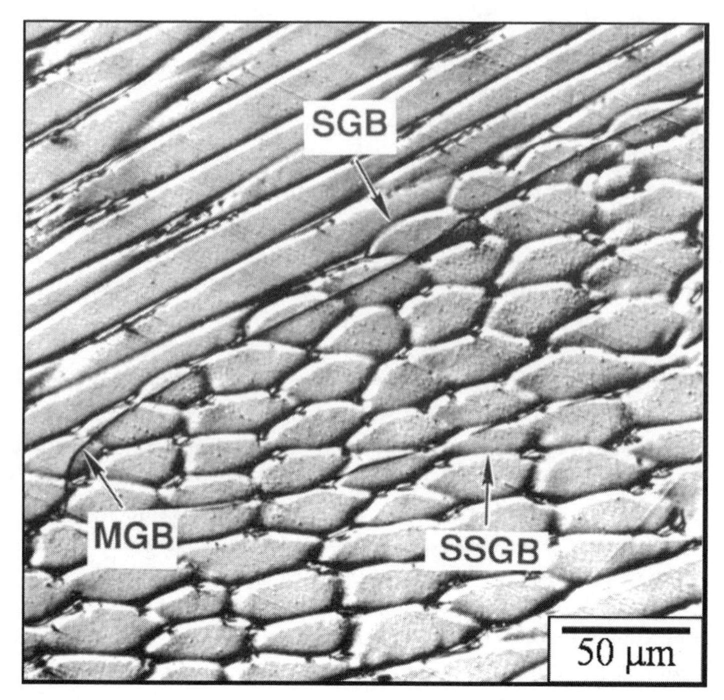

Figure 6.16 Fusion zone microstructure of Type 304L that has undergone A-type solidification. A solidification subgrain boundary (SSGB), solidification grain boundary (SGB), and migrated grain boundary (MGB) are indicated.

MGB does not form because the high-angle crystallographic boundary cannot migrate. An example of these interface types in a single-phase Type 304L fusion-zone microstructure is shown in Figure 6.16. Note that the actual migration distance of the MGB is only 5 to 10 μm away from the SGB and that it cuts through the center of the solidification subgrains.

When solidification occurs in the FA and F modes, SSGBs and SGBs also form but they are typically not apparent in the microstructure. This results for three reasons: (1) segregation during ferrite solidification is not as pronounced as during austenite solidification, (2) elevated temperature diffusion is much more rapid in ferrite relative to austenite (perhaps by 100 times), and (3) the ferrite-to-austenite transformation tends to mask any solidification segregation in the ferrite. MGBs are also typically, not observed in weld metals that solidify in FA or F modes. Although, technically, they must be present, they are virtually impossible to distinguish from the ferrite–austenite interface.

6.3.3 Heat-Affected Zone

The nature of the heat-affected zone (HAZ) in austenitic stainless steels depends on the composition and microstructure of the base metal. The following metallurgical reactions may occur in the HAZ of austenitic alloys.

6.3.3.1 Grain Growth Most stainless steels are welded in the solution-annealed or hot-rolled condition, so grain growth is usually restricted unless weld heat inputs are extremely high. Some grain coarsening can usually be observed, but in most cases it is not dramatic. In base metals that have been strengthened by cold working, recrystallization and grain growth can result in significant HAZ softening. In this case, a distinct HAZ results and the grain size is clearly larger than that of the base metal.

6.3.3.2 Ferrite Formation As shown in Figures 6.2 and 6.6, alloys whose compositions are to the right of the fully austenitic solidification range will form ferrite when heated to temperatures just below the solidus temperature. The higher the Cr_{eq}/Ni_{eq} ratio of the alloy, the more likely ferrite formation will be. When ferrite forms, it is usually along the grain boundary, as shown in Figure 6.17. Formation of ferrite along HAZ grain boundaries will restrict grain growth and also minimize the susceptibility to HAZ liquation cracking. The latter effect is discussed in more detail in Section 6.5.2. The degree of ferrite formation is usually low since the austenite-to-ferrite transformation is relatively sluggish and the HAZ thermal cycle is normally quite rapid. It is also possible that some of the ferrite that forms during elevated temperature exposure transforms back to austenite on cooling.

6.3.3.3 Precipitation Since the HAZ is heated to temperatures approaching the solidus temperature of the alloy, many of the precipitates that are present in the base metal may dissolve. This can lead to a supersaturation of the austenite matrix during

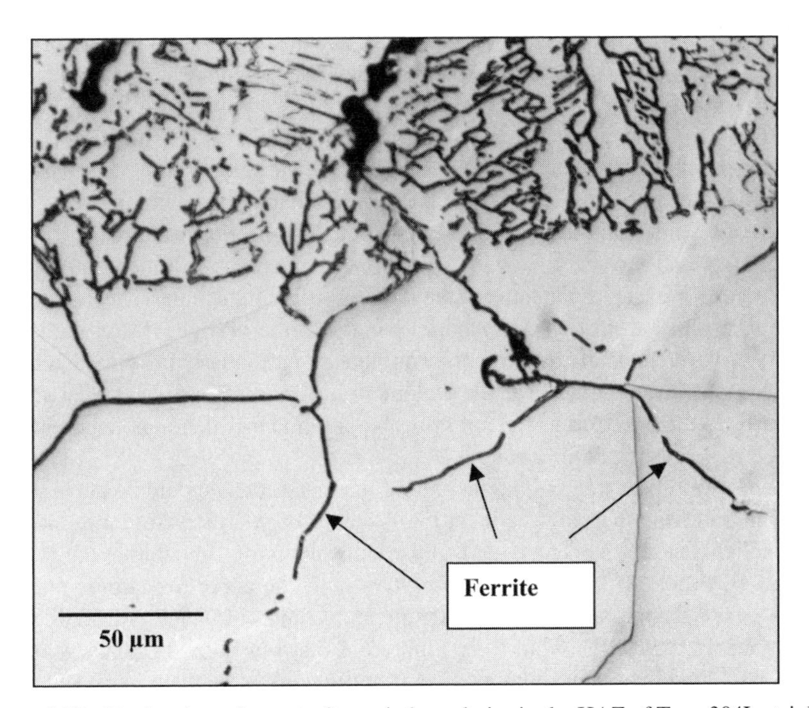

Figure 6.17 Ferrite along the austenite grain boundaries in the HAZ of Type 304L stainless steel.

cooling, resulting in the formation of various precipitates. Carbides and nitrides are the most likely precipitates to form in the HAZ of austenitic stainless steels. They will usually form along grain boundaries or at the ferrite–austenite interface (if ferrite is present). Reference to the calculated phase diagrams in Figure 6.3 shows the temperature ranges over which $M_{23}C_6$ carbides and Cr_2N nitrides are stable. Even though not apparent metallographically, it is likely that these precipitates will be present in the HAZ of most austenitic alloys. Their size, distribution, and morphology are dependent on the alloy composition and the HAZ thermal cycle. Extensive chromium-rich carbide precipitation can lead to degradation in corrosion resistance, as described in Section 6.6.

6.3.3.4 Grain Boundary Liquation Local melting along the austenite grain boundaries may also occur. This usually results from the segregation of impurity elements that reduce the grain boundary melting temperature. Alloys that contain titanium and niobium and that form MC carbides rich in these elements may undergo constitutional liquation, a phenomenon that can lead to HAZ liquation cracking. This is discussed in Section 6.5.2. Segregation of impurity elements, particularly sulfur and phosphorus, to the grain boundary can also promote liquation.

6.3.4 Preheat and Interpass Temperature and Postweld Heat Treatment

In general, preheat and interpass temperature control is not required for austenitic stainless steels as it is for transformable (martensitic) steels. High preheat and interpass temperatures will slow the cooling rate, but this will have little effect on the ferrite-to-austenite transformation, since this transformation occurs at very high temperatures, where preheat and interpass temperature control has very little effect on cooling rate. In cases where carbide precipitation leading to sensitization is a concern, interpass temperature may need to be controlled below some maximum since slow cooling through the carbide precipitation range may be damaging.

Postweld heat treatment is often required in thick-section weldments to relieve residual stresses. Because the coefficient of thermal expansion (and contraction) is much larger for austenitic stainless steels than for ferritic stainless steels, the magnitude of residual stress can be much larger in the former. Postweld stress relief may be necessary to reduce distortion in the component, particularly if postweld machining operations are required or if the weldment must maintain dimensional stability in service. Reduction of residual stress is also important if there is the potential for stress corrosion cracking in service.

The postweld heat treatment temperature that is selected should be dependent on the intention of the thermal treatment (stress relief or microstructure modification). Stress-relief heat treatments are generally conducted in the temperature range 550 to 650°C (1020 to 1200°F). This is below the nose of the curve for carbide precipitation (Figure 6.4) and also below the temperature range in which embrittling compounds form (Table 6.3). Note from Figure 6.4 that sensitization is possible if the stress-relief heat treatment requires several hours, as may be the case in very heavy section weldments. In this situation, the use of low-carbon base and filler metals or stabilized grades is recommended.

Higher postweld heat treatments may be desirable in some cases to more effectively relieve residual stresses or to modify the as-welded microstructure. Special care should be taken in the range 650 to 900°C (1200 to 1650°F), since both $M_{23}C_6$ carbides and sigma phase form rapidly in this temperature range. The former reaction can sensitize the material, while the latter can lead to embrittlement and loss of toughness (Figure 6.5). Postweld heat treatment in this range may be possible if the weld metal is fully austenitic and if the base and filler metals are low-carbon grades. As described in the following section, weld metals containing ferrite form sigma phase rapidly.

Heat treatment in the range 950 to 1100°C (1740 to 2010°F) will completely relieve residual stress and also result in modification of the as-welded microstructure without forming carbides or sigma phase. Heating above 950°C (1740°F) followed by rapid quenching will remove any carbides in the original microstructure. Heating to temperatures approaching 1100°C (2010°F) will dissolve some or all of the ferrite, depending on the hold time at temperature, composition of the weld metals, and the as-welded ferrite content. If this extreme heat treatment is used, rapid cooling to room temperature by water quenching is generally required since carbide precipitation may occur during slow cooling.

6.3.4.1 Intermediate-Temperature Embrittlement Austenitic stainless steel base metals and weld metals are susceptible to embrittlement by the formation of sigma phase.

This Cr-rich phase, nominally FeCr, is hard and brittle, and when present in large volume fraction can reduce toughness (Figure 6.5) and ductility. In a fully austenitic (no ferrite) microstructure, sigma phase precipitation is relatively sluggish and normally requires extended times (hundreds to thousands of hours) at elevated temperature. It may form in service or during postweld heat treatment. Its formation range is 600 to 900°C (1110 to 1650°F), and it forms most rapidly in austenitic–ferritic weld metals at around 750°C (1380°F) [20]. Besides Cr, additions of Mo, Nb, Si, W, V, Ti, and Zr promote sigma phase formation, whereas C and N tend to retard its growth. The presence of ferrite in the microstructure, as controlled by the balance of these ferrite- and austenite-promoting elements, significantly accelerates sigma phase formation.

Because ferrite is higher in Cr than austenite, the presence of ferrite greatly accelerates sigma phase formation, and as a result, weld metals containing retained ferrite are most susceptible to embrittlement. The use of filler metals to produce fully austenitic or low-ferrite weld metals is most effective in reducing the chances of sigma phase embrittlement. Care must be taken, however, not to induce solidification cracking in an effort to avoid sigma phase embrittlement. Studies by Vitek and David

Sigma Phase

- Equilibrium FeCr phase
- Precipitation range of 600 to 900°C (1110 to 1650°F)
- Forms rapidly in weld metals that contain ferrite
- May form during postweld heat treatment of large structures
- Reduces corrosion resistance, ductility, and toughness

[21,22] have shown that aging of Type 308 weld metals in the range 650 to 750°C (1200 to 1380°F) results in a dissolution of the ferrite phase, leading initially to the formation of Cr-rich $M_{23}C_6$ and finally to sigma phase nucleation. They found that nucleation is the rate-limiting step. Once nucleation occurs, sigma phase growth proceeds quite rapidly. Cold work will accelerate the nucleation of sigma. In Type 308 welds containing ferrite, sigma phase may form within less than 100 hours in the range 650 to 750°C (1200 to 1380°F).

Since sigma phase must be continuous or nearly continuous in the microstructure to cause significant losses in toughness and ductility, maintaining the weld metal FN in the range 3 to 8 FN is usually sufficient to avoid embrittlement. This is because the transformation from ferrite to sigma is not volumetrically equivalent since sigma has a higher Cr concentration than ferrite. Weld metals with 8 FN will transform to about 4 vol% sigma. This level is not sufficient to embrittle the weld, although some degradation in toughness can be expected (see Figure 6.5).

Studies by Alexander et al. [23] have also investigated the 475°C (885°F) embrittlement phenomenon in Type 308 weld metal containing ferrite (FN 11). This embrittlement phenomenon was discussed in Chapter 5. Since the ferrite in austenitic stainless steel weld metals is essentially a ferritic stainless steel (25 to 30 wt% Cr, 4 to 5 wt% Ni) embedded in an austenite matrix, it is not surprising that embrittlement of the ferrite by alpha-prime formation is observed. They found that aging in the range 475 to 550°C (885 to 1020°F) for up to 5000 hours reduced toughness significantly, as measured by an increase in the ductile-to-brittle fracture transition temperature and a lowering of the upper shelf energy. At 475°C (885°F), the reduction in toughness was associated with both the formation of alpha-prime and G-phase (see Table 6.3), while at 550°C (1020°F) it was associated with both carbide and sigma phase formation. In all cases, precipitation of the embrittling phases was associated with the chromium-enriched ferrite, and nucleation typically occurred at the ferrite–austenite interface.

6.4 MECHANICAL PROPERTIES OF WELDMENTS

The minimum mechanical properties for common welding consumables used to weld austenitic stainless steels are provided in Table 6.6. Austenitic stainless steels are usually welded in the annealed, hot-rolled, or cold-worked condition. In all cases some softening will occur in the HAZ, due either to grain growth in the case of hot-rolled materials or to recrystallization and grain growth in cold-worked material. Thus, when transverse tensile tests of weld samples are conducted, failure often occurs in the HAZ rather than in the weld metal. The presence of ferrite in the weld metal acts as a second phase-strengthening agent and increases the strength level relative to the base metal and HAZ.

Actual weld metal strength levels of Type 308L, Nitronic™ 40 (Type 219), and Type 312 weld metals are provided in Table 6.7. These data were generated from all weld metal tensile samples that were machined from weld metal buildups, both transverse and longitudinal to the welding direction. Note that the yield strength of the Type 308L weld deposit is significantly higher than that of Type 304 base metal,

TABLE 6.6 Minimum Mechanical Properties of Common Austenitic Stainless Steel Weld Metals[a]

Type	Tensile Strength		Elongation
	MPa	ksi	(%)
219	620	90	15
308	550	80	35
308H	550	80	35
308L	520	75	35
309	550	80	30
309L	520	75	30
310	550	80	30
316	520	75	30
316H	520	75	30
316L	480	70	30
317	550	80	30
317L	520	75	30
330	520	75	25
347	520	75	30

[a]Properties apply to all-weld metal deposits from covered electrodes of AWS A5.4 and flux-cored wires of AWS A5.22. Mechanical properties are not specified for deposits from bare wires, bare rods, tubular metal-cored wires, or strips of AWS A5.9, but are expected to be similar to those of corresponding alloy types in the covered electrode and flux-cored electrode standards. Also, the higher-silicon filler metals of AWS A5.9 can be expected to produce properties that do not differ from those of the lower-silicon grades.

TABLE 6.7 Experimental Weld Metal Tensile Properties

Material[a]	Orientation or Condition	Yield Strength		Tensile Strength		Elongation (%)	Reduction in Area (%)
		MPa	ksi	MPa	ksi		
308L	Transverse	452	65.6	605	87.7	55.5	75.3
308L	Longitudinal	450	65.3	595	86.3	59.8	73.7
219	Transverse	617	89.5	807	117.0	45.1	62.3
219	Longitudinal	600	87.0	811	117.6	48.4	61.5
312	Transverse	592	85.8	752	109.0	14.6	23.1
312	Longitudinal	607	88.0	774	112.2	24.9	31.0
304	Annealed plate	241	35	565	82	60	70

[a]Type 308L = FN 12, Nitronic 40 (Type 219) = FN 4, Type 312 = FN 30. Weld metal pads made with GTAW, cold wire feed process.

while the ductility (elongation and reduction in area) are comparable. Nitronic™ 40 and Type 312 weld deposits are considerably stronger than the Type 308L deposit. In the former case, this results from the presence of nitrogen (0.15 wt%) as an alloying addition, while in the latter the higher ferrite level and carbon content results in higher strength and lower ductility.

The effect of ferrite content on mechanical properties over a range of temperatures for E308-16 weld deposits has been studied in some detail by Hauser and Van Echo [24]. They evaluated extra low (FN 2), low (FN 6), medium (FN 10), and high (FN 16) ferrite weld deposits over the temperature range 25 to 650°C (80 to 1200°F). Their results are summarized in Table 6.8. Note that the increase in weld metal

TABLE 6.8 Properties of Type 308 Weld Metal as a Function of Temperature, Ferrite Content, and Orientation[a]

Temperature				Yield Strength		Tensile Strength			
°C	°F	FN	Orientation	MPa	ksi	MPa	ksi	Elongation (%)	Reduction in Area (%)
27	80	2	L	434	62.9	605	87.7	40.0	50.9
			T	472	68.4	628	91.0	35.8	40.7
		6	L	425	61.6	596	86.3	48.0	51.1
			T	490	71.0	642	93.1	40.8	44.9
		10	L	438	63.4	622	90.2	48.5	53.4
			T	458	66.3	628	91.0	49.3	46.3
		16	L	470	68.1	660	95.7	42.0	42.7
			T	529	76.7	689	99.8	41.0	48.2
260	500	2	T	368	53.4	485	70.3	22.8	40.1
		6	T	373	54.0	501	72.6	25.3	46.4
		10	T	385	55.8	504	73.0	25.5	48.8
		16	T	406	58.9	541	78.4	24.3	45.4
482	900	2	T	339	49.1	465	67.4	27.3	44.1
		6	T	323	46.8	467	67.7	25.3	39.8
		10	T	339	49.1	471	68.3	27.5	40.3
		16	T	351	50.9	505	73.2	24.8	38.9
593	1100	2	L	278	40.3	382	55.3	26.3	51.0
			T	288	41.7	382	55.3	24.3	39.2
		6	L	275	39.8	362	52.4	29.3	50.7
			T	295	42.8	382	55.4	22.8	47.5
		10	L	277	40.1	348	50.4	28.3	54.1
			T	293	42.5	381	55.2	23.8	47.7
		16	L	295	42.7	366	53.0	27.5	48.1
			T	297	43.0	376	54.5	23.0	40.9
649	1200	2	T	255	37.0	324	46.9	29.0	44.1
		6	T	255	37.0	324	46.9	29.0	44.1
		10	T	251	36.4	296	42.9	29.7	54.1
		16	T	273	39.6	329	47.7	29.3	43.5

Source: Hauser and Van Echo [24].

[a]Welding process, SMAW; electrode, E308-16. Samples machined from weld metal buildups.

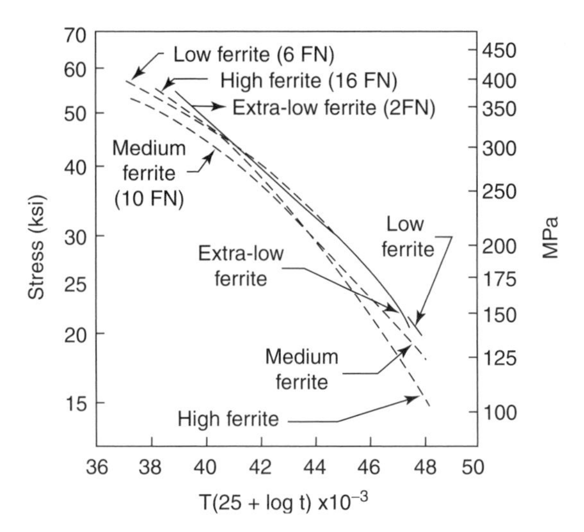

Figure 6.18 Stress rupture properties of Type 308 weld metal as a function of ferrite content. (From Hauser and Van Echo [24]. Courtesy of the American Welding Society.)

ferrite content results in an increase in strength at room temperature, and to a lesser extent at elevated temperature. They also studied the effect of ferrite content in E308-16 weld deposits on stress rupture properties at 540, 590, and 650°C (1000, 1100, and 1200°F). These data are summarized in Figure 6.18 using a Larson–Miller parameter that allows temperature and time to be combined on the same axis. Note that medium and high ferrite levels tend to reduce rupture life. This observation is similar to that of Thomas [25], who observed in Type 316 deposits that continuous networks of ferrite (FN 10) resulted in more rapid creep failure, due to early crack initiation at the ferrite–austenite interface. Thomas proposed that 5 FN in the weld deposit was ideal since a continuous ferrite network would not exist and this level would ensure resistance to weld solidification cracking.

Austenitic stainless steels are excellent engineering materials at cryogenic temperatures since they exhibit good strength, ductility, and toughness at those temperatures. The effect of weld metal ferrite content on cryogenic properties has been the subject of considerable research, since the presence of ferrite tends to reduce toughness [26,27]. This effect is shown in Figure 6.19 and Table 6.9 for austenitic stainless steel weld metals tested at liquid helium temperature (4 K) [7,28]. Note that both the ferrite content and welding process influence the cryogenic fracture toughness. For a given process and consumable, such as SMAW with 316L, an increase in deposit ferrite content clearly decreases fracture toughness. For non-flux-shielded processes such as GTAW and GMAW, the fracture toughness is superior to SMAW and SAW for equivalent ferrite content, due to reduced oxygen in the deposits from inert-gas shielded processes.

As shown in Figure 6.19, a 50% decrease in fracture toughness occurs when the weld metal ferrite content increases from FN 0 to FN 10. Also note that there is considerable scatter in the FN 0 welds. Lippold et al. [7] concluded that this variation

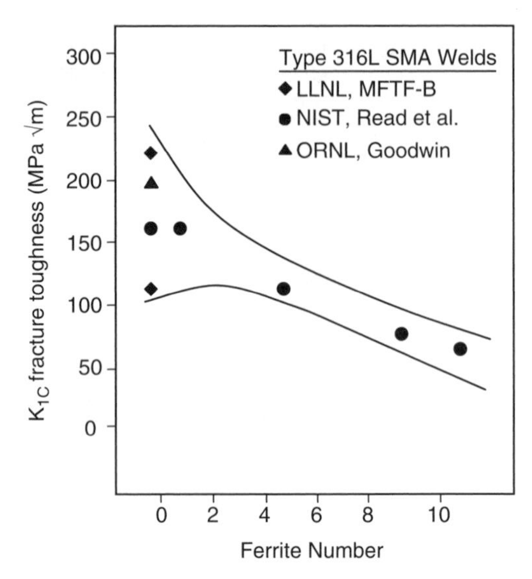

Figure 6.19 Effect of ferrite number on cryogenic fracture toughness of Type 316L shielded metal arc welds.

TABLE 6.9 Fracture Toughness at 4 K (−492°F) for Austenitic Stainless Steel Weld and Base Metals

Filler or Base Metal	Welding Process[a]	Ferrite Number[b]	K_{1C} Fracture Toughness	
			MPa · m$^{1/2}$	ksi · in$^{1/2}$
304L	—	—	211	192
316LN	—	—	224	204
316L	SMAW	0.1	179	162
316L	SMAW	0.8	177	161
316L	SMAW	4.1	141	128
316L	SMAW	8.5	108	98
316L	SMAW	10.1	98	90
316L	SAW	4.7	132	121
316L	GMAW	NR	163	148
316L	GTAW	5.0	272	247
308L	GMAW	NR	167	152
308L	GMAW	NR	133	121
308L	SMAW	NR	156	142
308L	FCAW	8.2	79	72

Source: Goodwin [28].

[a]SMAW, shielded metal arc welding; SAW, submerged arc welding; GMAW, gas-metal arc welding; GTAW, gas-tungsten arc welding; FCAW, flux-cored arc welding.

[b]NR, not reported.

was associated with microsegregation in the weld metal and its effect on martensite formation during cryogenic testing. In particular, the segregation of manganese in high-Mn filler metals was found to promote austenite stability in the interdendritic regions of the weld metal, resulting in lower fracture resistance relative to weld metals where martensite transformation was more complete.

6.5 WELDABILITY

Although the austenitic alloys are generally considered to be very weldable, they are subject to a number of weldability problems if proper precautions are not taken. Weld solidification and liquation cracking may occur depending on the composition of the base and filler metal and the level of impurities, particularly S and P. Solid-state cracking, including ductility dip, reheat (stress relief), and Cu contamination, has also been encountered in these alloys. Despite the good general corrosion resistance of austenitic stainless steels, they may be subject to localized forms of corrosion at grain boundaries in the HAZ (IGA and IGSCC) or at stress concentrations in and around the weld. Because many of the weld metals contain ferrite, intermediate temperature embrittlement due to sigma phase and carbide formation may also occur. As with the ferritic alloys, the sigma phase precipitation reaction is relatively sluggish and embrittlement by sigma is usually a service-related rather than a fabrication-related problem. However, as discussed previously, it may occur during the postweld heat treatment of large structures or thick sections when cooling rates from the postweld heat treatment temperature are extremely slow.

6.5.1 Weld Solidification Cracking

Weld solidification cracking can be a formidable problem with the austenitic stainless steels. Cracking susceptibility is primarily a function of composition. Weld metals that solidify in the A mode and are fully austenitic (contain no ferrite) tend to be the most susceptible. Those that solidify in the FA mode tend to be very resistant to solidification cracking. High impurity levels, particularly sulfur and phosphorus, tend to increase the susceptibility in alloys that solidify in the A or AF mode. Examples of weld solidification cracks in weld metals that solidify in the A and FA modes are shown in Figure 6.20. Weld restraint conditions and weld shape also influence cracking susceptibility, particularly when solidification occurs as primary austenite (A or AF). Welding conditions that impose high levels of restraint on the solidifying weld metal tend to increase cracking susceptibility. High heat inputs resulting in large weld beads or excessive travel speeds that promote teardrop-shaped weld pools are most problematic with respect to cracking. Concave bead shape and underfilled craters at weld stops also promote solidification cracking.

Weld solidification cracking is a strong function of composition, as shown by the schematic representation of cracking susceptibility versus Cr_{eq}/Ni_{eq} (WRC-1992 equivalents) in Figure 6.21. Note that compositions which result in primary austenite solidification (A and AF) are most susceptible to cracking, while the FA mode offers

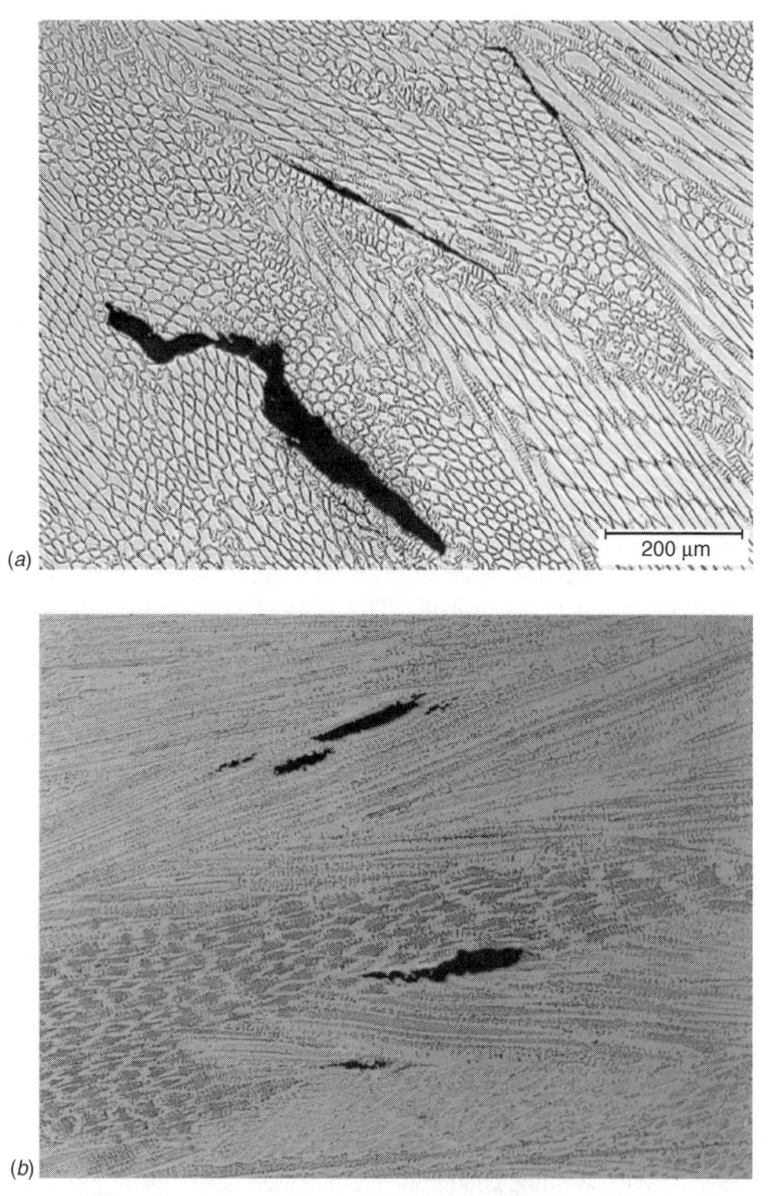

Figure 6.20 Weld solidification cracks in Varestraint samples tested at 5% strain (*a*) fully austenitic weld metal (FN 0), and (*b*) weld metal with FN 6–FA solidification mode.

the greatest resistance to solidification cracking. The F mode is more susceptible to cracking than FA but superior to A and AF. Thus, composition can be used very effectively to control weld solidification cracking. Solidification as primary ferrite in the FA mode has been shown to ensure superior resistance to weld solidification cracking over alloys that solidify as austenite. The principal reason for this superior resistance is the presence of a two-phase austenite + ferrite mixture along SGBs at the

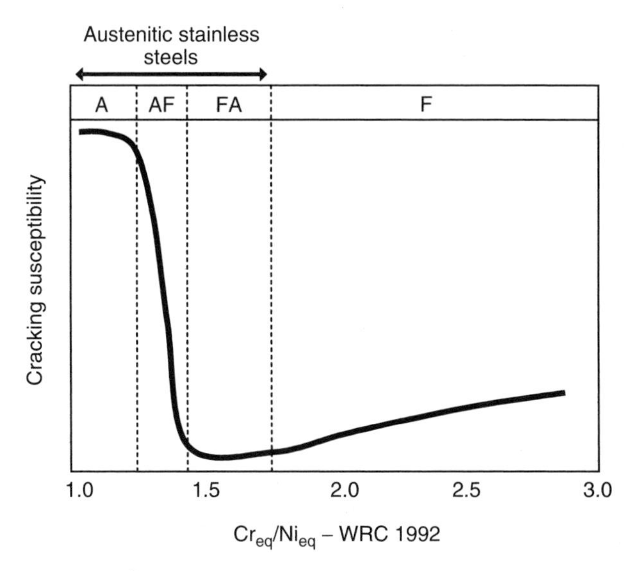

Figure 6.21 Weld solidification cracking susceptibility as a function of composition based on Varestraint data.

end of solidification that resists wetting by liquid films and presents a tortuous (not straight and smooth) boundary along which cracks must propagate.

Room-temperature weld metal ferrite content can be used as an approximation of the solidification behavior. If the FN is 0, the alloy is presumed to have solidified in the A mode. Between FN 0 and 3, solidification probably occurred as AF. Above 3 FN, but less than 20 FN, solidification is most likely in the FA mode. The latter range has been shown to be extremely resistant to weld solidification cracking. Note, however, in the WRC-1992 diagram (Figure 3.14) that the boundary which separates AF from FA solidification is not parallel to any nearby isoferrite line. Lean alloy compositions such as the AWS A5.4 16-8-2 filler metal (16% Cr, 8% Ni, 2% Mo) are predicted by the diagram to solidify as primary ferrite even at less than 2 FN. These alloys are known to have high resistance to solidification cracking. Conversely, rich alloys such as 317LM and 209 can solidify as primary austenite at 5 FN or more and are sensitive to solidification cracking at 3 or 4 FN, or even higher [29].

6.5.1.1 Beneficial Effects of Primary Ferrite Solidification

Historically, a number of factors have been used to explain the beneficial effect of ferrite, or ferrite solidification, on solidification cracking resistance in austenitic stainless steels. These are summarized in Table 6.10. Many of these have subsequently been found to have little or no effect on susceptibility, as indicated by the "N" in the right-hand column. Ferrite certainly has higher solubility for impurities such as sulfur and phosphorus, which restricts the partitioning of these elements to interdendritic regions during primary ferrite solidification. The most important factors, however, are the

TABLE 6.10 Proposed Beneficial Effects of Ferrite for Preventing Weld Solidification Cracking

Factor	Effect
High solubility of impurity elements	Some
Better high-temperature ductility than austenite	Negligible
Lower CTE than austenite	Negligible
Smaller solidification temperature range	Negligible
Less partitioning during solidification	Some
Less wetting at F–F and F–A boundaries	Strong
More difficult crack propagation along tortuous F–A boundaries at the end of solidification	Strong

nature of boundary wetting and the inherent boundary tortuosity that occurs when ferrite and austenite are both present at the end of solidification.

In the FA mode, a ferrite–austenite boundary is present at the end of solidification that is both difficult for liquid films to wet and presents a very nonplanar crack path. Thus, once a crack is initiated, it becomes very difficult for it to propagate along this tortuous boundary. Both austenite–austenite (Type A) and ferrite–ferrite (Type F) boundaries are much straighter, since no secondary solidification product is present. This makes crack propagation much easier. In the AF mode, some ferrite is present along a relatively smooth A–A boundary resulting in some improvement over fully austenitic (Type A) solidification.

The effect of boundary tortuosity is shown schematically in Figure 6.22. Weld solidification cracks occur preferentially along solidification grain boundaries (SGBs). Under Type A solidification, these boundaries are very straight, contain no residual ferrite, and offer little resistance to crack propagation if a liquid film wets the boundary. By contrast, an SGB under type FA solidification contains a mixture

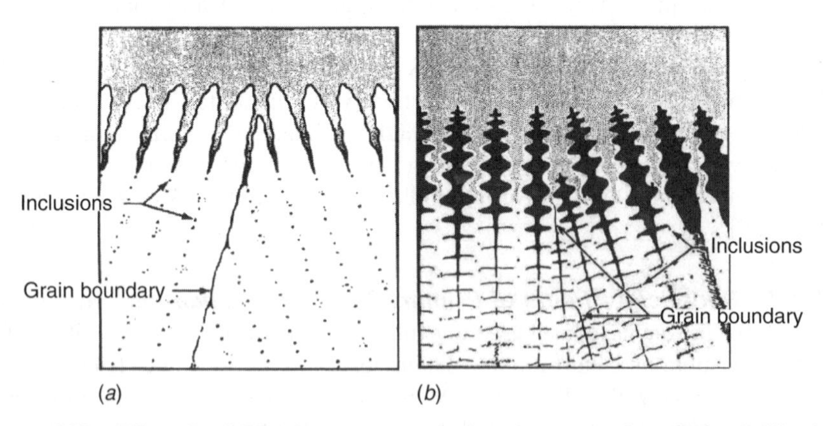

Figure 6.22 Effect of solidification type on grain boundary tortuosity: (*a*) A solidification; (*b*) FA solidification with skeletal ferrite. (From Brooks et al. [12]. Courtesy of the American Welding Society.)

of ferrite and austenite that mitigates liquid film wetting and complicates crack propagation, since the crack must follow a very tortuous austenite–ferrite interface.

6.5.1.2 *Use of Predictive Diagrams* A number of diagrams have been developed to predict cracking susceptibility, based on composition. One of the earliest of these, introduced in the 1980s, was developed by Kujanpää and Suutala (30). It is generally referred to as the *Suutala diagram*. This diagram, shown in Figure 6.23, was developed by evaluating a wide range of published austenitic stainless steel weld metal cracking studies. The equivalents used are those developed by Hammar and Svenson [31]. This diagram demonstrates the importance of composition on cracking susceptibility in austenitic stainless steel weld metals. As the Cr_{eq}/Ni_{eq} increases above a critical level, the resistance to cracking shows a dramatic increase, irrespective of the impurity level. This sharp increase results from the shift in solidification behavior from primary austenite to primary ferrite.

At extremely low S + P contents, cracking resistance is high across the range of compositions. Achieving these low levels of impurities is not generally economical using conventional melting practices. AOD melting can effectively reduce sulfur contents but has no effect on phosphorus. Even in extremely "clean" steels, P + S contents can be expected to exceed 0.02 wt% (200 ppm). As a result, solidification cracking is best controlled by controlling solidification behavior.

The WRC-1992 diagram (Figure 3.14) can also be used to estimate both ferrite content (in terms of FN) and solidification behavior. As described in Chapter 3, this

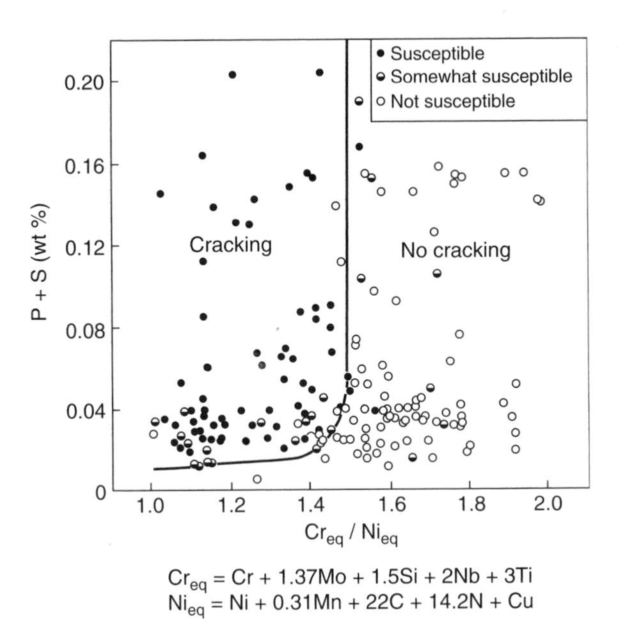

Figure 6.23 Suutala diagram for predicting weld solidification cracking from weld metal composition. (From Kujanpää et al. [30].)

diagram was developed through the efforts of the Stainless Steel Subcommittee of the Welding Research Council (WRC). Hundreds of welds were analyzed to determine the Ferrite Number (FN) and solidification mode as a function of composition. Based on the compositions of commercial austenitic stainless steels and filler metals, ferrite contents in the range 0 to 20 FN can be expected. Within the specification range of a given alloy, solidification behavior can range from A to FA or even F with a corresponding range in FN. Thus, the solidification cracking susceptibility of a weld metal can be predicted by plotting its composition on the WRC-1992 diagram and determining the type of solidification that it will undergo. As described previously, compositions that lie in the A or AF regions will be more susceptible to cracking than those that lie in the FA region.

If the composition of the base and filler metal are known, the FN and solidification behavior can be estimated, as shown in Figure 6.24. In this example, a fully austenitic base metal is being welded using a filler metal with FN 10. The composition of any weld made between these two materials must lie along the tie line connecting them. The position along the tie line is determined by the dilution of the filler metal by the base metal. If the dilution is 50% (case 1), the weld metal will solidify in the AF mode and have FN 1. This weld metal may potentially be susceptible to weld solidification cracking if the restraint level is sufficiently high. If the dilution is reduced to 20% (case 2), which may be typical of low-heat-input welds, the solidification mode will shift to FA with a resultant ferrite content of FN 6. This weld metal would be expected to be quite resistant to cracking, even under high restraint conditions.

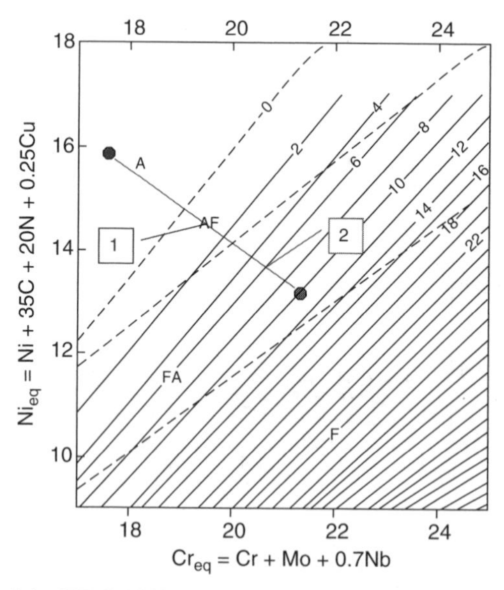

Figure 6.24 Use of the WRC-1992 diagram to predict solidification behavior and Ferrite Number.

6.5.1.3 Effect of Impurity Elements It is well recognized that impurity elements, particularly phosphorus and sulfur, promote solidification cracking in steels. Even low concentrations of these elements can promote cracking in austenitic stainless steel welds if solidification occurs as primary austenite (A or AF). This is shown quite clearly on the Suutala diagram (Figure 6.23), where P + S levels as low as 0.02 wt% are enough to promote cracking when the Cr_{eq}/Ni_{eq} value is below 1.48.

In stainless steels, typical P + S levels are in the range 0.02 to 0.05 wt%, depending on the steel type and specification. Removing sulfur from stainless steels is easily accomplished using argon–oxygen decarburization (AOD) melt practice. Using this practice, an Ar–O_2 mixture is blown into the molten steel to reduce the carbon level through the formation and release of CO and CO_2. At the same time the oxygen combines with sulfur, forming SO_2, which also escapes from the melt. In AOD-processed steels it is possible to achieve sulfur levels as low as 0.001 wt% (10 ppm). Unfortunately, removing phosphorus is much more difficult and levels below 0.02 wt% are achieved only by careful control of the starting stock.

Work by Arata et al. [32] and Ogawa and Tsunetomi [33] has shown that phosphorus may in fact be more damaging than sulfur with regard to promoting solidification cracking. However, they recommend that both S and P should be reduced to levels below 0.002 wt% to eliminate solidification cracking in fully austenitic weld metal. The effect of these impurity elements on solidification cracking in 25Cr–20Ni fully austenitic weld metal evaluated using the Varestraint test is shown in Figure 6.25. From a practical standpoint, achieving impurity levels low enough to minimize or eliminate cracking in fully austenitic weld metals is not practical using commercial steel production methods. Thus, control of solidification behavior is paramount.

As a result of a study of GTA welding of specially produced very low ferrite heats of 308, Li and Messler [34], contend that phosphorus is more potent than sulfur in promoting solidification cracking in the fusion zone. However, they also found that sulfur is more potent than phosphorus in promoting liquation cracking in the HAZ. A complicating factor in reducing sulfur to very low levels is the beneficial effect of sulfur on weld penetration. Heiple and Roper [35] found that autogenous GTA welds had very poor penetration when the base metal sulfur content was below 50 ppm (0.005%). They attributed this to surface-tension-driven convection (Marangoni effect) in the weld pool. When sulfur is too low, the surface tension of the weld pool decreases with increasing temperature, so that the hottest metal, directly under the arc, is pulled toward the weld pool edges. The result is a wide but shallow penetration pattern. On the other hand, when the sulfur is higher, the surface tension of the pool increases with increasing temperature. Then the liquid metal is pulled by surface tension along the surface toward the weld center, where it has nowhere to go but down. The hottest metal is driven downward, resulting in a deep, narrow penetration pattern. So although it is desirable from a cracking viewpoint to reduce sulfur to very low levels, it is not always beneficial from a productivity standpoint.

If solidification occurs in the FA mode, resistance to cracking is very high, irrespective of impurity level. As shown on the Suutala diagram, if solidification occurs in the FA mode (above 1.48 Cr_{eq}/Ni_{eq} in Figure 6.23), very high levels of S + P can

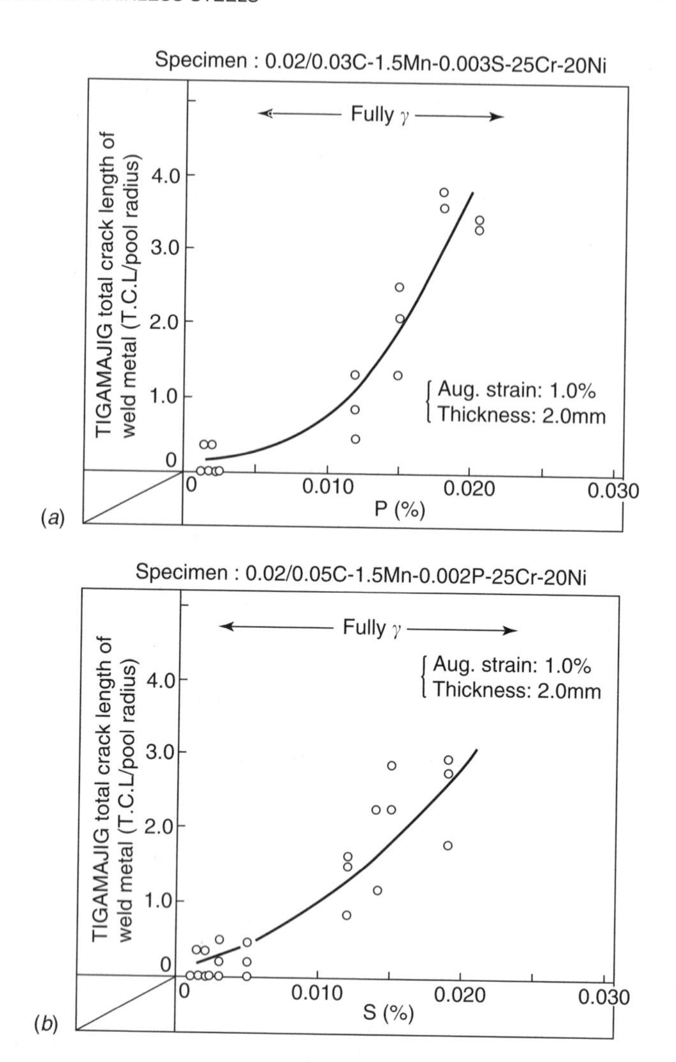

Figure 6.25 Weld solidification cracking susceptibility of fully austenitic 25Cr–20Ni weld metal based on Varestraint testing: (*a*) effect of phosphorus; (*b*) effect of sulfur. (From Ogawa and Tsunetomi [33]. Courtesy of the American Welding Society.)

be tolerated with no cracking. This has been verified by Lundin et al. [36] and Brooks et al. [37], who have shown that high-sulfur, free-machining steels (Type 303S) can be welded without cracking if solidification is maintained in the FA mode. This is shown in Figure 6.26 for both GTA and pulsed laser welds.

Note that a Cr_{eq}/Ni_{eq} range exists in which cracking does not occur for either process: 1.55 to 1.9 for GTA welds and above 1.7 for pulsed laser welds. The lower bound for GTA welds is slightly higher than that originally proposed by the Suutala diagram (1.48), but the transition in susceptibility was again consistent in that it reflected the change in solidification from primary austenite to primary ferrite. The

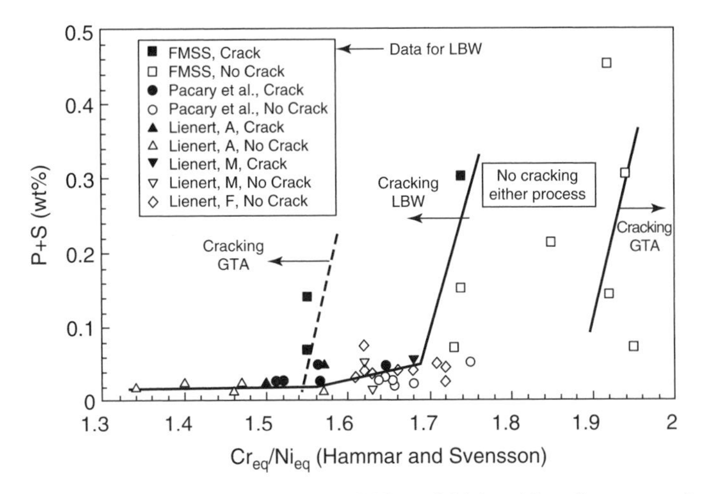

Figure 6.26 Solidification cracking susceptibility of high-sulfur alloys as a function of Cr_{eq}/Ni_{eq} for both GTA and pulsed laser welds. (From Brooks et al. [37] Courtesy of the American Welding Society.)

increase in cracking susceptibility at high Cr_{eq}/Ni_{eq} values (approximately 1.9) reflects the shift from FA to F type solidification. The presence of ferrite–ferrite boundaries at the end of solidification allows better wetting by sulfur-rich liquid films than the ferrite–austenite boundaries that are present with FA solidification.

6.5.1.4 Ferrite Measurement As should be apparent from the previous discussion, the ferrite content of the weld metal is important since it is an indicator of weld solidification behavior and associated resistance to cracking. The use of composition alone may not be sufficient to predict solidification behavior and ferrite content, since small variations in composition obtained by various analytical techniques can lead to large variations in predicted behavior. This is particularly true for carbon and nitrogen, which are potent austenite-promoting elements. Thus, it is critical to be able to accurately measure the ferrite content of weld metals. It is possible to use metallographic measurement techniques that require multiple sections and specialized characterization methods. These techniques have three major disadvantages. First, they are destructive, requiring sections to be removed from an actual weldment. This precludes their use for field examination of welds. Second, they are time consuming and inherently inaccurate unless many sections in various locations and orientations are examined. Third, the reproducibility, among a number of laboratories, of ferrite percent determination by metallographic means has been shown to be poor [38].

For these reasons, magnetic techniques have been widely adopted and standardized. These techniques take advantage of the fact that ferrite is ferromagnetic at room temperature, while austenite is not. One of the most widely used techniques involves relating the force required to pull a small magnet from the weld surface (called *tearing force*) to the ferrite content. A number of instruments have been designed using this principle, most recognizably the MagneGage and the Severn Gage. An AWS standard, AWS A4.2-98, has been developed to allow calibration of the MagneGage using secondary weld metal samples or primary coating thickness standards. The same principles and calibration procedure are reflected in the international standard ISO 8249.

Another technique uses an eddy current probe to measure ferrite content. Instruments of this type, such as the Fischer FeritScope™, are very useful in the field, since they are portable and can access small spaces. Such instruments can only be calibrated with secondary weld metal standards but produce results identical to those obtained with a MagneGage. All of the instruments can be calibrated to measure ferrite content in terms of a Ferrite Number. FN does not correlate exactly with volume percent ferrite and ranges from 0 to 140 FN, or more, depending on the composition of the ferrite. At values below 8, FN is often considered to be approximately equal to volume percent. At higher FN levels found in common austenitic and duplex stainless steel weld metals, volume percent ferrite may be about 70% of the FN value [39].

6.5.1.5 Effect of Rapid Solidification The Suutala and WRC-1992 diagrams are usually sufficient for predicting the solidification behavior and weld cracking susceptibility of austenitic stainless steels. Under rapid solidification conditions, however, a shift in solidification behavior may occur and these diagrams may not accurately predict either solidification behavior or ferrite content. This behavior has been studied by a number of investigators [40–45]. It is generally agreed that the shift in solidification mode from primary ferrite to primary austenite is associated with dendrite tip undercooling, as described by Kurz and Fisher [46]. This effect in austenitic stainless steel welds has been described in some detail by Lippold [41] and Brooks and Baskes [44]. Under rapid solidification conditions, the dendrite tip undercooling increases the stability of austenite relative to ferrite as the primary phase of solidification. This is shown schematically in Figure 6.27. The preferred

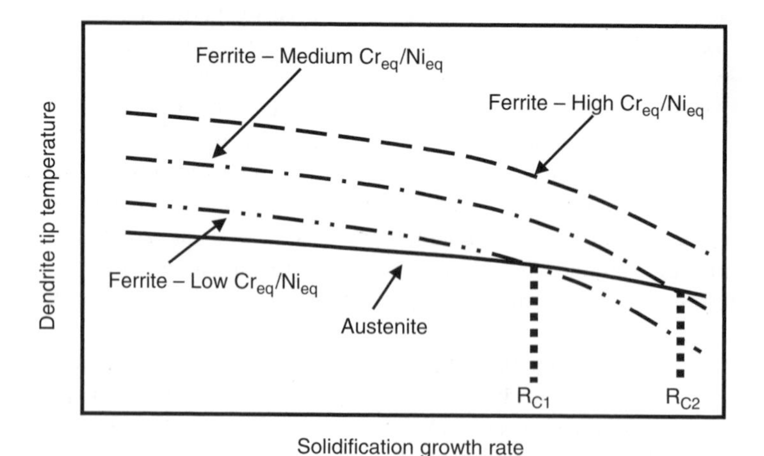

Figure 6.27 Effect of rapid solidification on dendrite tip undercooling.

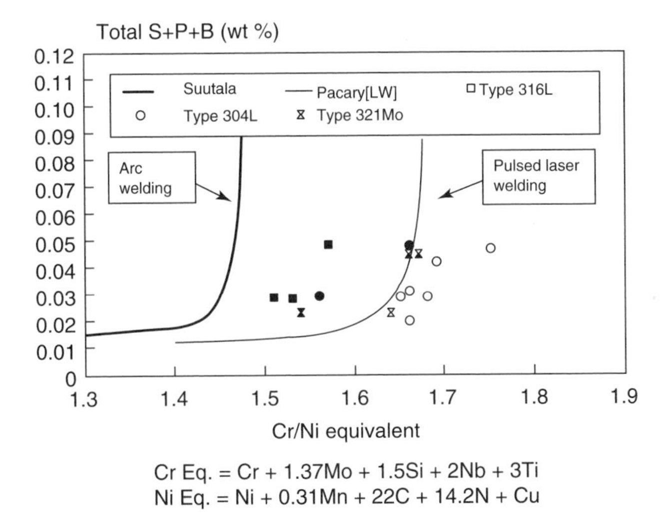

Figure 6.28 Modified Suutala diagram for rapid solidification conditions. Solid symbols, cracking; open symbols, no cracking. (From Pacary et al. [47]. Courtesy of the American Welding Society.)

solidification phase will be that with the highest dendrite tip temperature at a given growth rate. Thus, as growth rate increases, austenite becomes favored over ferrite. In alloys that would normally solidify as ferrite, but with low Cr_{eq}/Ni_{eq} values, the critical growth rate for austenite solidification is indicated as R_{C1} in Figure 6.27. As the Cr_{eq}/Ni_{eq} increases, higher growth rates are required for austenite solidification, as shown by the value R_{C2}.

Recognizing this effect and its influence on solidification cracking, Pacary et al. [47] developed a modified version of the Suutala diagram based on rapid solidification in pulsed laser beam welds (Figure 6.28). This diagram is also applicable when using other processes that produce extreme solidification rates, such as electron-beam welding or gas-tungsten arc welding performed at high weld travel speeds. Note that relative to the original Suutala demarcation between cracking/no cracking developed for arc welds, the critical equivalency ratio (based on Hammar and Svennson equivalents [31]) for pulsed laser welds has shifted to a higher value (from 1.48 to 1.68). As a consequence, alloys that are crack resistant under "normal" weld solidification conditions may be crack susceptible if the solidification rate is extremely rapid.

As discussed previously, the dramatic shift in the critical Cr_{eq}/Ni_{eq} value for rapid solidification results from a shift in solidification behavior at high solidification rates. Some alloys that would normally solidify as primary ferrite, $(Cr_{eq}/Ni_{eq})_{WRC}$ = 1.35 to 1.55, shift to primary austenite solidification (A). As a result, these alloys may be susceptible to weld solidification cracking. Figure 6.29 shows the transition from primary austenite solidification to primary ferrite solidification under rapid solidification conditions, plotted using the WRC-1992 equivalency relationships.

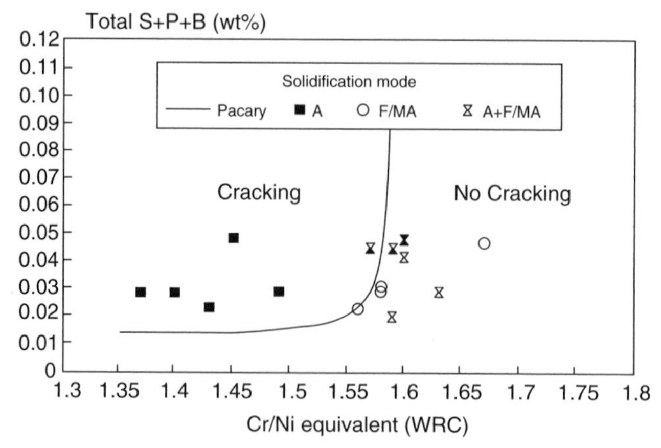

Figure 6.29 Transition from primary austenite to primary ferrite solidification mode for rapidly solidified stainless steel welds. (From Lippold [41]. Courtesy of the American Welding Society.)

Above the WRC-1992 equivalency ratio 1.55, solidification occurs as ferrite, but this ferrite may transform completely to austenite upon cooling in the solid state by a massive transformation (F/MA). This is thought to be a diffusionless, massive transformation, since no remnants of the ferrite are observed in the microstructure, being replaced by a massive austenite (MA) structure [41,44,49]. In the transition region, austenite formed during solidification and that formed by a massive transformation from ferrite may coexist (A + F/MA). Weld microstructures from pulsed laser welds representative of A and F/MA solidification are shown in Figure 6.30. More recently, Lienert and Lippold [48] have investigated the transition in solidification behavior under rapid solidification conditions using a wider range of alloys. This has resulted in a slightly modified diagram relative to Figure 6.28 that shows a distinct transition region between A- and F-type solidification, as shown in Figure 6.31.

The behavior described above is not unique to pulsed laser welds. Electron-beam welds made at very high speeds may also exhibit this behavior, as was demonstrated by Elmer et al. [49,50]. Lippold [51] observed a shift in solidification behavior along the centerline of electron-beam welds in Type 304L. This shift, as shown in Figure 6.32, was associated with the higher solidification rate at the weld centerline relative to elsewhere in the weld. Similar behavior was observed by Kou and Le [45] in gas-tungsten arc welds made at very high speeds, resulting in a teardrop-shaped weld pool. A shift from primary ferrite to primary austenite solidification was observed along the weld centerline.

To combine the effects of composition and solidification rate, Lippold [41] proposed the microstructure diagram shown in Figure 6.33. A similar diagram based on electron-beam welds in ternary Fe–Cr–Ni alloys was also proposed by Elmer et al. [49] and is shown in Figure 6.34. Note that in Figure 6.33 for values of Cr_{eq}/Ni_{eq} between 1.3 and 1.6, an increase in solidification rate results in a transition from AF

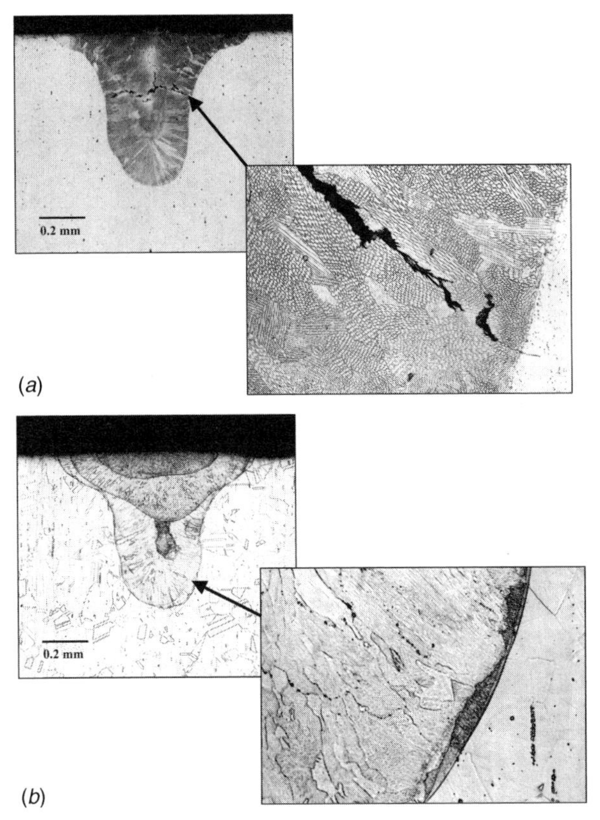

Figure 6.30 Microstructure of pulsed laser welds: (*a*) A solidification; (*b*) F/MA solidification. Both structures are fully austenitic.

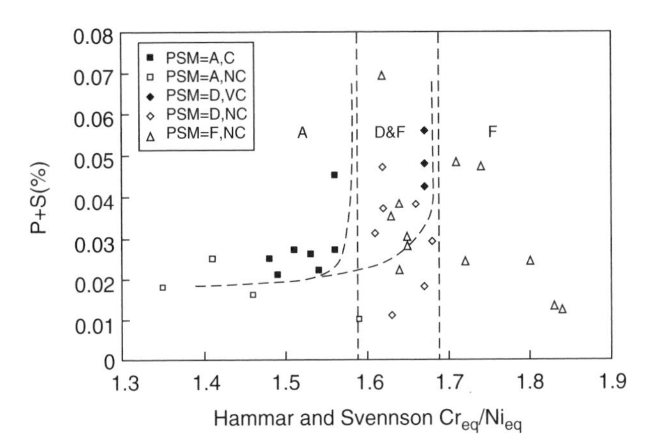

Figure 6.31 Primary solidification mode (PSM) diagram. A, austenite solidification; D, dual (A + F) solidification; F, ferrite solidification; C, cracking; NC, no cracking; VC, variable cracking. (From Lienert and Lippold [48]. Courtesy of *Science and Technology of Welding and Joining*.)

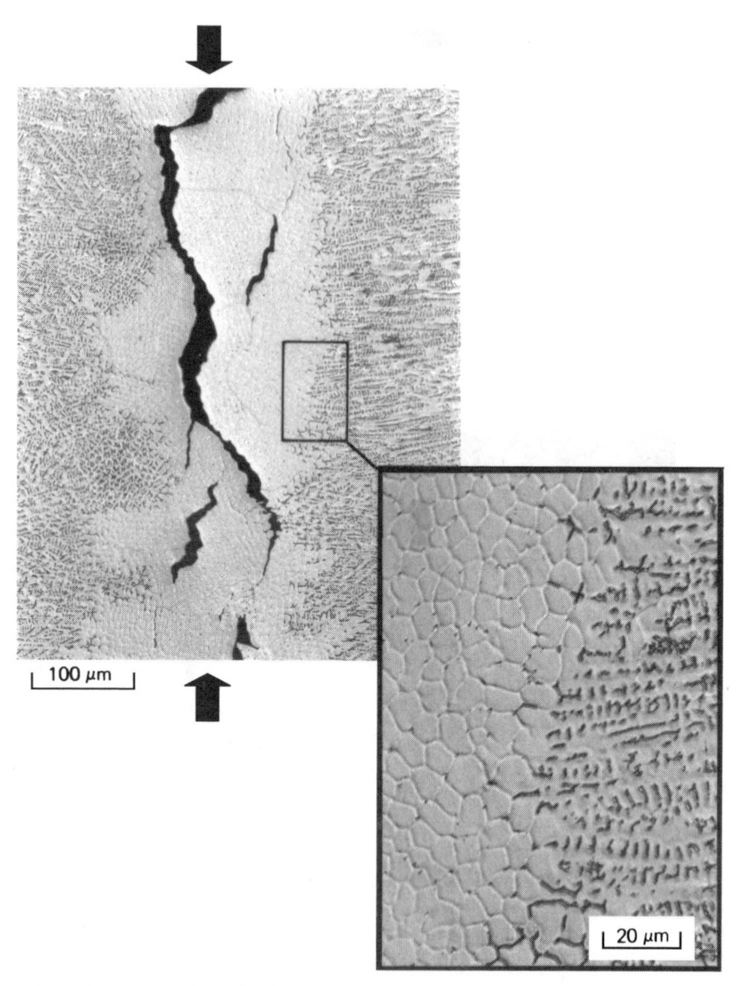

Figure 6.32 Transition in solidification mode along the centerline of an electron-beam weld in Type 304L. Transverse cross section, arrows indicate location of weld centerline. (From Lippold [51]. Courtesy of the American Welding Society.)

and FA solidification to Type A. This can result in a corresponding increase in solidification cracking susceptibility.

6.5.1.6 Solidification Cracking Fracture Morphology Because solidification cracking is associated with liquid films along solidification grain boundaries or solidification subgrain boundaries (see Figure 6.15), the fracture surface normally exhibits a dendritic appearance. A representative solidification crack surface as viewed in the scanning electron microscope (SEM) is shown in Figure 6.35. The "eggcrate" appearance of this fracture surface results from the separation of opposing dendrites due to the presence of liquid films along solidification grain or subgrain boundaries (SGBs or SSGBs). In fully austenitic weld metals, a transition from

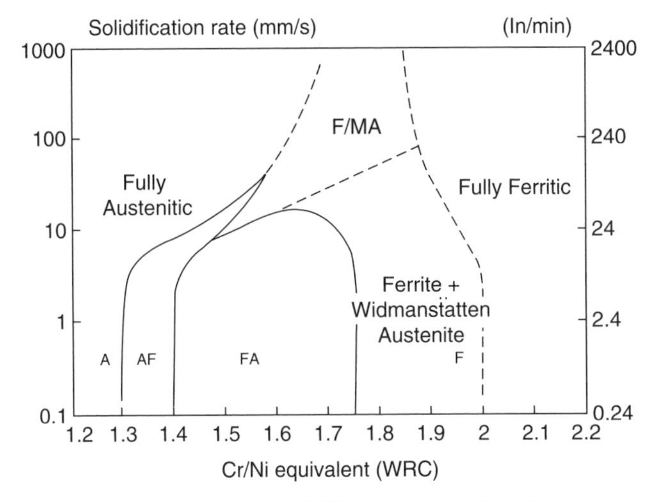

Figure 6.33 Effect of composition and solidification rate on the microstructure of austenitic stainless steels. (From Lippold [41]. Courtesy of the American Welding Society.)

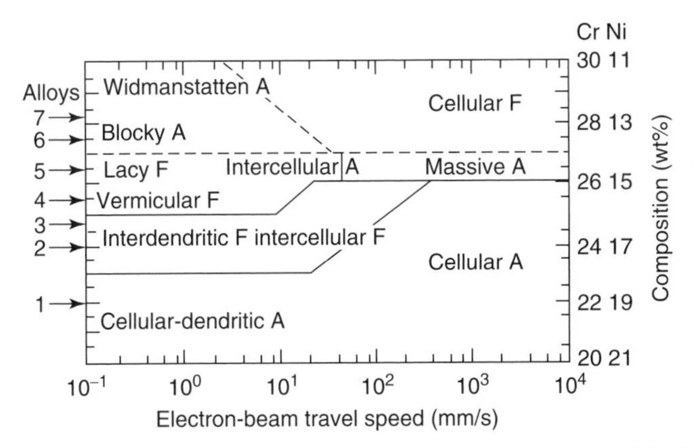

Figure 6.34 Effect of electron-beam travel speed on the microstructure of Fe–Cr–Ni alloys. (From Elmer et al. [49].)

dendritic to flat fracture may occur. This is shown in the schematic in Figure 6.36. This occurs because the solidification grain boundary begins to solidify and the dendritic character is lost. Cracking progresses along what is probably then a migrated grain boundary.

In general, weld solidification crack surfaces are nearly entirely dendritic in character. The presence of the morphology shown in Figure 6.35 is usually a good indication that cracking occurred during solidification rather than in the solid state. Other elevated temperature cracking phenomena, such as ductility-dip cracking, occur along migrated grain boundaries in the weld metal and do not exhibit a clear dendritic appearance.

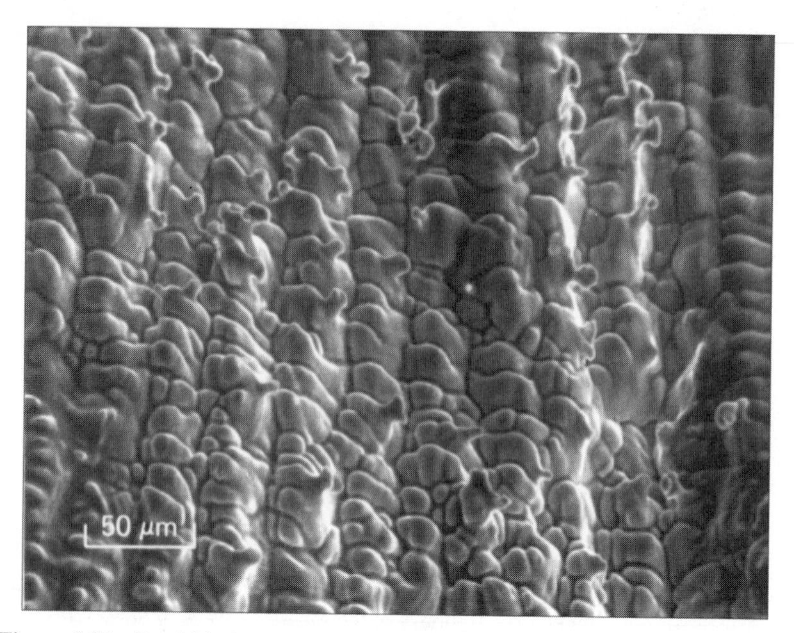

Figure 6.35 Dendritic fracture surface characteristic of a weld solidification crack.

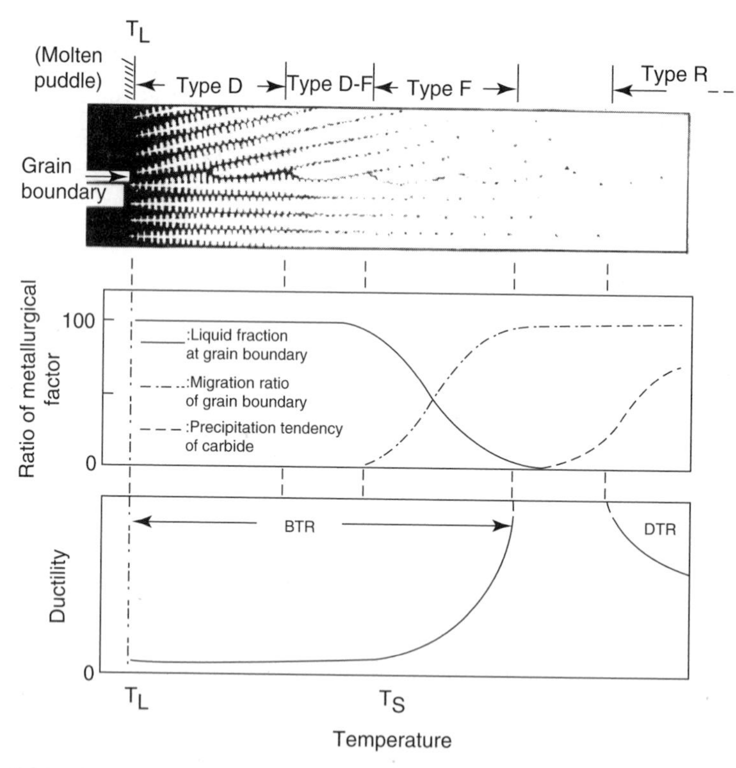

Figure 6.36 Schematic of fracture morphology in fully austenitic weld metal. D, dendritic; F, flat; BTR, brittle temperature range; DTR, ductility-dip temperature range. (From Arata et al. [13]. Courtesy of the Japanese Welding Research Institute.)

6.5.1.7 Preventing Weld Solidification Cracking

Avoiding or minimizing weld solidification cracking in austenitic stainless steels is accomplished simply and most effectively by controlling the composition of base and filler materials. By assuring solidification as primary ferrite, the potential for cracking will be effectively nil. For most austenitic stainless steel weld metals, this means that the composition should be controlled to achieve the FA solidification mode, resulting in the presence of FN 3-20 in the weld deposit. As shown in Figure 6.24, the WRC-1992 diagram can be used quite effectively to accomplish this if the composition of the base and filler metals are known.

In systems where the base and filler compositions preclude FA solidification [i.e., solidification occurs as primary austenite (A or AF)], the potential for cracking will be much higher. The most effective way to avoid cracking in these weld metals is to reduce impurity content (see Figure 6.23, Suutala diagram) and/or minimize the weld restraint. High-purity fully austenitic weld metals can be quite resistant to weld cracking under conditions of low to moderate restraint. Convex bead shape and filled weld stops (craters) are also helpful.

Depending on the application and/or service conditions, some care should be taken when prescribing a weld metal ferrite content. While the presence of ferrite levels in the range FN 3 to 20 is almost certain to avoid solidification cracking, ferrite above FN 10 may in fact compromise mechanical properties if the weldment is to be stress-relieved or the structure is put in service at either cryogenic temperatures or elevated temperatures. Loss of cryogenic fracture toughness at ferrite levels as low as FN 3 has been shown previously (Figure 6.19).

Service temperatures from 425 to 870°C (800 to 1600°F) can lead to embrittlement due to formation of alpha-prime and sigma phase, both of which form preferentially at the ferrite–austenite interface. For weld metals above FN 10, the formation of these phases can severely reduce toughness and ductility. High ferrite contents have also been shown to reduce elevated temperature stress–rupture properties (Figure 6.18). Thus, while it is tempting to use the WRC-1992 diagram simply to control composition to produce FA solidification and a "safe" level of weld metal ferrite relative to weld solidification cracking, the engineer must be aware of the implications of high weld metal ferrite contents. When laser and electron-beam welding processes are used, the solidification behavior and ferrite content predicted by the WRC-1992 diagram may not be valid. In this case, the precautions and predictive diagrams in Section 6.5.1.5 should be used.

6.5.2 HAZ Liquation Cracking

Austenitic stainless steels can be susceptible to various forms of cracking in the HAZ, as reviewed by Thomas [52,53]. Liquation cracking in the HAZ occurs due to formation of liquid films along grain boundaries in the partially melted zone adjacent to the fusion boundary. This liquation can occur due to segregation of impurities to grain boundaries at elevated temperatures or by the constitutional liquation of NbC (Type 347) and TiC (Type 321). As noted previously, sulfur has been found to be more harmful than phosphorus with respect to HAZ liquation cracking [34].

HAZ liquation cracking can be controlled by managing the composition of the base metal. Base metals that have a *Ferrite Potential** (from WRC-1992 diagram) of 1 or higher will form some ferrite along HAZ/PMZ boundaries (see Figure 6.17) and effectively inhibit liquation cracking. This occurs because the ferrite–austenite boundaries that are present are not easily wet by liquid films. The formation of ferrite at the grain boundaries also restricts grain growth. This also has a beneficial effect on reducing cracking susceptibility.

In alloys where the HAZ is fully austenitic (no grain boundary ferrite), liquation cracking can be minimized by restricting impurity levels and grain size. Lowering the heat input will result in steeper temperature gradients in the surrounding HAZ and restrict the distance over which liquation occurs. Grain size also has an important effect on liquation cracking susceptibility, with smaller grains improving the cracking resistance. Since failure occurs along grain boundaries, increasing the grain boundary area reduces segregation and local stress on individual boundaries resulting in higher total stress to initiate cracking.

An example of cracking in the HAZ of austenitic stainless steels is shown in Figure 6.37. These microstructures were produced using the spot Varestraint test on alloys with different Ferrite Potentials. When the Ferrite Potential is zero, cracking occurs at relatively low strains and the cracks propagate some distance from the fusion boundary. When the Ferrite Potential exceeds 1, some ferrite forms along the austenite grain boundary in the high-temperature HAZ. The presence of this ferrite reduces the cracking susceptibility. Results of the spot Varestraint testing for Type 304 and 304L alloys and A286 are shown in Figure 6.38.

6.5.3 Weld Metal Liquation Cracking

Weld metal liquation cracking occurs in multipass welds along SGBs or MGBs (see Figures 6.7 and 6.8). Fully austenitic welds are the most susceptible, due to the presence of a primary austenite solidification structures (A or AF) that exhibit significant segregation. Weld metals that contain sufficient ferrite (greater than FN 2 to 6) are generally resistant to WM liquation cracking. These defects are also referred to as *microfissures* since they tend to be small and buried within the weld deposit. An example of a weld metal liquation crack in fully austenitic weld metal is shown in Figure 6.39. Note that this crack lies in the HAZ of the pass above and resides along an MGB. It is relatively short in the cross-section dimension of the weld since it exists only where remelting of the MGB occurs during reheating. Detection of such cracks is difficult because they are generally short [1 to 2 mm (0.04 to 0.08 in.) in length] and tight. Metallographic examination or a fissure bend test (discussed in Chapter 10) is necessary to find them. If the top surface of the weld is ground and dye penetrant inspection is performed, they can sometimes be located. Because they are very short and the dye penetrant tends to "bleed" around the crack, they often appear as spots rather than cracks. The fact that they are aligned along a fusion boundary in reheated weld or base metal can cause them to be mistakenly identified as "linear porosity."

*Ferrite Potential is determined the same way as Ferrite Number in the weld metal. It indicates the tendency for forming ferrite in the base metal HAZ.

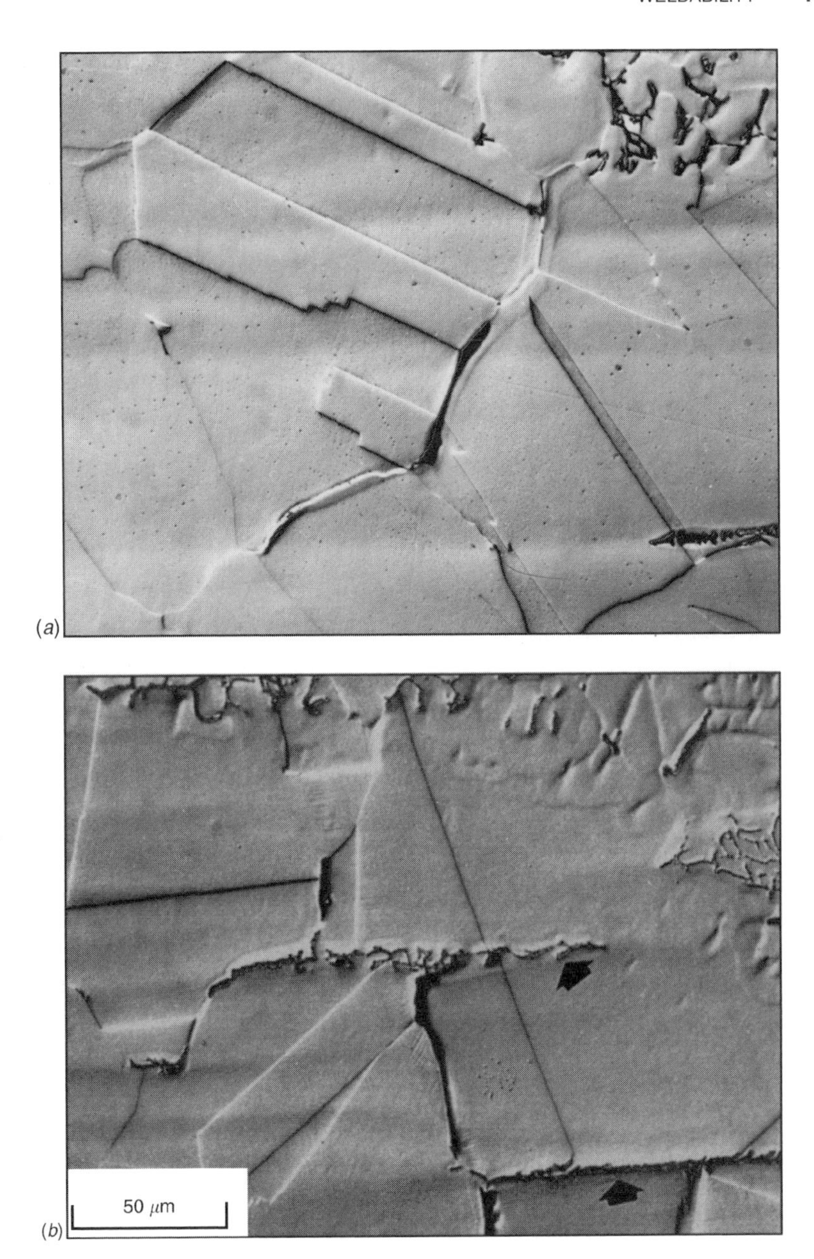

Figure 6.37 HAZ liquation cracking in the HAZ of austenitic stainless steels: (*a*) Type 304L with FP 0; (*b*) Type 304 with FP 1. (From Lippold et al. [54]. Courtesy of the American Welding Society.)

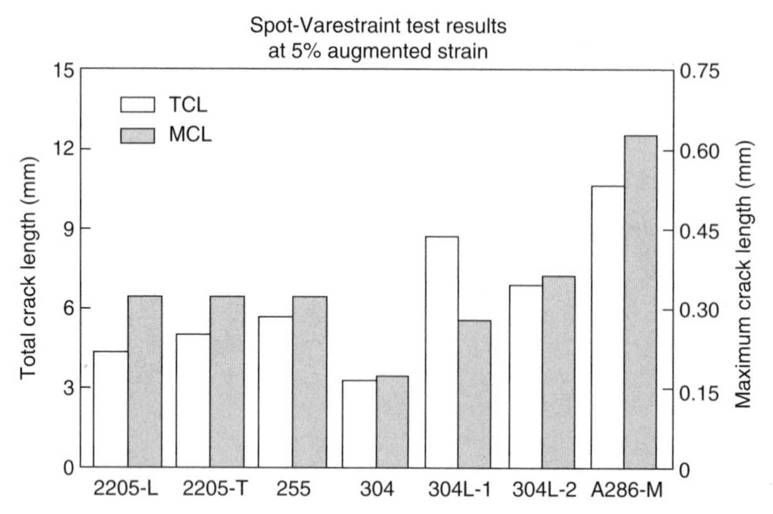

Figure 6.38 Results of spot Varestraint tests of austenitic and duplex stainless steels. Ferrite potentials: 2205 and 255 = FP > 50, 304 = FP 8, 304L-1 = FP 1, 304L-2 = FP 0, A286 = FP 0. (From Lippold et al. [54]. Courtesy of the American Welding Society.)

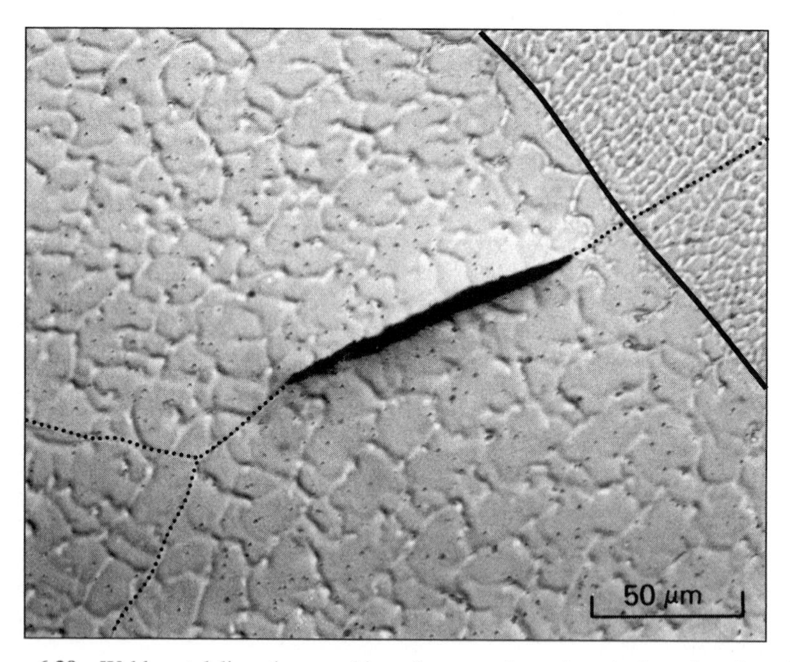

Figure 6.39 Weld metal liquation cracking along a migrated grain boundary in a fully austenitic multipass weld. Solid line represents fusion boundary of subsequent pass. Dotted lines show location of migrated grain boundaries.

Weld metal liquation cracking is best controlled by adjusting the deposit composition such that some ferrite is present in the deposit. In fully austenitic deposits, control of impurity levels and minimizing weld heat input can reduce or eliminate this form of cracking. Some filler metals have been developed with increased Mn to minimize cracking in fully austenitic weld metals [55,56]. Weld metal liquation cracking has been studied in some detail by Lundin et al. [57–59]. Their results showed that control of weld metal ferrite content is also extremely important for avoiding cracking in multipass welds. They used the Fissure Bend test on multipass weld pads to determine the critical level of ferrite necessary to eliminate cracking. This test is described in Chapter 10.

Based on this test, they provided the Ferrite Number guidelines listed in Table 6.11 for preventing weld metal liquation cracking. They also discussed the mechanism for cracking and defined a critical region in the weld metal HAZ where cracking occurs. This region, termed the "Hazard HAZ," results from dissolution of the original weld metal ferrite in the temperature range from 1095 to 1290°C (2000 to 2350°F). This effect is shown in Figure 6.40. Note that for weld metals that originally contain FN 5, dissolution in the HAZ can reduce the ferrite level by up to 80%, to a level of

TABLE 6.11 Minimum Ferrite Number Needed to Avoid Weld Metal Liquation Cracking

Weld Metal	Minimum FN
316	1.5
308	2.0
316	2.5
308L	3.0
309	4.0
347	6.0

Source: Lundin and Chou [59].

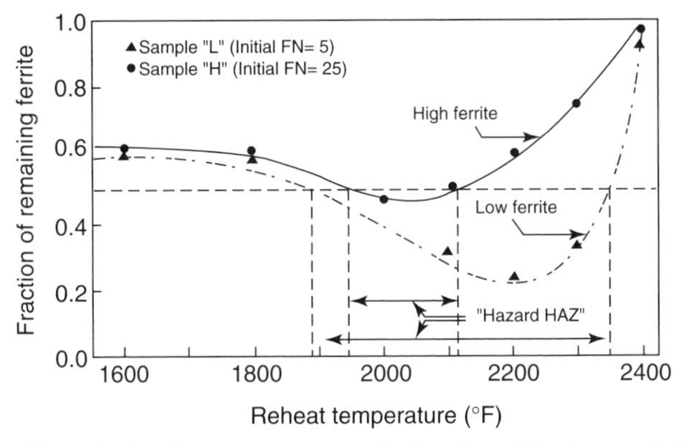

Figure 6.40 Effect of reheating temperature on the fraction of ferrite remaining in the weld metal HAZ. (From Lundin and Chou [59]. Courtesy of the American Welding Society.)

FN 1. This region then becomes susceptible to liquation cracking. When weld deposits with higher ferrite levels are reheated, the reduction in HAZ ferrite will not be sufficient to promote cracking.

6.5.4 Ductility-Dip Cracking

Ductility-dip cracking (DDC) occurs in many alloys that have an austenitic (FCC) microstructure, including austenitic stainless steels, Ni-base alloys, and Cu-base alloys. It is associated with a precipitous drop in elevated temperature ductility above approximately one-half the melting temperature of the material. A schematic ductility versus temperature curve for both normal elevated-temperature ductility and a material that exhibits a ductility dip is shown in Figure 6.41. Note that this ductility dip is separate and distinct from the brittle temperature range (BTR) in which solidification and liquation cracking occur, although the high-ductility region that separates them may be 200°C (360°F) or less (see Figure 6.36). It is likely that DDC has been mistakenly identified as liquation cracking by many investigators because of this small temperature differential. It is also possible that liquation cracks and DDC may actually link to form a single crack, although by two different mechanisms.

DDC has been observed in both the weld metal and the HAZ of austenitic stainless steels [60,61]. It is normally associated with large grain size and high restraint conditions, such as those generated in thick-section weldments. In weld metals, DDC occurs along MGBs. An example of weld metal DDC from the work of Nissley and Lippold [62] is shown in Figure 6.42. Note that the grain size is quite large and that cracking is clearly along the MGBs in the microstructure. DDC is now thought to be a form of elevated temperature creep failure that occurs very rapidly at temperatures well above the normal creep regime for stainless steels [63,64]. Cracking is most prevalent in weld metals where the MGBs are very straight and grain boundary sliding is facilitated. Increasing the boundary *tortuosity* by grain boundary pinning

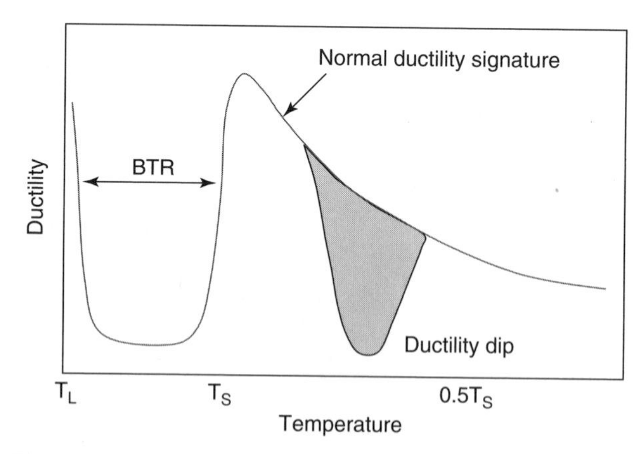

Figure 6.41 Elevated temperature ductility including the brittle temperature range (BTR) during solidification. BTR = brittle temperature range, T_L = liquidus temperature, T_S = solidus temperature.

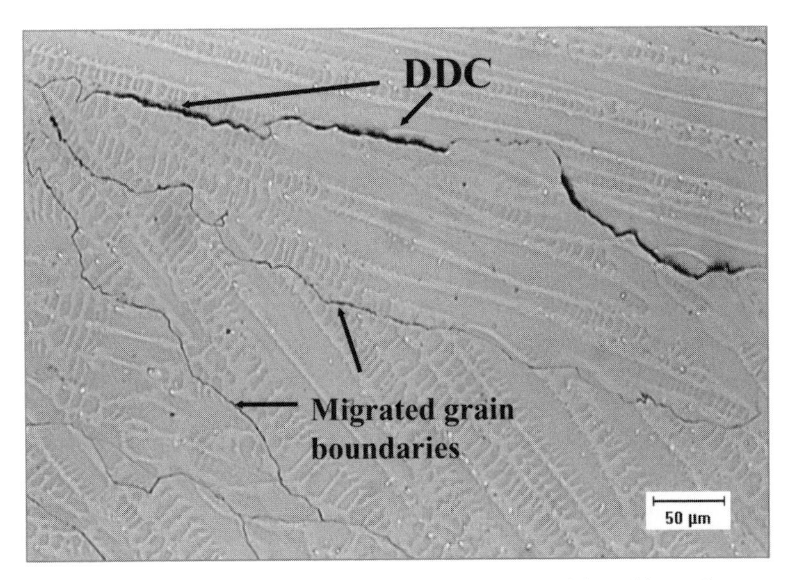

Figure 6.42 Ductility-dip cracking in fully austenitic weld metal.

effects will reduce susceptibility to DDC. In austenitic stainless steel welds, AF or FA solidification result in the presence of ferrite in the austenitic microstructure. This ferrite is effective in pinning the MGBs, resulting in very tortuous, crack-resistant boundaries (see Figures 6.11 and 6.22). The tortuosity of the boundary imposes a mechanical locking effect that opposes the elevated temperature sliding that leads to cracking.

Susceptibility to DDC can be quantified using the strain-to-fracture test [65]. Using this test, the minimum strain to cause solid-state, grain boundary cracking over a range of temperatures can be determined. This test method is described in more detail in Chapter 10. An example of strain-to-fracture results for autogenous welds in three austenitic stainless steels is shown in Figure 6.43. Note that Type 310 shows a minimum strain to fracture of 5% or below over a temperature range from approximately 750 to 1000°C (1380 to 1830°F) while Type 304 has a much higher minimum strain for cracking. The Type 310 weld metal was fully austenitic, while Type 304 solidified in the FA mode and exhibited FN 4. The presence of ferrite in the Type 304 weld metal results in very tortuous MGBs, since the grain boundary is pinned at the ferrite–austenite interface. The superaustenitic alloy, AL6XN, shows a low minimum strain to fracture between 900 and 950°C (1650 and 1740°F) but relatively high resistance to DDC outside this range. The MGBs in AL6XN were also quite tortuous, due to the presence of a eutectic constituent that forms at the end of solidification.

The mechanism for DDC is still not fully understood. In austenitic stainless steel weld metals, the presence of ferrite significantly reduces the risk of DDC by creating grain boundary paths that are very tortuous and resistant to crack initiation and propagation. Multipass, fully austenitic weld metals in thick-section, highly restrained joints are the most susceptible to this form of cracking.

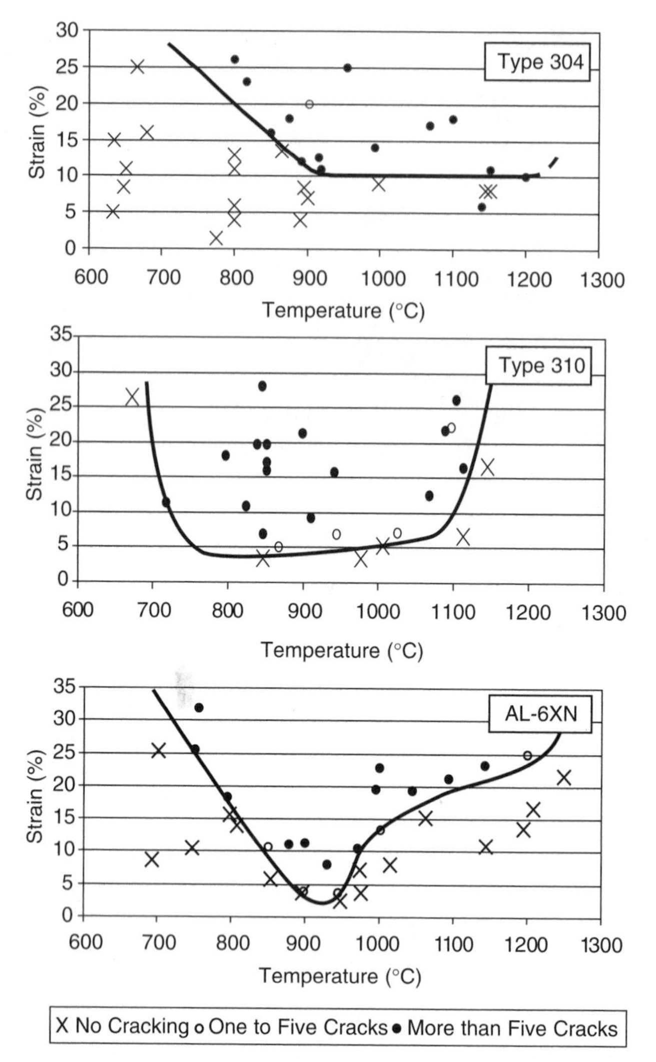

Figure 6.43 Ductility-dip cracking susceptibility as determined by the strain-to-fracture test for three austenitic stainless steels. (From Nissley et al. [66].)

6.5.5 Reheat Cracking

Reheat, or *stress-relief, cracking* is unusual in standard grades of austenitic stainless steels but may occur in alloys that form MC-type carbides during the stress-relief thermal cycle. Type 347, which contains Nb and forms NbC, is known to be susceptible to this form of cracking [52,67]. Higher-carbon heat-resisting alloys such as 304H and 316H may also be susceptible to reheat cracking, as described in Section 6.7. The mechanism is similar to that in low-alloy steels containing Cr, Mo, and V. During welding, alloy carbides dissolve in the high-temperature HAZ adjacent to the fusion boundary. The weld metal also contains carbon and carbide-forming alloy elements in solution. When the weldment is reheated during a postweld stress relief, carbides precipitate in the grain interiors and strengthen these regions relative to the grain boundaries. If significant stress relaxation occurs in this same temperature range, failure can occur preferentially along the grain boundaries. This phenomenon has been observed in both the HAZ and weld metal.

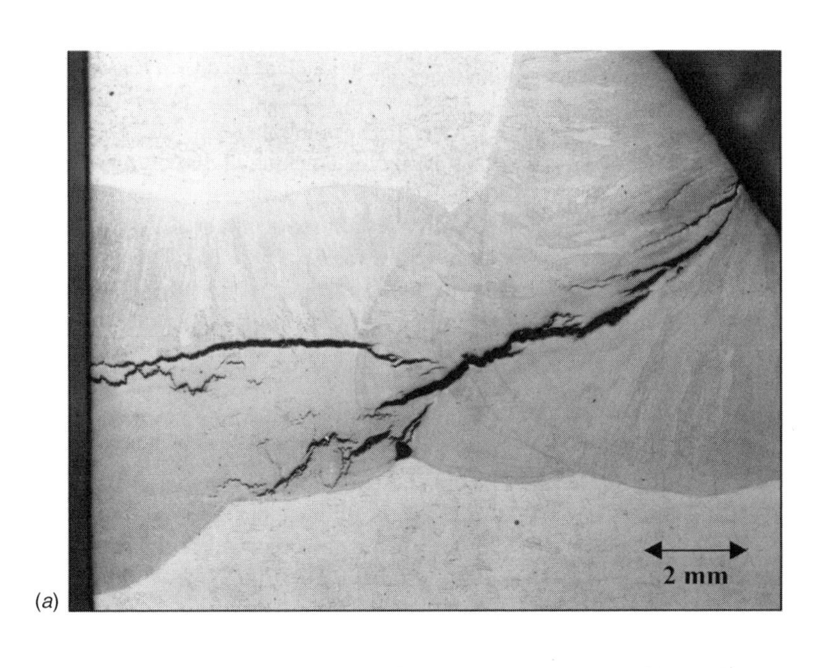

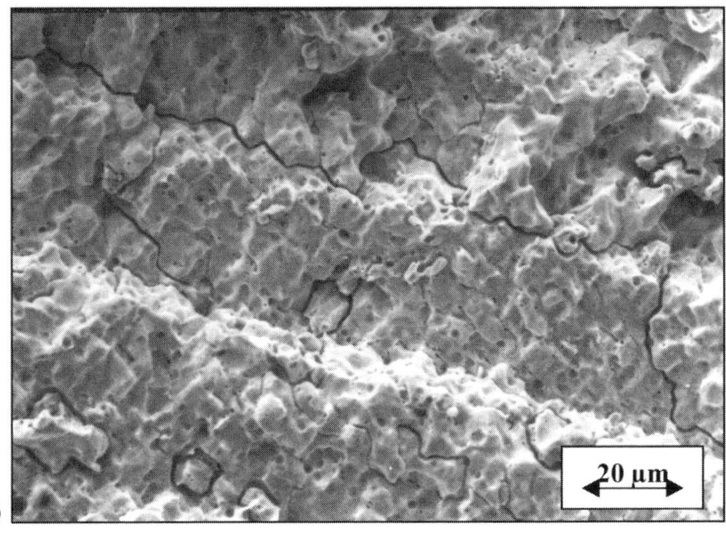

Figure 6.44 Reheat cracking in Type 347 weld metal: (*a*) macrograph; (*b*) SEM fractograph. (From Lin et al. [68].)

An example of reheat cracking in Type 347 weld metal is shown in Figure 6.44. This was a very highly restrained weld that required postweld stress relief at 900°C (1650°F). Note that this cracking shows a branched appearance. Upon closer examination it was found that the crack paths were along crystallographic grain boundaries. A SEM fractograph of a typical fracture surface is shown in Figure 6.44*b*. Note that remnants of the dendritic solidification structure are evident. This is consistent with the fact that the original weld microstructure solidified in the FA mode and had a ferrite level of approximately FN 8. This means that the crystallographic boundary would not have migrated away from the solidification grain boundary and therefore still retains a dendritic character. Much of the ferrite dissolved during the stress-relief heat treatment, resulting in a final ferrite level of FN 2. The combination of stress relaxation and NbC precipitation in the grain interiors promoted cracking in this weld metal.

Reheat cracking normally exhibits a "C-curve" temperature–time relationship. This behavior is shown in Figure 6.45 from the work of Lin et al. on the same Type 347 stainless steel weld metals discussed previously [68]. The data in Figure 6.45 were generated using a Gleeble thermo-mechanical simulator. Weld metal samples were heated to various postweld heat treatment temperatures and then loaded to 75 or 100% of their elevated temperature yield strength. Samples were held under load until failure occurred (open symbols in Figure 6.45). This testing resulted in a C-curve cracking response. The temperature range over which failure occurs, 700 to 1050°C (1290 to 1920°F), represents the range over which NbC precipitates form. Precipitation is most rapid in the range 800 to 1000°C (1470 to 1830°F). This represents the danger zone for postweld stress relief, since precipitation can occur within about 20 minutes. It is not surprising that the weldment in Figure 6.44 exhibits cracking since the postweld stress-relief temperature [about 900°C (1650°F)] is almost precisely at the nose of the reheat cracking C-curve.

From a practical standpoint, reheat cracking occurs when the precipitation-strengthening mechanism, as described by Figure 6.45, overlaps the temperature range in which residual stresses in the weld begin to relax. Thus, for large fabrications

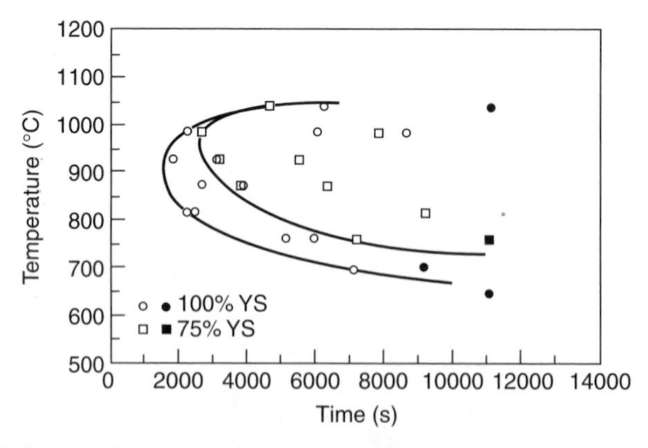

Figure 6.45 Reheat cracking susceptibility for Type 347 weld metal. Open symbols, failure; solid symbols, test stopped with no failure. (From Lin et al. [68].)

where heating times to the stress-relief temperature range 650 to 900°C (1200 to 1650°F) may require several hours, there is sufficient time for the reheat cracking mechanism to be operative. Heating into the low end of this range will delay precipitation but may not sufficiently relieve the residual stresses.

6.5.6 Copper Contamination Cracking

Copper contamination cracking (CCC) is a well-known phenomenon in both austenitic stainless steels and structural steels. It occurs by a liquid metal embrittlement mechanism, whereby molten copper penetrates austenite grain boundaries. Since molten copper (or a copper alloy) is a prerequisite, it can only occur above the melting point of copper, approximately 1083°C (1981°F). Welding-related failures due to CCC almost always result from the abrasion of copper on the parts to be welded from fixturing, contact tips, or other parts or tools made from copper. Copper added to the material as an alloying element does *not* promote CCC. A schematic of CCC is shown in Figure 6.46.

Because copper wets the grain boundaries most effectively at temperatures close to 1100°C [69], CCC is usually observed at some distance (a few millimeters) from the fusion boundary. The degree of cracking depends on the weld thermal cycle and applied stress. Residual and/or external stresses tend to accelerate CCC. It is generally quite straightforward to determine if the cracking mechanism is CCC. Copper along an austenite grain boundary is very apparent in an as-polished metallographic section, since the copper color will be very distinct from the stainless steel. Examination of the fracture surface in an SEM equipped with composition analysis capabilities (EDS or EDAX) will also readily reveal the presence of copper. Once identified, eliminating the source of the copper may be more challenging. It is usually associated with copper fixturing, tooling, or shielding systems that come in contact with the pieces to be welded. In some situations, copper fixtures are plated with chromium, nickel, or other metals to isolate the copper from the parts to be welded.

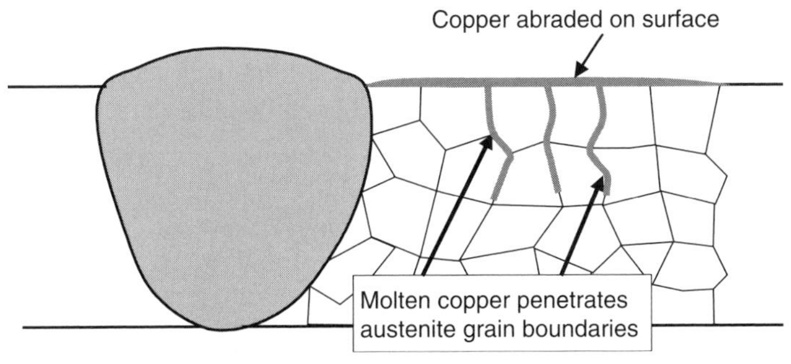

Figure 6.46 Copper-contamination cracking resulting from copper abrasion on the piece to be welded.

6.5.7 Zinc Contamination Cracking

Zinc contamination cracking (ZCC) is very similar to copper contamination cracking, but the means by which zinc enters the picture is different. Zinc melts at a temperature of 419.5°C (787°F), which is much lower than copper. Zinc boils at 906°C (1663°F), which may also be germane to the problem. ZCC occurs in welding galvanized steel to austenitic stainless steel [70], regardless of the filler metal used. The zinc from the galvanized steel reaches the HAZ of the stainless steel base metal, possibly by evaporation and condensation. Then it penetrates the grain boundaries of the stainless steel, just as does the copper in copper contamination cracking. The result is the same—cracking along grain boundaries in the HAZ of the austenitic stainless steel, a few millimeters from the fusion boundary. It is best prevented by removing the galvanizing from the mild steel in the joint area by acid dissolution before welding.

6.5.8 Helium-Induced Cracking

Another unusual form of embrittlement was encountered in the mid-1980s during repair of nuclear fuel reactor vessels at the Savannah River plant in South Carolina [71]. Upon remote GTA repair of irradiated Type 304L stainless steel, significant cracking was observed in the HAZ of the repair welds. This cracking was associated with helium bubble formation along the Type 304L base metal grain boundaries that forms as a by-product of neutron irradiation. Helium has very low solubility in steel and thus tends to form small bubbles along grain boundaries and other defect sites. In the HAZ, these bubbles grow rapidly by diffusion of helium along the grain boundary. If sufficient restraint is present, cracking will occur along the grain boundary, normally in very close proximity to the fusion boundary. This phenomenon has been studied in some detail by Goods and Karfs [72]. Neutron capture by tramp boron atoms in the steel converts the most common isotope of boron, ${}^{11}_{5}B$, into ${}^{12}_{5}B$, which has a very short half-life and decays by emitting an alpha particle (helium nucleus). An isotope of nickel can also be formed by neutron capture, and the resulting isotope (${}^{59}_{28}Ni$) decays by emitting an alpha particle. The alpha particle becomes a helium atom in the steel.

Cracking is very difficult to avoid since it is nearly impossible to remove the helium from its defect sites within the steel. Some success in repair welding was achieved by first applying low heat input weld overlays and then making repair weld attachments to the overlay material, effectively a form of "buttering" the substrate. The low heat input overlay welds minimized grain boundary cracking by reducing the overall heat input and weld restraint.

6.6 CORROSION RESISTANCE

Although austenitic stainless steels are often selected because of their corrosion resistance, some precautions are required when these alloys are welded and exposed to certain environments. The general atmospheric corrosion resistance of austenitic stainless steels is very good. At room temperature, atmospheric corrosion is essentially nil and the material will remain stainless indefinitely. At elevated temperature, general corrosion rates increase and some degradation and material loss will occur with time. In freshwater marine environments, general corrosion rates are also quite low, on the order of 2.5×10^{-5} mm/yr or less. In addition to general corrosion, austenitic stainless steels may experience the following forms of corrosion: pitting, intergranular, stress-assisted, crevice, galvanic, erosion corrosion, and microbiologically induced corrosion [3,73]. An excellent review on the corrosion behavior of welded stainless steels has been published by Gooch [74].

Welding can produce metallurgical modifications that can increase susceptibility to corrosion attack. In combination with the residual stresses that are present following welding, these modifications can result in accelerated attack of the weld region. Two forms of welding-related corrosion have been studied extensively in austenitic stainless steels because of the possibility of compromising the engineering usefulness of the welded structure. These forms are *intergranular corrosion* (IGC), often called *intergranular attack* (IGA), in the HAZ, and *stress corrosion cracking* (SCC). SCC can occur both inter- and intragranularly (or transgranularly), depending on microstructure and stress state. When it occurs intergranularly, it is called *intergranular stress corrosion cracking* (IGSCC). A number of forms of corrosion that effect austenitic stainless steel weldments are described in the following sections.

6.6.1 Intergranular Corrosion

Figure 6.47 represents the appearance of a weld that has undergone intergranular attack in the HAZ. On the surface of the weld exposed to the corrosive environment, there often appears a linear area of attack that parallels the fusion boundary. These are sometimes called "wagon tracks" because they are symmetric and parallel on either side of the weld. In cross section, severe attack (or weld "decay") can be observed along a sensitized band in the HAZ. Note that this band is at some distance from the fusion boundary. This is due to the fact that the carbide precipitation that leads to sensitization occurs in the temperature range from about 600 to 850°C (1110 to 1560°F). Above this temperature range, carbides go back into solution and thus the region adjacent to the fusion boundary is relatively free of carbides (assuming that cooling rates are rapid enough to suppress carbide precipitation during cooling).

In the HAZ of most austenitic stainless steels, Cr-rich $M_{23}C_6$ carbides form preferentially along grain boundaries, as shown in Figure 6.48. This results in a chromium-depleted zone along the grain boundary that is "sensitive" to corrosive attack. Hence, the term *sensitization* is often used to describe the metallurgical condition leading to intergranular attack. The exception to this is the stabilized grades of stainless steel containing Nb and/or Ti (such as Types 347 and 321). In these steels the Nb and Ti ties up carbon in the form of stable MC-type carbides and minimizes the formation of $M_{23}C_6$ carbides at the grain boundaries.

Intergranular corrosion results from the localized precipitation of Cr-rich carbides, or carbonitrides, at the grain boundary. This precipitation requires short-range

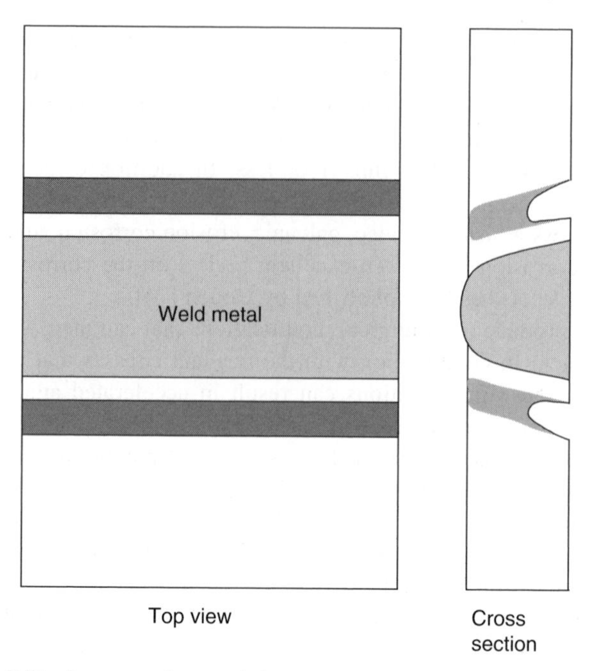

Figure 6.47 Intergranular attack in the HAZ of an austenitic stainless steel.

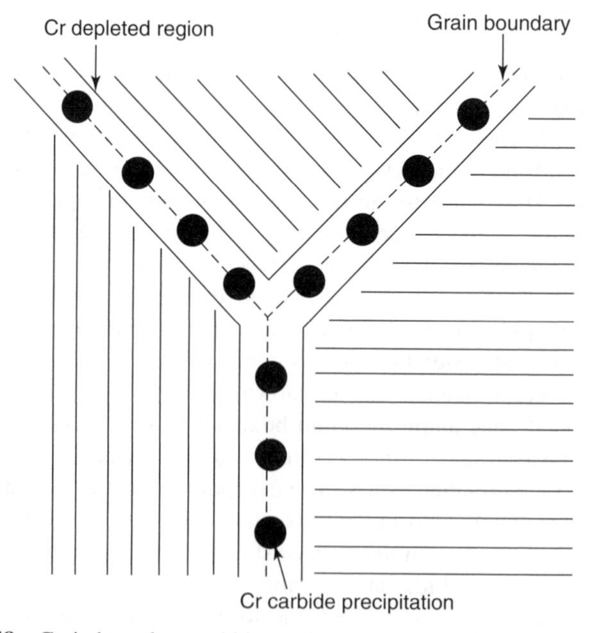

Figure 6.48 Grain boundary carbide precipitation and local chromium depletion.

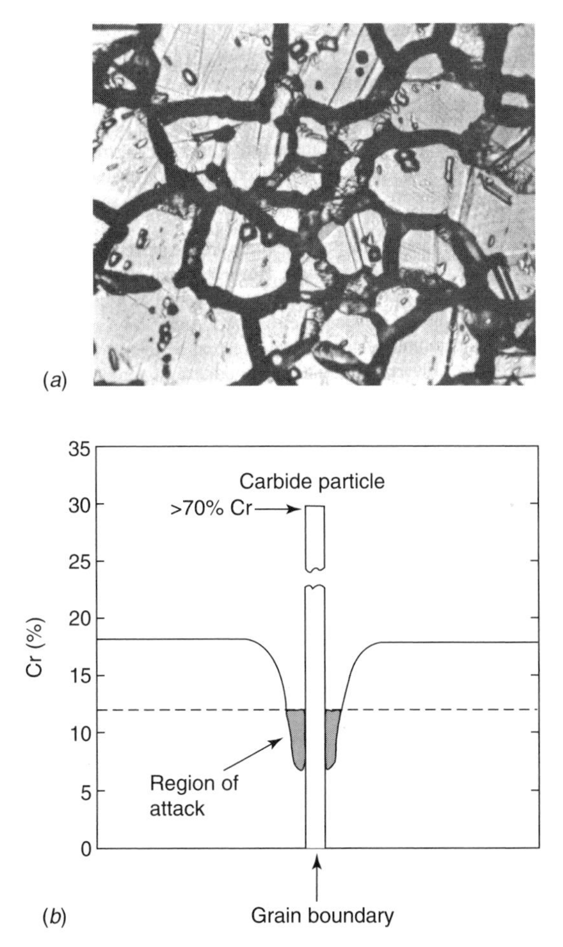

Figure 6.49 Intergranular corrosion: (*a*) grain boundary attack in the HAZ of Type 304 (C = 0.06 wt%); (*b*) Cr depletion adjacent to the grain boundary carbide.

diffusion of Cr from the adjacent matrix and produces a Cr-depleted region surrounding the precipitate, as shown in Figure 6.49. This reduces the local corrosion resistance of the microstructure and promotes rapid attack of the grain boundary region. In certain corrosive environments the effect is a local "ditching" at the grain boundary, as shown in the metallographic section in Figure 6.49. In extreme cases, the grains will actually drop out of the structure because of complete grain boundary attack and dissolution.

Carbon content has the most profound influence on susceptibility to IGC in austenitic stainless steels. The use of low-carbon (L grade) alloys minimizes the risk of sensitization by slowing down the carbide precipitation reaction. The time–temperature–precipitation curves shown in Figure 6.50 demonstrate the effect of carbon content on the time to precipitation. Note that at low carbon contents ($C < 0.04$ wt%), the nose of the curve is beyond 1 hour, while for carbon levels

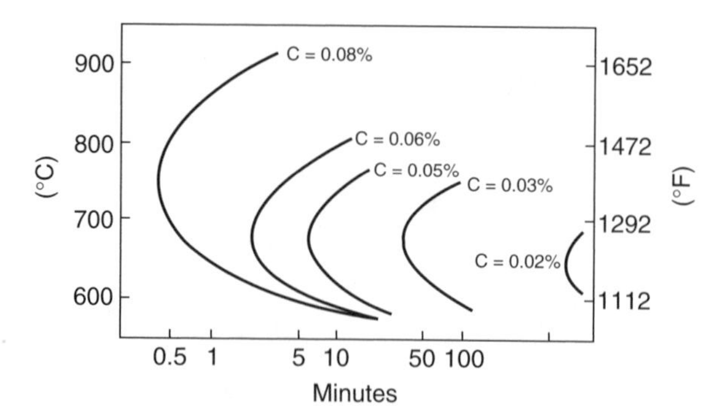

Figure 6.50 $M_{23}C_6$ time–temperature–precipitation curves for 18Cr–8Ni alloys with variable carbon content. (From Peckner and Bernstein [3]. Courtesy of McGraw-Hill.)

from 0.06 to 0.08 wt% the time for precipitation may be less than a minute. This difference demonstrates the benefit of the low-carbon alloys (L grades) for reducing or eliminating HAZ grain boundary sensitization during welding. The presence of residual stresses in the HAZ may also serve to accelerate the precipitation reaction.

In most cases, sensitization occurs in the HAZ as a direct result of the weld thermal cycle. It should be noted, however, that the stress relief temperature range for most austenitic stainless steels overlaps the carbide precipitation range. Care must be taken not to sensitize the entire structure during PWHT. This is a particular concern with alloys containing more than 0.04 wt% C.

In general, weld metals such as 308 and 316 are less likely to be sensitized than corresponding 304 and 316 base metals. The ferrite that is normally found in the weld metal is richer in Cr than the austenite, and Cr diffuses much more rapidly in ferrite than in austenite, which helps overcome any Cr depletion. $M_{23}C_6$ carbides tend to precipitate at the tortuous ferrite–austenite boundaries instead of at usually much straighter austenite–austenite boundaries. All of these factors greatly limit the tendency for sensitization in austenitic stainless steel weld metals containing ferrite [74]. So, except in fully austenitic stainless steel weld metals, sensitization is largely a HAZ problem, not a weld metal problem.

6.6.1.1 Preventing Sensitization It is possible to minimize or eliminate intergranular corrosion in austenitic stainless steel welds by the following methods.

- Select base and filler metals with as low a carbon content as possible (L grades such as 304L and 316L).
- Use base metals that are "stabilized" by additions of niobium (Nb) and titanium (Ti). These elements are more potent carbide formers than chromium and thus tie up the carbon, minimizing the formation of Cr-rich grain boundary carbides.
- Use annealed base material or anneal prior to welding to remove any prior cold work (cold work accelerates carbide precipitation).

- Use low weld heat inputs and low interpass temperatures to increase weld cooling rates, thereby minimizing the time in the sensitization temperature range.
- In pipe welding, water cool the inside of the pipe after the root pass. This will help to eliminate sensitization of the ID resulting from subsequent passes.
- Solution heat treat after welding. Heating the structure into the temperature range 900 to 1100°C (1650 to 2010°F) dissolves any carbides that may have formed along grain boundaries in the HAZ. The structure is then quenched from this temperature to prevent carbide precipitation during cooling. Note, however, that there are a number of practical considerations that tend to limit the usefulness of the latter approach. Distortion during quenching is a serious problem for plate structures. Inability to quench complex pipe weldments is also a limiting factor.

6.6.1.2 Knifeline Attack Intergranular attack can also occur in certain situations in the stabilized grades, such as Types 347 and 321. This attack, shown schematically in Figure 6.51, normally occurs in a very narrow region just adjacent to the fusion boundary. It is sometimes called *knifeline attack* because the weld appears as if it were cut out with a knife. This type of attack occurs when the stabilized carbides (NbC or TiC) dissolve at elevated temperatures in the region just adjacent to the fusion zone. Upon cooling, Cr-rich carbides will form faster than the NbC or TiC, resulting in a narrow sensitized region. Farther from the fusion boundary, NbC and TiC do not dissolve and sensitization does not occur.

6.6.1.3 Low-Temperature Sensitization In the 1970s and 1980s, it was observed that sensitization can actually occur after long exposures at low temperatures [$<$300°C (570°F)] following an initial high-temperature thermal cycle such as that experienced in the HAZ. This has come to be known as *low-temperature sensitization* (LTS) and has been a problem with stainless steel piping used in the power generation industry [75,76]. LTS occurs because carbide "embryos" form during the original welding process and then grow to form carbide precipitates at low temperature. Figure 6.52 shows this phenomenon schematically. The long-term effect is the sensitization of the grain boundaries and the potential for IGA or IGSCC, even though the C-curve for sensitization under normal conditions is not crossed during

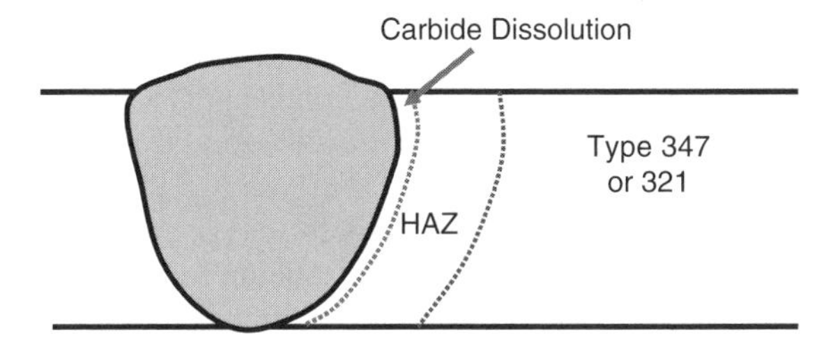

Figure 6.51 Location of knifeline attack in stabilized grades of austenitic stainless steels. Knifeline attack occurs in the MC carbide dissolution region, due to the reformation of $M_{23}C_6$ during cooling.

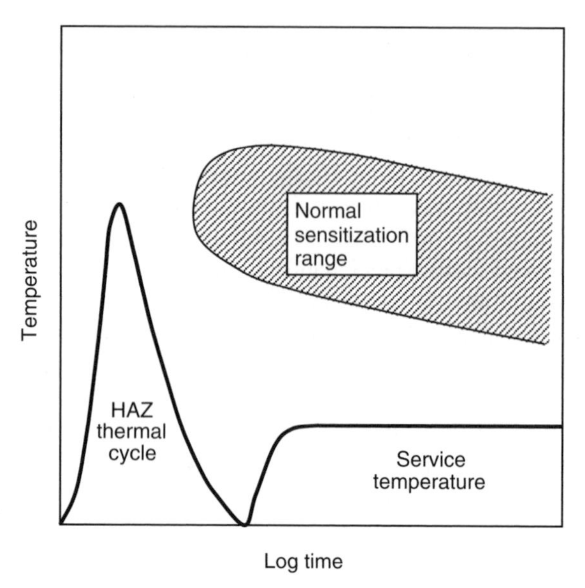

Figure 6.52 Thermal history leading to low-temperature sensitization.

the thermal history of the weldment. This phenomenon has occurred in L-grade alloys but does not seem to be a problem with stabilized grades such as Type 347.

6.6.2 Stress Corrosion Cracking

Many of the austenitic stainless steels are inherently susceptible to SCC, particularly in Cl-bearing environments (such as seawater). The Copson curve [77], shown in Figure 6.53, plots resistance to SCC in boiling magnesium–chloride as a function of nickel content. The use of this aggressive environment is intended to accelerate the corrosion processes that would occur in other Cl-bearing environments (such as seawater). Note that the trough of the resistance curve occurs in the range 8 to 12% Ni. This is precisely the nickel range of many popular austenitic alloys, such as 304 and 316.

SCC can be avoided by selecting alloys with either higher (>20%) or lower (<5%) nickel content. In the former case, the use of superaustenitics or Ni-base alloys is common. In the low-Ni case, ferritic or duplex stainless steels are often selected. SCC has also been observed in caustic (high-pH) environments, such as in pulp and paper mills. It appears that the same rules apply in these environments as with Cl-bearing environments with respect to alloy selection to avoid caustic-induced SCC.

An example of severe transgranular SCC in a Type 316 tubesheet after exposure to a caustic solution of sodium hydroxide in a pulp and paper mill is shown in Figure 6.54. This structure was exposed to the caustic solution for less than a year prior to failure. The residual stresses resulting from the weld in addition to imposed operating stresses led to the severe cracking seen in Figure 6.54. The Type 316 alloy was replaced with a duplex grade, Alloy 2205. This alloy has not exhibited any cracking after several years of service.

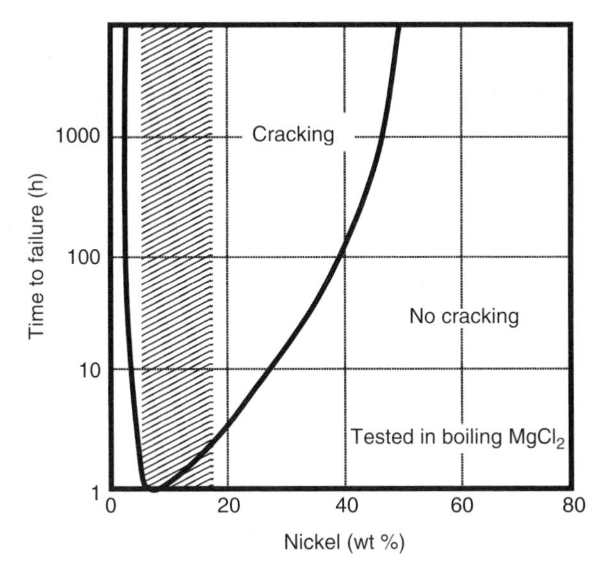

Figure 6.53 Copson curve for SCC in stainless steels. Shaded region represents nickel range of many austenitic stainless steel. (Redrawn from Copson [77].)

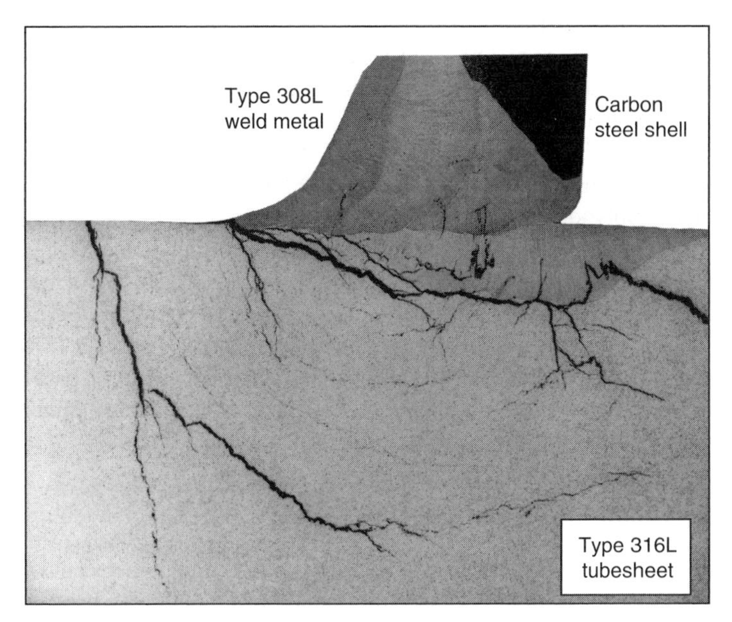

Figure 6.54 Transgranular SCC near the weld in Type 316L after exposure to a caustic sodium hydroxide solution.

Stress corrosion cracking is best avoided by proper alloy selection. The use of duplex and ferritic stainless steels in applications where austenitic grades would otherwise be selected can avoid SCC. Welding may exacerbate SCC in alloy systems that are otherwise resistant due to changes in microstructure and residual stress. Sensitization can promote IGSCC in both austenitic and ferritic grades of stainless steel. Weld designs or conditions that generate high residual stress or create stress concentrations can also promote SCC. Postweld stress relief can sometimes be used to reduce these stresses and minimize susceptibility to SCC. But as noted above, postweld stress relief needs to be done with care to avoid sensitization.

6.6.3 Pitting and Crevice Corrosion

Pitting and crevice corrosion are related phenomena. Halide element ions, most commonly Cl^-, are very often present in aqueous solutions. A penetration of the passive film on the stainless steel surface allows the metal beneath to become active. The penetration site may be an inclusion exposed to the corrosive media [78]. A crevice, as might exist at the edge of a gasket or sealing ring at a flanged connection, can also serve as a pit initiation site. Pitting and crevice corrosion resistance can be measured by the ASTM G48 ferric chloride test. A critical pitting temperature (CPT) can be determined by Method C of this standard. A critical crevice temperature (CCT) can be determined by Method D of this standard. The test solution consists of 6% ferric chloride and 1% HCl in water. For the CPT test, the sample is simply immersed, held for a prescribed time period, and then examined for pits. For the CCT test, TFE-fluorocarbon washers are pressed against the sample to provide crevice sites. Various temperatures are evaluated in each test method, with a view to finding the minimum temperature at which pitting will occur (CPT) or the minimum temperature at which crevice corrosion will occur. The higher the CPT or CCT, the more resistant is the alloy under test. Testing temperatures from 0 to 85°C (32–185°F) can be used.

Chromium, molybdenum, tungsten and nitrogen are alloy elements that improve pitting and crevice corrosion resistance. Garner [79] determined the CPT for a number of stainless steel base metals and their autogenous welds. He found that the CPT increases with increasing Mo and that weld metal was invariably inferior to base metal of the same composition. Figure 6.55 presents Garner's findings. The reasons for the inferiority of the weld metal as compared to the corresponding base metal are related to segregation during solidification, which leaves local areas of molybdenum depletion, normally at the center of the cells or dendrites.

6.6.4 Microbiologically Induced Corrosion

Microbiologically induced corrosion (MIC) occurs in certain aqueous environments where aerobic bacteria literally attack the metal. MIC manifests itself as pitting, whereby the metal surface exhibits a small hole, or pit, and rapid attack occurs subsurface. In the past, MIC was probably often misinterpreted as normal pitting corrosion. However, it can be distinguished by the presence of "tubercules" of biological residue and corrosion products over the pit. This form of corrosion has been

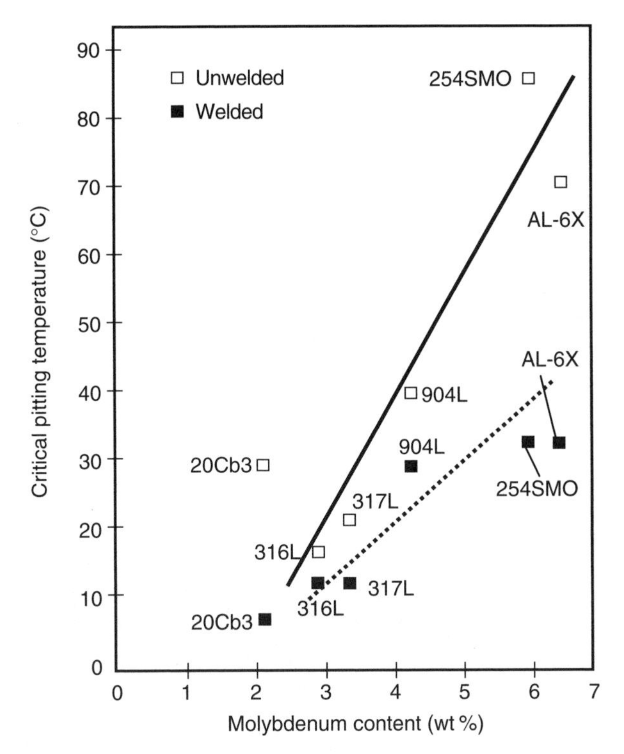

Figure 6.55 CPT of commercial high-Mo austenitic stainless steels and their weld fusion zones. Open symbols, base metal values; solid symbols, autogenous welds; solid line, average base metal behavior; dotted line, weld metal behavior. (Redrawn from Garner [79]. Courtesy of the American Welding Society.)

observed in both fresh water and seawater and requires oxygenated water to support the metal dissolution reaction, as shown in Figure 6.56. The presence of specific "metal-munching" bacteria, a warm-water oxygenated environment, and specific materials are required to support MIC. Austenitic stainless steels seem to be particularly susceptible and the presence of a two-phase austenite + ferrite microstructure, such as that present in many weld metals, seems to increase susceptibility.

An example of MIC that occurred during the construction of a storage tank is shown in Figure 6.57. This is a shielded metal-arc weld using E308-16 and joining two Type 304 plates in a fillet weld configuration. During construction, water was allowed to accumulate in the bottom of the tank, covering the weld shown in Figure 6.57. This created the environment for MIC, as evidenced by the severe attack of the weld metal.

6.6.5 Selective Ferrite Attack

In general, nominally austenitic weld metals that contain a small amount of ferrite have corrosion resistance equivalent to their corresponding ferrite-free base metals

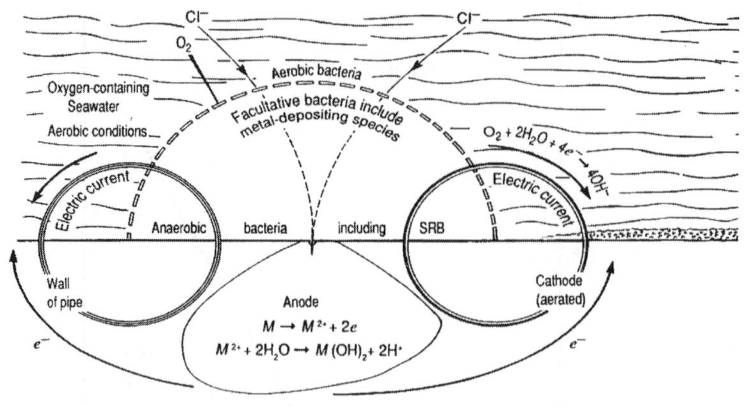

Figure 6.56 Microbiologically induced corrosion (MIC). (From Little et al. [80].)

Figure 6.57 Microbiologically induced corrosion (MIC) in Type 308 weld metal. (Sample courtesy of Christopher Hayes.)

in most corrosive environments. However, there are certain environments in the manufacture of urea, or organic acids such as pteraphthalic acid, which selectively attack ferrite in nominally austenitic stainless steel welds and overlays [74]. If ferrite is present in any appreciable amount (generally greater than 5FN), corrosion will proceed along ferrite networks and can effectively destroy the weld metal. Type 316L or similar alloys are normally used in these applications, and special welding filler metals similar to 316L, but virtually ferrite-free, are available. The AWS filler metal specifications do not provide for filler metals of this type, but the ISO 3581 and EN 1600 standards provide for 18% Cr–15% Ni–3% Mo filler metals without nitrogen alloying (18-15-3 L) and with nitrogen alloying (18-16-5N L) that have proven successful in these environments. These are generally high in manganese (>2%) to improve resistance to hot cracking, but it is prudent to produce small convex beads, with filled craters, for maximum solidification cracking resistance.

6.7 SPECIALTY ALLOYS

A number of specialty austenitic stainless steels have been developed to meet exceptional service environments. These alloys can generally be broken into two groups, heat-resistant alloys that are used for elevated temperature service and nitrogen-alloyed steels that are used in demanding corrosion environments. The latter group includes alloys that have been referred to as superaustenitics.

6.7.1 Heat-Resistant Alloys

These alloys are used in elevated-temperature applications in the power generation and petrochemical industries. Most heat-resisting alloys have high carbon contents, since carbon provides superior elevated-temperature strength. Some alloys are based on standard 300 series alloys, with additional carbon added, such as 304H and 316H. The "H" designation indicates carbon in the high range of the general specification, 0.04 to 0.10 wt%. The cast alloys in this group, including HK40 and HP45Nb, contain carbon contents of 0.40 wt% or higher. A list of wrought and cast heat-resistant alloys is provided in Table 6.12.

Note that most of these alloys have high chromium contents for improved elevated-temperature corrosion resistance and high nickel content to stabilize the austenite phase to prevent embrittlement. Many alloys also contain carbide formers, including Nb, Ti, Mo, and W to improve creep resistance. Criteria used by the materials engineer to select from this group of materials includes creep resistance, thermal fatigue resistance, and resistance to different forms of high-temperature corrosion such as oxidation, sulfidation, and carburization.

Although mechanical properties, corrosion resistance, availability, fabricability, and cost are all important when selecting materials for these demanding applications, it is the stability of the microstructure and the properties of the material following extended service that dictate material selection and ultimate performance. The ability to repair the material after extended service is also important, but is often not factored into the material selection process. The microstructure and properties of the materials used at elevated temperature may degrade from a variety of causes. As a consequence of this degradation, the materials become more susceptible to cracking in service, during plant shutdowns, or when repair welding is attempted. The cracking and embrittlement phenomena of corrosion- and heat-resistant alloys include reheat cracking, sigma-phase formation, and aging embrittlement. A comprehensive review of the characteristics and weldability of these materials after service exposure was conducted recently by Zhang et al. [81]. Ebert [82] provided a practical approach to field repair welding of service-embrittled HK40 which

TABLE 6.12 Austenitic Stainless Steels for Elevated-Temperature Service

	Nominal Composition (%)							
Alloy	Cr	Ni	Co	C	Si	Mn	Al	Other
				Established Heat-Resistant Wrought Alloys				
304H	19	9	—	0.07	0.75	1	—	—
316H	17	12	—	0.07	0.75	2	—	Mo: 2.5
321H	18	10.5	—	0.07	0.75	2	—	Ti: 10 × (C+N)
347H	18	10.5	—	0.07	0.75	2	—	Nb: 12 × C
309H	23	13.5	—	0.07	0.75	2	—	—
310H	25	20.5	—	0.07	0.75	2	—	—
85H	19	15	—	0.20	3.5	0.8	1	—
253MA	21	11	0.04	0.08	1.7	0.6	—	N: 0.17
330	19	35	—	0.05	1.2	1.5	—	—
800H	20	31	—	0.08	0.3	0.8	0.3	Ti: 0.3
				New-Generation Wrought Alloys				
803	27	34	—	0.08	0.3	0.4	0.4	—
HK4M	25	25	0.75	0.25	0.4	0.4	—	B: 0.004
HPM	25	38	1.7	0.15	0.4	—	—	Mo: 2; Zr: 0.05; B: 0.01
HR120	25	37	1	0.05	0.6	0.7	0.1	W: 2; Mo: 2; Nb: 0.7; B: 0.004 N: 0.2
				Cast Alloys				
HK40	25	20	—	0.4	—	—	—	—
HP-45Nb	25	35	—	0.45	—	—	—	Nb: 1.5
HP-45NbMA	25	35	—	0.45	—	—	—	Nb: 1.5; Ti, Zr
HP-45NbW	25	35	—	0.45	—	—	—	Nb: 1.5; W: 1.5
HP-45W	25	35	—	0.45	—	—	—	Nb: 1.5; W: 1.5
HP-45Mo	25	35	—	0.45	—	—	—	Mo: 1.5
HP-15Nb	25	35	—	0.15	—	—	—	Nb: 1.5

involves local annealing to dissolve carbides to provide for enough ductility in the weld area so that cracking does not occur beside the repair welds.

Reheat cracking is usually associated with the HAZ of restrained welded joints during either postweld heat treatment or elevated temperature service (where it is often called *relaxation cracking*). Reheat cracking occurs in a range of materials, including austenitic and ferritic steels and nickel-base alloys. Historically, reheat cracking was first observed in austenitic stainless steels by the power generation industry. Metallurgical investigations have shown that the cracking is associated with the formation of fine intragranular precipitates. Essentially, this precipitation strengthens the grain interiors and transfers the strain necessary for the relief of residual welding and system stresses to the grain boundaries. The overall effect is a reduction in creep ductility leading to intergranular failure. This mode of cracking poses a considerable practical problem, particularly because the microstructural and compositional factors affecting its occurrence have not been defined unambiguously. Examples of this can be found in a paper by Dhooge [83] which reviewed cracking in a variety of high-temperature applications.

Relaxation cracking is associated with long-term exposure of heavy-section or highly restrained weldments in the temperature range 500 to 700°C (930 to 1290°F). This form of cracking occurs by the same mechanism as reheat cracking, except that time to failure is typically in the range 10,000 to 100,000 hours. This has been a particular problem in large components constructed of high-carbon alloys (304H, 316H, 321H, 800H) that cannot be given a postweld stress-relief heat treatment. As a result, stress relaxation and carbide precipitation occurs simultaneously, leading to intergranular cracking, normally in the HAZ of circumferential or longitudinal welds [84].

AWS classifies only a limited number of filler metals to match these high-temperature base metals. These are listed in Table 6.13. Except for 308H and 316H compositions, primary ferrite solidification is impossible in these filler metals. One of the filler metals (NiCrCoMo-1) is considered to be a nickel-base alloy. Again, the procedure of producing small convex beads and filling craters is essential for producing welds without solidification cracking when primary ferrite solidification cannot be obtained.

TABLE 6.13 Austenitic Filler Metals for Elevated-Temperature Service

	Nominal Composition (%)							
Alloy	Cr	Ni	Co	C	Si	Mn	Al	Other
				Covered Electrodes				
308H	19	10	—	0.06	0.4	1	—	—
316H	18	12	—	0.06	0.4	2	—	Mo: 2.5
310	26	21	—	0.14	0.4	2	—	—
310H	26	21	—	0.40	0.4	2	—	—
330	15.5	35	—	0.22	0.4	2	—	—
330H	15.5	35	—	0.40	0.4	2	—	—
NiCrCoMo-1	23	51	12	0.10	0.3	1.5	—	Mo: 9.0
				Bare Wires				
19-10H	19	10	—	0.06	0.4	1.5	—	—
308H	20	10	—	0.06	0.4	1.5	—	—
316H	19	12	—	0.06	0.4	1.5	—	2,2
310	26	21	—	0.12	0.4	1.5	—	—
330	16	35	—	0.22	0.4	2	—	—
NiCrCoMo-1	22	55	12	0.10	0.4	0.5	1.2	Mo: 9.0

6.7.2 High-Nitrogen Alloys

Nitrogen is added to austenitic stainless steels to increase strength, improve pitting corrosion resistance, or both. These alloys include standard 300 series alloys, such as Types 304L and 316L, that contain nitrogen additions in the range 0.10 to 0.16 wt% and are designated 304LN and 316LN. Another group includes 200 series alloys that are often designated by trade names such as Nitronic™ and Gall-Tough™. These alloys contain high levels of manganese, up to 15 wt%, since Mn increases the solubility of nitrogen in the austenite matrix. This prevents chromium nitride precipitation that can compromise mechanical and corrosion properties. These alloys can contain as much as 0.40 wt% nitrogen. The third group of alloys, the superaustenitics, are formulated to provide superior SCC and pitting resistance compared to standard austenitic grades. Common alloys include AL-6XN and 254SMo. These alloys contain nominally 20 wt% chromium, 18 to 25 wt% nickel, 6 to 7 wt% Mo, and 0.15 to 0.25 wt% nitrogen. The compositions of the more common nitrogen-alloyed stainless steels are provided in Table 6.14.

These alloys pose some special challenges with respect to welding. Because of the high levels of N and Mn, the earlier constitution diagrams (such as Schaeffler and DeLong) make serious errors in predicting microstructure and/or cracking susceptibility. Typically, these diagrams will predict fully austenitic weld metal when, in fact, the weld metal contains enough ferrite for FA solidification and freedom from solidification cracking. A special diagram, developed by Espy as a modification to the Schaeffler diagram, has been found to be effective, although it makes predictions in ferrite percent [85]. This diagram is shown in Figure 6.58. More recently, Kotecki, using the data of Espy, demonstrated that the WRC-1992 diagram provides good FN predictions for these alloys, except in the case of the very high silicon Nitronic 60 and Gall-Tough alloys [86]. For the latter two alloys, the Espy diagram is still the best predictive tool available.

In general, the same rules and precautions apply to the high-nitrogen alloys as to those for standard 300 series alloys. Solidification and ferrite control are essential to avoid weld solidification cracking. A study by Robino et al. [87] on two high-nitrogen steels (Gall-Tough and Nitronic 60) showed the importance of maintaining FA solidification in order to minimize susceptibility to solidification cracking. Due to their high Si content, these alloys are especially prone to solidification cracking if the solidification mode is primary austenite (AF or A). Some problems may occur with these alloys due to porosity formation during electron-beam and laser welding. The rapid solidification rates in these processes result in a supersaturation of nitrogen in the molten pool, which, upon solidification, results in nitrogen pore nucleation [88].

Matching filler metals are available for only a few of the high-Mn, high-N steels. Table 6.15 lists high-nitrogen covered electrode and bare wire classifications found in AWS specifications. For other compositions, the manufacturer of the base metal needs to be contacted for guidance. In some cases it is common to shear strips of the base metal into pieces suitable for filler metal for GTA welding. However, the base metal may not have sufficient ferrite potential to provide for primary ferrite

TABLE 6.14 Nitrogen-Alloyed Austenitic Stainless Steels

Alloy	Nominal Composition (%)								PRE_N^a
	C	Mn	Si	Cr	Ni	Mo	N	Other	
				Standard, High N					
304LN	0.02	1.0	0.4	19.0	10.0	—	0.13	—	21
316LN	0.02	1.0	0.4	17.0	12.0	2.2	0.13	—	26
317LN	0.02	1.0	0.4	19.0	13.0	3.3	0.15	—	32
317LMN	0.02	1.0	0.4	18.5	15.5	4.5	0.16	—	36
				High Mn, High N					
Nitronic 30	0.02	8.0	0.5	16.0	2.25	—	0.23	—	20
Nitronic 32	0.08	18.0	0.5	18.0	—	1.0	0.50	Cu: 1.0	25
Nitronic 33	0.04	13.0	0.4	18.0	3.0	—	0.30	—	23
Nitronic 40	0.04	9.0	0.5	20.0	6.5	—	0.28	—	24
Nitronic 50	0.04	5.0	0.4	22.0	12.5	2.25	0.30	Nb: 0.20	34
Nitronic 60	0.05	8.0	4.0	17.0	8.5	—	0.13	—	19
Gall-Tough	0.15	5.0	3.5	16.5	5.0	3.5	0.15	—	30
				Super-Austenitic					
254SMo	0.01	0.5	0.4	20.0	18.0	6.25	0.20	Cu: 0.75	44
AL-6XN	0.02	1.0	0.5	21.0	24.5	6.5	0.22	—	46

[a] $PRE_N = Cr + 3.3(Mo + 0.5W) + 16N$.

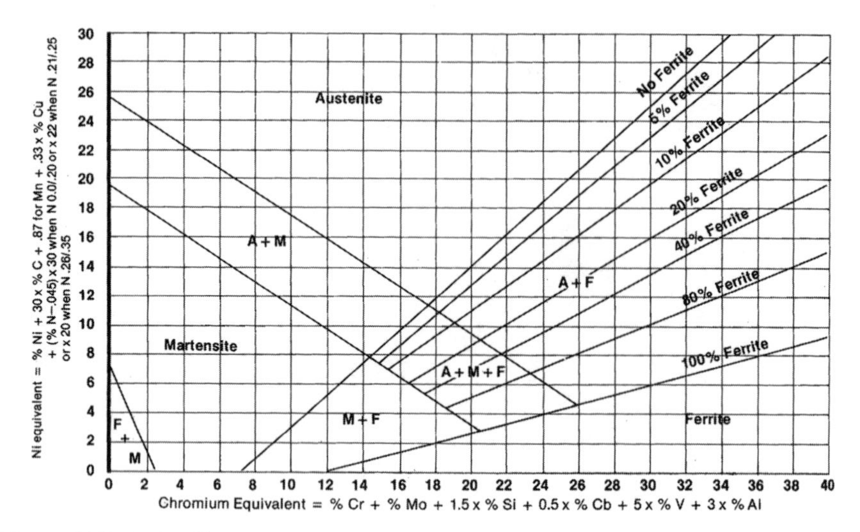

Figure 6.58 Schaeffler diagram for high-N, high-Mn stainless steels as modified by Espy. (From Espy [85]. Courtesy of the American Welding Society.)

solidification, while commercial filler metal wires or electrodes are generally designed for primary ferrite solidification.

Super-austenitic stainless steels exhibit superior stress corrosion cracking and pitting corrosion resistance relative to the standard 18-8 austenitic grades and are often using in demanding applications where corrosion resistance is paramount. This is achieved by increased levels of nickel, molybdenum, and nitrogen. Most of the super-austenitic grades lie within the following composition limits (wt%): 20 to 25% Cr, 15 to 25% Ni, 4 to 8% Mo, 0.01 to 0.03% C, and 0.2 to 0.6% N. The pitting resistance equivalent (PRE_N), a common measure of pitting corrosion, is typically greater than 45, where

with W = usually "PREW"

$$PRE_N = Cr + 3(Mo + 0.5W) + 16N \tag{6.1}$$

As seen in Table 6.14, these alloys typically have quite high nickel content, which in combination with the high nitrogen levels, results in austenite solidification and a weld metal microstructure consisting of austenite and a Mo-rich eutectic phase. Thus solidification cracking is a potential problem, although the impurity levels in these alloys are usually extremely low.

Using ThermoCalc [2], Perricone and DuPont [89] developed a solidification diagram for Fe–Ni–Cr–Mo alloys that shows the solidification behavior of superaustenitic stainless steels (Figure 6.59). According to this diagram, Mo segregates strongly during austenite solidification, while there is little tendency for Cr segregation. When the solidification path reaches the line of twofold saturation between the austenite (γ) and sigma (σ), a eutectic reaction occurs and sigma will form along cell and dendrite boundaries. (Note that this is not the same as the sigma phase that forms in the solid state, resulting in embrittlement.)

See also p 261

TABLE 6.15 AWS Classifications for High-Manganese, High-Nitrogen Austenitic Stainless Steel Filler Metals

AWS Type	Matching Alloy	Nominal Composition (%)								PRE_N^a
		C	Mn	Si	Cr	Ni	Mo	N	Other	
					Covered Electrode					
E240-XX	Nitronic 33	0.04	12.0	0.4	18.0	5.0	—	0.20	—	21
E219-XX	Nitronic 40	0.04	9.0	0.5	20.0	6.0	—	0.20	—	23
E209-XX	Nitronic 50	0.04	5.5	0.4	22.0	11.5	2.25	0.20	V: 0.20	33
					Bare Wire					
ER240	Nitronic 33	0.03	12.0	0.4	18.0	5.0	—	0.20	—	21
ER219	Nitronic 40	0.03	9.0	0.4	20.0	6.0	—	0.20	—	23
ER209	Nitronic 50	0.03	5.5	0.4	22.0	10.5	2.25	0.20	V: 0.20	33
ER218	Nitronic 60	0.06	8.0	4.0	17.0	8.5	—	0.13	—	19

[a] $PRE_N = Cr + 3.3(Mo + 0.5W) + 16N$.

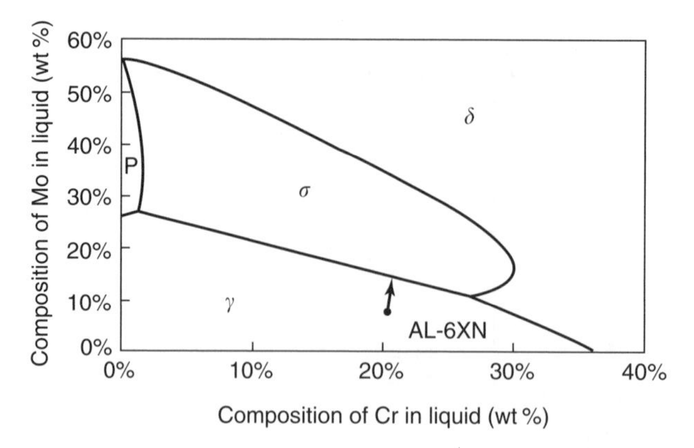

Figure 6.59 Liquidus projection for the Fe–Ni–Cr–Mo system showing the solidification path for AL6XN. (From Perricone and Dupont [89]. Courtesy of ASM International.)

Figure 6.60 Pitting of a super-austenitic stainless steel weld metal associated with local Mo depletion during solidification. (From Woolin [90]. Courtesy of the Welding Institute.)

A reduced resistance to corrosion has been observed in welded super-austenitic alloys due to the segregation of Mo during weld solidification. Since the partition coefficient (k) for Mo is less than 1, Mo segregates to the solidification subgrain and grain boundaries and leaves the center of the subgrains (dendrite or cell cores) depleted in Mo. Because of this, local pitting corrosion attack can occur more easily in the microstructure, because pits nucleate at the low-Mo dendrite or cell cores. This behavior is shown in Figure 6.60, where severe pitting has occurred in the weld metal. Pit initiation occurs at the dendrite core, where the Mo content is below 4 wt%, and then pit growth progresses, resulting in the widespread attack shown in Figure 6.60.

One method employed to avoid this attack is the use of high-Mo Ni-base filler metals. At least 9% Mo is needed in the filler metal to match the pitting resistance of the 6% Mo base metals. Table 6.16 lists AWS classifications for such filler metals.

TABLE 6.16 AWS Classifications for Nickel-Base Alloy Filler Metals That Are Applied to 6% Mo Super-Austenitic Stainless Steel Base Metals

Alloy	Nominal Composition (%)								PRE_N^a
	C	Mn	Si	Cr	Mo	Fe	N	Other	
Covered Electrode									
ENiCrMo-3	0.05	0.5	0.4	21.5	9.0	3.5	—	Nb: 3.6	51
ENiCrMo-4	0.01	0.5	0.1	15.5	16.0	5.5	—	W: 3.75	74
ENiCrMo-10	0.01	0.5	0.1	21.25	13.5	4.0	—	W: 3.0	71
Bare Wire									
ERNiCrMo-3	0.04	0.5	0.3	21.5	9.0	1.0	—	Nb: 3.6	51
ERNiCrMo-4	0.01	0.5	0.05	15.5	16.0	5.5	—	W: 3.75	74
ERNiCrMo-10	0.01	0.3	0.05	21.0	13.5	4.0	—	W: 3.0	70

[a] $PRE_N = Cr + 3.3(Mo + 0.5W) + 16N$.

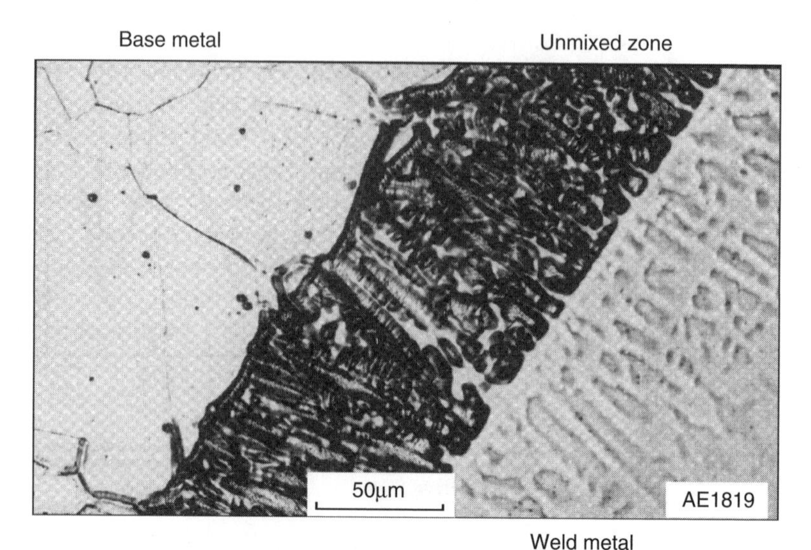

Figure 6.61 Unmixed zone that forms at the fusion boundary between a super-austenitic alloy and a Ni-base filler metal. (From Gooch [74]. Courtesy of the American Welding Society.)

This approach has been somewhat problematic since an unmixed zone (UMZ) can form at the edge of the fusion zone. This UMZ, as shown in Figure 6.61, represents a narrow region where the super-austenitic alloy melts and resolidifies without mixing with the Ni-base filler metal. As a result, the microstructure that forms is essentially an autogenous weld that is subject to the same Mo-depletion susceptibility noted previously. As a consequence of this UMZ, local pitting attack may occur at the fusion boundary.

In general, the superaustenitic alloys have good weldability and perform well in moderately aggressive environments. It is expected that their application will continue to expand, particularly if the corrosion problems described here can be overcome.

6.8 CASE STUDY: SELECTING THE RIGHT FILLER METAL

This case study is a good example of how the WRC-1992 diagram can be used to predict solidification behavior in austenitic stainless steels and avoid weld solidification cracking. A welding engineer is attempting to weld Type 320 stainless steel to Type 316L. Initially, the welding engineer selects a Type 316L filler metal but encounters severe centerline solidification cracking. What filler metal should be used with this combination to prevent cracking?

Type 320 stainless is a very high nickel stainless steel. The weld produced with 316L filler metal contained no ferrite in the root pass, due to nickel pickup from the 320 base metal. The absence of ferrite makes the weld metal sensitive to solidification cracking. Table 6.17 lists typical compositions of 320 stainless steel base metal, 316L

TABLE 6.17 Typical Compositions of 320 and 316L Base Metals, E316L Covered Electrode, and Root-Pass Weld Metal

	Nominal Composition (%)									
Metal	C	Cr	Ni	Mo	Nb	Cu	N	Cr_{eq}	Ni_{eq}	FN
320	0.02	20.0	34.0	2.5	0.30	3.5	0.02	22.7	36.0	0
316L	0.02	17.0	12.0	2.3	—	0.2	0.02	19.3	13.1	2.9
E316L	0.03	18.5	12.0	2.3	—	0.2	0.06	20.8	14.3	4.3
Root pass	0.027	18.5	15.3	2.33	0.045	0.7	0.048	20.9	17.4	0

base metal, and 316L covered electrode filler metal. With covered electrodes, dilution of about 30% is normally expected, with half of that dilution coming from each side of the joint. So the root pass composition can be expected to consist of 15% 320 base metal, 15% 316L base metal, and 70% 316L filler metal. This calculated root-pass composition is included in Table 6.17. The chromium equivalent (Cr_{eq}) and nickel equivalent (Ni_{eq}), and the calculated Ferrite Number for each composition, according to the WRC-1992 diagram, are also included in Table 6.17.

Use of 320 filler metal, which cannot contain any ferrite due to its high nickel content, will be no better than using of 316L filler metal. The selection of Type 320 (or Type 320LR) filler metal is not a good solution. Several other filler metals could be considered. The most likely would be austenitic grades 309L, 309LMo (formerly 309MoL), and 312, or the duplex alloy 2209. Of these, 312 is a good choice for many dissimilar metal combinations, but because both sides of this joint are very low carbon while 312 typically contains on the order of 0.08 wt% C (which is not low), that selection should probably be avoided out of concern for inferior corrosion resistance in the joint. The other three candidate filler metals are interesting to examine using the same sort of analysis as in Table 6.17. This is shown in Table 6.18. Again, 30% dilution is assumed, half of the dilution coming from the 316L side and half coming from the 320 side of the joint.

It can be seen from Table 6.18 that the 309L root pass has insufficient ferrite (1.6 FN), so solidification cracking could be expected. The 309LMo root (3.5 FN), at first glance, looks acceptable, but this is not the case. To understand that, it is necessary to refer to the WRC-1992 diagram (Figure 6.62) to see where each of the root-pass weld compositions lie. The 320 composition is entirely off the diagram, well above the maximum 18 Ni_{eq} of the diagram. The 309L root pass and the 309MoL root pass are both slightly above the maximum 18 Ni_{eq} of the diagram also, but close enough to 18 to get a pretty good idea of the situation.

There is no single "magic" FN that separates compositions on the WRC-1992 diagram that are sensitive to solidification cracking from those that are not sensitive. However, the diagram includes a heavy dotted line that separates the compositions into those that solidify as primary austenite (AF region) and are therefore sensitive to solidification cracking, and those that solidify as primary ferrite (FA region) and are therefore not sensitive to cracking. That dotted line runs at a slight angle to the

TABLE 6.18 Typical Compositions of 309L, 309MoL, and 2209 Filler Metals and Typical Root-Pass Weld Compositions in Joining 320 to 316L Stainless with Those Electrodes

	Nominal Composition (%)									
Metal	C	Cr	Ni	Mo	Nb	Cu	N	Cr_{eq}	Ni_{eq}	FN
E309L	0.03	23.5	13.5	0.2	—	0.2	0.06	23.8	15.7	10.5
E309L root	0.027	22.07	16.28	0.86	0.045	0.7	0.048	23.0	18.4	1.6
E309MoL	0.03	23.0	13.5	2.2	—	0.2	0.06	25.2	15.7	16.8
E309MoL root	0.027	21.65	16.28	2.26	0.045	0.7	0.048	23.9	18.4	3.5
E2209	0.03	22.5	9.0	3.0	—	0.1	0.15	25.5	13.1	35.4
E2209 root	0.027	21.3	13.2	2.82	0.045	0.6	0.111	24.2	16.5	9.0

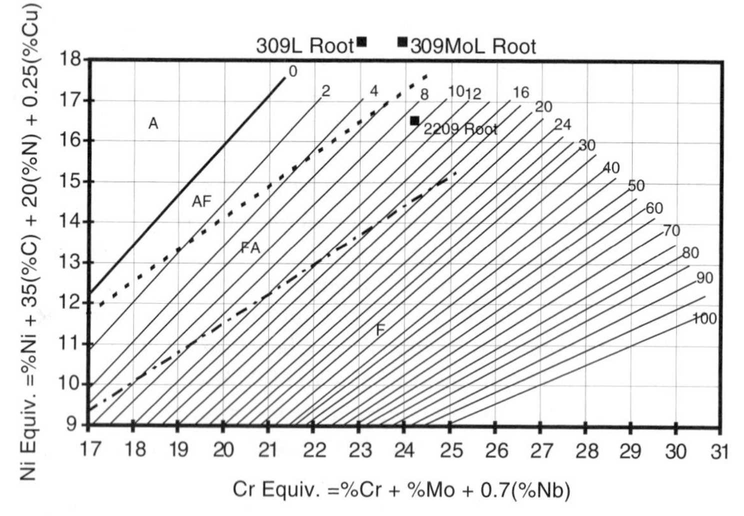

Figure 6.62 WRC-1992 diagram with root-pass compositions plotted for the filler metals indicated.

nearby isoferrite lines. At the lower left end of the heavy dotted line, it is near 1 FN. At the upper right end of the heavy dotted line, it is above 5 FN.

It is common to use, as an initial design assumption, the "rule of thumb" that a FN greater than 3 or 4 will probably be free from solidification cracking. However, for higher alloy compositions, more than 5 FN may be necessary. This feature of the WRC-1992 diagram agrees well with experience. For example, it is well known that the 16-8-2 composition of AWS A5.4, which, at less than 2 FN, plots in the FA region near the lower left end of the heavy dotted line of the WRC-1992 diagram, is not troubled by solidification cracking. But the 317LM composition, which, at a little over 4 FN, plots in the AF region near the upper right end of the heavy dotted line, is sensitive to cracking—more than 5 FN is necessary to make that filler metal insensitive.

So it can be concluded from this analysis that neither 309L nor 309MoL are safe choices for the dissimilar combination of 320 and 316L stainless steel base metals. But the 2209 duplex stainless steel filler metal is a different story. Its high ferrite content results in a root pass estimated at 9 FN, which seems safely on the FA side of the heavy dotted line in the WRC-1992 diagram. The 2209 root pass matches, or slightly overmatches the molybdenum content of the two base metals, which also makes it attractive. The corrosion resistance of 2209 filler metal is generally better than that of 316L base metal, and its tensile strength is higher than that of 316L.

The welding engineer would be well advised to consider 2209 filler metal for the joining of 320 to 316L. The only consideration about that selection is if the weldment were intended for service where ferrite is detrimental: (1) urea manufacture where ferrite is selectively attacked, (2) elevated-temperature service where sigma phase embrittlement may occur, or (3) cryogenic temperature service where toughness would be compromised.

6.9 CASE STUDY: WHAT'S WRONG WITH MY SWIMMING POOL?

Every summer the community of German Village in Columbus, Ohio, holds a house and garden tour, allowing people to visit many of the old historic homes in that region. While on one of these visits, I (J.C.L.) toured a house with an exquisite garden containing an L-shaped lap pool. Upon closer inspection of the pool, I discovered it was of welded stainless steel construction. But there was something peculiar about the pool. Periodically along the length of the pool there would be two parallel "brown" lines. The owner of the house happened to be present at the time of my inspection and expressed some frustration that his beautiful stainless steel pool had these unsightly brown lines. He mentioned that the contractor had been back on two different occasions to "buff out" these lines, but within a matter of weeks they would return. What was going on?

This was a classic case of intergranular attack in the HAZ of the welds. The material of construction turned out to be Type 304, and during welding of the sheet metal sections, sensitization of the HAZ microstructure occurred. The chlorinated pool water provided the Cl^- ions that exacerbated the intergranular attack. The manifestation of this was the local "rusting" that occurred on opposite sides of the weld and the wagon track appearance described in Section 6.6.1.

So what was the unfortunate homeowner to do? Because permanent metallurgical damage in the form of sensitization was the root cause of the problem, there are no remedial options short of replacing the pool with a low-carbon grade of stainless steel (304L or 316L) or more resistant alloy altogether. In other situations, it may be possible to reverse the damage by a solutionizing heat treatment. This, of course, is not appropriate for an in-ground 50-m-long lap pool. Contractors using austenitic stainless steel for such applications should be aware of the potential for localized corrosion resulting from sensitization and select base metals that are resistant to this form of corrosion, particularly in situations where exposure to chloride-containing environments (such as seawater) are possible.

6.10 CASE STUDY: CRACKING IN THE HEAT-AFFECTED ZONE

A sheet metal fabricator who manufacturers commercial kitchen equipment from various grades of stainless steel suddenly began to encounter cracking problems in the HAZ of autogenous GTA welds in 2-mm-thick Type 304L. The welding process was completely automated and the procedure used had made miles of welds without problems. The HAZ cracking was sporadic but occurred over a period of several months and was associated with multiple heat compositions of Type 304L. What could be the possible nature of the cracking?

In considering the possible cracking mechanisms, there are three possibilities: HAZ liquation cracking, ductility-dip cracking, and Cu-contamination cracking. Examination of welds that showed HAZ cracking revealed that the cracks were always located at a distance of 2 to 3 mm from the fusion boundary. The cracking was not continuous but appeared sporadically along the length of the weld. Analysis of the many heat compositions revealed no linkage between impurity level (P + S content) or Ferrite Potential (based on the WRC-1992 diagram).

The location of the cracks at some distance from the fusion boundary and the lack of heat-to-heat variation in susceptibility eliminated HAZ liquation cracking as a possibility. HAZ liquation cracks are characteristically located immediately adjacent to the fusion boundary and are associated with materials that have a Ferrite Potential of zero and high impurity levels (see Section 6.5.2 for more details). While ductility-dip cracking (DDC) is a possibility, this form of cracking is quite rare in thin sheet materials because the restraint levels are generally not high enough. In addition, the lack of heat-to-heat variation in cracking susceptibility would suggest that DDC is not the culprit. This leaves only Cu-contamination cracking (CCC) as a feasible mechanism.

To determine if CCC was indeed the root cause, samples containing cracks were obtained for metallographic and fracture surface analysis. As noted previously, the cracks were located at a distance of approximately 2 mm from the fusion boundary. In this region there was only a small amount of grain growth due to the HAZ thermal excursion, but the cracks were clearly intergranular. Examination of the fracture surface in the scanning electron microscope (SEM) confirmed the intergranular nature of the fracture and composition analysis with the EDAX system clearly showed the presence of Cu on the fracture surface. Additionally, examination of metallographic samples in the as-polished (unetched) condition clearly showed the presence of a yellowish-gold constituent coating the austenite grain boundaries. The CCC mechanism was confirmed (see Section 6.5.6 for more details).

The problem now became one of finding the source of the copper that was being deposited on the Type 304L sheet. In many cases the copper comes from the fixturing or copper components in the welding system. This was not the case here. Rather it appeared that copper was somehow being plated onto the sheet, but the source of that could not immediately be determined. Abrading the sheet in the vicinity of the weld was successful in removing copper from the surface and preventing CCC from occurring, but the mystery continues.

REFERENCES

1. Lippold, J. C., and Savage, W. F. 1979. Solidification of austenitic stainless steel weldments, 1: a proposed mechanism, *Welding Journal*, 58(12):362s–374s.
2. Sundman, B., Jansson, B., and Andersson, J.-O. 1985. *Calphad*, 6:153–190.
3. Peckner, D., and Bernstein, I. M. 1977. *Handbook of Stainless Steels*, McGraw-Hill, New York.
4. Lacombe, P., Baroux, B., and Beranger, G. 1993. *Stainless Steels*, Les Éditions de Physique, Les Ulis, France.
5. Cihal, V. 1968. *Protection of Metals (USSR)*, 4(6):563.
6. Talbot, A. M., and Furman, D. E. 1953. *Transactions of the American Society for Metals*, 45:429–440.
7. Lippold, J. C., Juhas, M. C., and Dalder, E. N. C. 1985. The relationship between microstructure and fracture behavior of fully austenitic Type 316L weld filler materials at 4.2K, *Metallurgical Transactions*, 16A:1835–1848.
8. David, S. A., Goodwin, G. M., and Braski, D. N. 1979. Solidification behavior of austenitic stainless steel filler metals, *Welding Journal*, 58(11):330s–336s.
9. David, S. A. 1981. Ferrite morphology and variations in ferrite content in austenitic stainless steel welds, *Welding Journal*, 60(4):63s–71s.
10. Lippold, J. C., and Savage, W. F. 1981. Modelling solute redistribution during solidification of austenitic stainless steel weldments, in *Modelling of Casting and Welding Processes*, H. D. Brody and D. Apelian, eds., Metallurgical Society of AIME, Warrendale, PA, pp. 443–458.
11. Lippold, J. C., and Savage, W. F. 1980. Solidification of austenitic stainless steel weldments, 2: the effect of alloy composition on ferrite morphology, *Welding Journal*, 59(2):48s–58s.
12. Brooks, J. A., Thompson, A. W., and Williams, J. C. 1984. A fundamental study of the beneficial effects of delta ferrite in reducing weld cracking, *Welding Journal*, 63(3):71s–83s.
13. Arata, Y., Matsuda, F., and Katayama, S. 1976. Solidification cracking susceptibility of fully austenitic stainless steels, report 1: fundamental investigation on solidification behavior of fully austenitic and duplex microstructures and effect of ferrite on microsegregation, *Transactions of JWRI*, 5(2):135.
14. Katayama, S., Fujimoto, T., and Matsunawa, A. 1985. Correlation among solidification process, microstructure, microsegregation and solidification cracking susceptibility in stainless steel weld metals, *Transactions of JWRI*, 14(1):123.
15. Leone, G. L., and Kerr, H. W. 1982. The ferrite to austenite transformation in stainless steels, *Welding Journal*, 61(1):13s–21s.
16. Fredriksson, H. 1972. Solidification sequence in an 18-8 stainless steel investigated by directional solidification, *Metallurgical Transactions*, 3(11):2989–2997.
17. Suutala, N., Takalo, T., and Moisio, T. 1980. Ferritic–austenitic solidification mode in austenitic stainless steel welds, *Metallurgical Transactions*, 11A(5):717–725.
18. Brooks, J. A., and Thompson, A. W. 1991. *International Materials Review*, 36(1):16–44.
19. Lippold, J. C., Clark, W. A. T., and Tumuluru, M. 1992. An investigation of weld metal interfaces, in *The Metal Science of Joining*, Metals, Minerals and Materials Society, Warrendale, PA, pp. 141–146.

20. Wegrzyn, J., and Klimpel, A. 1981. The effect of alloying elements on sigma phase formation in 18-8 weld metals, *Welding Journal*, 60(8):146s–154s.
21. Vitek, J. M., and David, S. A. 1984. The solidification and aging behavior of Types 308 and 308CRE stainless steel welds, *Welding Journal*, 63(8):246s–253s.
22. Vitek, J. M., and David, S. A. 1986. The sigma phase transformation in austenitic stainless steels, *Welding Journal*, 65(4):106s–111s.
23. Alexander, D. J., Vitek, J. M., and David, S. A. 1995. Long-term aging of Type 308 stainless steel welds: effects on properties and microstructure, in *Proceedings of the 4th International Conference on Trends in Welding Research*, ASM International, Materials Park, OH, pp. 557–561.
24. Hauser, D., and Van Echo, J. A. 1982. Effects of ferrite content in austenitic stainless steel welds, *Welding Journal*, 61(2):37s–44s.
25. Thomas, R. G. 1978. The effect of delta ferrite on the creep rupture properties of austenitic weld metals, *Welding Journal*, 57(3):81s–86s.
26. Szumachowski, E. R., and Reid, H. F. 1978. Cryogenic toughness of SMA austenitic stainless steel weld metals, 1: role of ferrite, *Welding Journal*, 57(11):325s–333s.
27. Read, D. T., McHenry, H. I., Steinmeyer, P. A., and Thomas, R. D., Jr. 1980. Metallurgical factors affecting the toughness of 316L SMA weldments at cryogenic temperatures, *Welding Journal*, 59(4):104s–113s.
28. Goodwin, G. M. 1984. *Fracture Toughness of Austenitic Stainless Steel Weld Metal at 4K*, ORNL/TM-9172, Oak Ridge National Laboratory, Oak Ridge, TN.
29. Kotecki, D. J. 2003. Stainless Q & A, *Welding Journal*, 82(11):80–81.
30. Kujanpää, V., Suutala, N., Takalo, T., and Moisio, T. 1979. Correlation between solidification cracking and microstructure in austenitic and austenitic–ferritic stainless steel welds, *Welding Research International*, 9(2):55.
31. Hammar, O., and Svennson, U. 1979. *Solidification and Casting of Metals*, Metal Society, London, pp. 401–410.
32. Arata, Y., Matsuda, F., Nakagawa, H., and Katayama, S. 1978. Solidification cracking susceptibility of fully austenitic stainless steels, report 4: effect of decreasing P and S on solidification cracking susceptibility of SUS 310 austenitic stainless steel weld metals, *Transactions of JWRI*, 7(2):169.
33. Ogawa,T., and Tsunetomi, E. 1982. Hot cracking susceptibility of austenitic stainless steels, *Welding Journal*, 61(3):82s–93s.
34. Li, L., and Messler, R. W., Jr. 1999. The effects of phosphorus and sulfur on susceptibility to weld hot cracking in austenitic stainless steels, *Welding Journal*, 78(12):387s–396s.
35. Heiple, C. R., and Roper, J. R. 1982. Mechanism for Minor Element Effect on GTA Fusion Zone Geometry, *Welding Journal*, 61(4):97s–102s.
36. Lundin, C. D., Lee, C. H., and Menon, R. 1988. Hot ductility and weldability of free machining austenitic stainless steel, *Welding Journal*, 67(6):119s–130s.
37. Brooks, J. A., Robino, C. V., Headley, T. J., and Michael, J. R. 2003. Weld solidification and cracking behavior of free-machining stainless steel, *Welding Journal*, 82(3):51s–64s.
38. AWS. 1997. *Standard Procedures for Calibrating Magnetic Instruments to Measure the Delta Ferrite Content of Austenitic and Duplex Ferritic–Austenitic Stainless Steel Weld Metals*, ANSI/AWS A4.2M/A4.2:1997, American Welding Society, Miami, FL, p. 13.
39. Kotecki, D. J. 1997. Ferrite determination in stainless steel welds: advances since 1974, *Welding Journal*, 76(1):24s–37s.

40. Nakao,Y., Nishimoto, K., and Zhang, W. 1988. Effects of rapid solidification by laser surface melting on solidification modes and microstructures of stainless steels, *Transactions of the Japan Welding Society*, 19:101.
41. Lippold, J. C. 1994. Solidification behavior and cracking susceptibility of pulsed-laser welds in austenitic stainless steels, *Welding Journal*, 73(6):129s–139s.
42. David, S. A., Vitek, J. M., and Hebble, T. M. 1987. Effect of rapid solidification on stainless steel weld metal microstructures and its implications on the Schaeffler diagram, *Welding Journal*, 66(10):289s–300s.
43. Elmer, J. W., Allen, S. M., and Eagar, T. W. 1990. The influence of cooling rate on the ferrite content of stainless steel alloys, in *Recent Trends in Welding Science and Technology*, S. A. David and J. M. Vitek, eds., ASM International, Materials Park, OH, pp. 165–170.
44. Brooks, J. A., and Baskes, M. I. 1990. Microsegregation modeling and transformation in rapidly solidified austenitic stainless steels welds, in *Recent Trends in Welding Science and Technology*, S. A. David and J. M. Vitek, eds., ASM International, Materials Park, OH, pp. 153–158.
45. Kou, S., and Le, Y. 1982. The effect of quenching on the solidification structure and transformation behavior of stainless steel welds, *Metallurgical Transactions*, 13A:1141–1152.
46. Kurz, W., and Fisher, D. J. 1981. Dendrite growth at the limit of stability: tip radius and spacing, *Acta Metallurgica*, 29:11.
47. Pacary, G., Moline, M., and Lippold, J.C. 1990. *A Diagram for Predicting the Weld Solidification Cracking Susceptibility of Pulsed-Laser Welds in Austenitic Stainless Steels*, EWI Research Brief B9008.
48. Lienert, T. J., and Lippold, J. C. 2003. Weldability and solidification mode diagrams for pulsed-laser welds in austenitic stainless steels, *Science and Technology of Welding and Joining*, 8(1):1–9.
49. Elmer, J. W., Allen, S. M., and Eagar, T. W. 1989. Microstructural development during solidification of stainless steels alloys, *Metallurgical Transactions*, 20A:2117.
50. Elmer, J. W., Allen, S. M., and Eagar, T. W. 1990. Single phase solidification during rapid resolidification of stainless steel alloys, in *Weldability of Materials*, R. A. Patterson and K. W. Mahin, eds., ASM International, Materials Park, OH, pp. 143–150.
51. Lippold, J. C. 1985. Centerline cracking in deep penetration electron beam welds in type 304L stainless steel, *Welding Journal*, 64(5):127s–136s.
52. Thomas, R. D., Jr. 1984. HAZ cracking in thick sections of austenitic stainless steels, 1, *Welding Journal*, 62(12):24–32.
53. Thomas, R. D., Jr. 1984. HAZ cracking in thick sections of austenitic stainless steels, 2, *Welding Journal*, 62(12):355s–368s.
54. Lippold, J. C., Varol, I., and Baeslack, W. A. 1992. An investigation of heat-affected zone liquation cracking in austenitic and duplex stainless steels, *Welding Journal*, 71(1):1s–14s.
55. Honeycombe, J., and Gooch, T. G. 1972. Effect of manganese on cracking and corrosion resistance of fully austenitic stainless steel weld metals, *Metal Construction and British Welding Journal*, 4(12):456.
56. Gooch, T. G., and Honeycombe, J. 1975. Microcracking in fully austenitic stainless steel weld metal, *Metal Construction*, 7(3):146.
57. Lundin, C. D., and Spond, D. F. 1976. The nature and morphology of fissures in austenitic stainless steel weld metals, *Welding Journal*, 55(11):356s–367s.

58. Lundin, C. D., DeLong, W. T., and Spond, D. F. 1975. Ferrite-fissuring relationship in austenitic stainless steel weld metals, *Welding Journal*, 54(8):241s–246s.
59. Lundin, C. D., and Chou, C. P. D. 1985. Fissuring in the "Hazard HAZ" region of austenitic stainless steel welds, *Welding Journal*, 64(4):113s–118s.
60. Hemsworth, B., Boniszewski, T., and Eaton, N. F. 1969. Classification and definition of high temperature welding cracks in alloys, *Metal Construction and British Welding Journal*, 1(2):5.
61. Hadrill, D. M., and Baker, R. G. 1965. Microcracking in austenitic weld metal, *British Welding Journal*, 12(8):411.
62. Nissley, N. E., and Lippold, J. C. 2003. Ductility-dip cracking susceptibility of austenitic alloys, in *Proceedings of the 6th International Conference on Trends in Welding Research*, ASM International, Materials Park, OH, pp. 64–69.
63. Ramirez, A. J., and Lippold, J. C. 2004. High temperature cracking in nickel-base weld metal, 1: ductility and fracture behavior, *Materials Science and Engineering A*, 380: 259–271.
64. Ramirez, A. J., and Lippold, J. C. 2004. High temperature cracking in nickel-base weld metal, 2: insight into the mechanism, *Materials Science and Engineering A*, 380: 245–258.
65. Nissley, N. E., and Lippold, J. C. 2003. Development of the strain-to-fracture test for evaluating ductility-dip cracking in austenitic alloys, *Welding Journal*, 82(12):355s–364s.
66. Nissley, N. E., Collins, M. G., Guaytima, G., and Lippold, J. C. 2002. *Development of the Strain-to-Fracture Test for Evaluating Ductility-Dip Cracking in Austenitic Stainless Steels and Ni-Base Alloys*, IIW Document IX-2050-02., International Institute of Welding, Paris.
67. Christoffel, R. J. 1962. Cracking in Type 347 heat-affected zone during stress relaxation, *Welding Journal*, 41(6):251s–256s.
68. Lin, W., Lippold, J. C., and Luke, S. 1994. Unpublished research performed at Edison Welding Institute.
69. Savage, W. F., Nippes, E. P., and Mushala, M. C. 1978. Copper-contamination cracking in the weld heat-affected zone, *Welding Journal*, 57(5):145s–152s.
70. Kotecki, D. J. 1999. Stainless Q & A, *Welding Journal*, 78(10):113.
71. Kanne, W. R., Jr. 1988. Remote reactor repair: GTA weld cracking caused by entrapped helium, *Welding Journal*, 67(8):33–39.
72. Goods, S. H., and Karfs, C. W. 1991. Helium-induced weld cracking in low heat input GMA weld overlays, *Welding Journal*, 70(5):123s–132s.
73. ASM. 1987. *ASM Metals Handbook*, 10th ed., Vol. 13, *Corrosion*, ASM International, Materials Park, OH.
74. Gooch, T. G. 1996. Corrosion behavior of welded stainless steel, *Welding Journal*, 75(5):135s–154s.
75. Povich, M. J. 1978. *Corrosion*, 34:60.
76. Povich, M. J., and Rao, P. 1978. *Corrosion*, 34:269.
77. Copson, H. R. 1959. Effect of composition on stress corrosion cracking of some alloys containing Ni, in *Physical Metallurgy of Stress Corrosion Fracture*, Interscience, New York, pp. 247–272.
78. Jones, D. A. 1995. *Principles and Prevention of Corrosion*, 2nd ed., Prentice Hall, Upper Saddle River, NJ.

79. Garner, A. 1983. Pitting corrosion of high alloy stainless steel weldments in oxidizing environments, *Welding Journal*, 62(1):27–34.
80. Little, B., Wagner, P., and Mansfeld, F. 1991. Microbiologically influenced corrosion of metals and alloys, *International Materials Review*, 36(6):253.
81. Zhang, H., Shi, S., Ramirez, J., and Lippold, J. C. 2004. *Review of Reheat Cracking and Elevated Temperature Embrittlement of Austenitic Materials*, EWI report.
82. Ebert, H. 1974. Solution annealing in the field, *Welding Journal*, 53(2):88–93.
83. Dhooge, A. 1997. *Survey on Reheat Cracking in Austenitic Steels and Ni-Base Alloys*, IIW Document IX-1876-97, International Institute of Welding, Paris.
84. Van Wortel, J. C. 1995. Relaxation cracking in austenitic welded joints: an underestimated problem, *Stainless Steel World*, pp. 47–49.
85. Espy, R. H. 1982. Weldability of nitrogen-strengthened stainless steels, *Welding Journal*, 61(5):149s–156s.
86. Kotecki, D. J. 2002. Stainless Q & A, *Welding Journal*, 81(11):86–87.
87. Robino, C. V., Michael, J. R., and Maguire, M. C. 1998. The solidification and weld metallurgy of galling resistant stainless steels, *Welding Journal*, 77(11):446s–457s.
88. Brooks, J. A. 1975. Weldability of high N, high Mn austenitic stainless steels, *Welding Journal*, 54(6):189s–195s.
89. Perricone, M. J., and DuPont, J. N. 2003. Laser welding of superaustenitic stainless steel, in *Proceedings of the 6th International Conference on Trends in Welding Research*, ASM International, Materials Park, OH, pp. 64–69.
90. Woolin, P. 1997. *Autogenous Welding of High Nitrogen Superaustenitic Stainless Steels*, TWI Research Report 593/1997.

CHAPTER 7

DUPLEX STAINLESS STEELS

Duplex ferritic–austenitic stainless steels derive their name from their normal room-temperature microstructure that is roughly half ferrite and half austenite. Duplex stainless steels have been known since the 1930s [1]. Progress in their development came in fits and starts, apparently receiving more emphasis during periodic shortages of nickel [1,2]. As recently as 1982, the *Welding Handbook* [3] did not recognize duplex stainless steels as a separate family of stainless steels. At that time, the *Welding Handbook* listed Type 329, one of the older duplex stainless steels, among the austenitic stainless steels. At least one manufacturer of Type 329 duplex stainless steel recommended against welding this alloy [4]. The cast alloy CD4MCu, developed during the 1950s, was known for its brittleness when welded [1]. The duplex stainless steels have evolved rapidly since the early 1980s, and this evolution is chronicled in the compiled papers of a series of international conferences on duplex stainless steels over this same time frame [5–10]. These steels are currently used in a wide range of applications, most requiring superior corrosion resistance. Significant improvements have been made in both the weldability and corrosion resistance of these alloys over this period, due primarily to realizing the critical role of nitrogen as an alloying element [1].

Duplex stainless steels are used in applications that take advantage of their superior corrosion resistance, strength, or both. Because they have a higher ferrite content than austenitic steels, they are more ferromagnetic and have higher thermal conductivity and lower thermal expansion. They are most often selected for corrosion resistance and have

Welding Metallurgy and Weldability of Stainless Steels, by John C. Lippold and Damian J. Kotecki
ISBN 0-471-47379-0

been substituted for austenitic alloys in many applications where stress corrosion cracking and pitting corrosion are concerns. They are also vastly superior to structural steels in most corrosive applications and can have comparable strength. For example, duplex stainless steels have been used widely for oil and gas pipelines, both on- and offshore.

Because duplex stainless steels form a number of embrittling precipitates at relatively low temperatures, they are not recommended for service applications exceeding about 280°C (535°F). They are more expensive than austenitic stainless steels, not so much due to the cost of alloy elements, but due primarily to the cost of processing the as-cast steel to finished plate, sheet, or tubular form. They offer distinct corrosion advantages and weight savings. They may be used in place of some Ni-base alloys in mildly aggressive environments, at a fraction of the material cost.

Duplex stainless steels are significantly stronger than the austenitic stainless steels. Typical yield strengths are above 425 MPa (60 ksi) compared to 210 MPa (30 ksi) for austenitic steels. Since the duplex stainless steels are stronger, they are also harder, possibly making them more attractive where abrasion is a concern, as well as corrosion. Most of today's duplex stainless steels have good toughness and ductility, however, they do undergo a ductile-to-brittle transition at low temperature, so that they are generally not suitable for service at cryogenic temperatures. The service temperature range of duplex stainless steels is generally limited to a range of about −40 to 280°C (−40 to 535°F).

The thermal expansion of the duplex alloys is close to that of carbon and low-alloy steels. As a result of this similarity, they may find application in situations, such as pressure vessels, where they are coupled with carbon steels. Stresses due to the differential thermal expansion would be reduced relative to austenitic alloys. However, due to the relatively low temperature precipitation reactions that occur in duplex stainless steels, their application in situations requiring thermal stress relief (postweld heat treatment) is generally inappropriate.

7.1 STANDARD ALLOYS AND CONSUMABLES

Table 7.1 lists the composition of a number of duplex stainless steel base metal alloys. It is noteworthy that two of these alloys, Type 329 wrought steel and Type CD4MCu, do not show a nitrogen requirement in ASTM A240 or ASTM A890, respectively. These two alloys were developed before the importance of nitrogen was appreciated, and they were considered to be very difficult to weld successfully. A nitrogen modification to each has since been added to the ASTM specifications (UNS S32950 and CD4MCuN, respectively), for improved weldability, as well as for improved corrosion resistance.

Welding consumables are selected to create the proper phase balance in the weld deposit and provide corrosion resistance at least equal to that of the base metal. Nickel contents are often "boosted" in nominally matching filler metals in order to promote austenite formation during the rapid cooling associated with welding. One such filler is AWS ER/E 2209, which contains nominally 9% Ni, typically used for base metals such as 2304 and 2205 that contain nominally 5% Ni. Table 7.2 lists a number of filler metal compositions from the AWS standards.

TABLE 7.1 Composition of Duplex Ferritic–Austenitic Stainless Steels

Type[b]	UNS No.[b]	Composition (wt%)[a]										
		C	Mn	P	S	Si	Cr	Ni	Mo	N	Cu	W
—	S32201	0.030	4.0–6.0	0.040	0.030	1.00	19.5–21.5	1.00–3.00	0.60	0.05–0.17	1.00	—
2304	S32304	0.030	2.50	0.040	0.030	1.00	21.5–24.5	3.0–5.5	0.05–0.60	0.05–0.20	0.05–0.60	—
2205[c]	S31803	0.030	2.00	0.030	0.020	1.00	21.0–23.0	4.5–6.5	2.5–3.5	0.08–0.20	—	—
2205[c]	S32205	0.030	2.00	0S.030	0.020	1.00	22.0–23.0	4.5–6.5	3.0–3.5	0.14–0.20	—	—
329	S32900	0.08	1.00	0.040	0.030	0.75	23.0–28.0	2.0–5.00	1.00–2.00	—	—	—
—	S32950	0.030	2.00	0.035	0.010	0.60	26.0–29.0	3.5–5.2	1.00–2.50	0.15–0.35	—	—
—	S31260	0.03	1.00	0.030	0.030	0.75	24.0–26.0	5.5–7.5	2.5–3.5	0.10–0.30	0.20–0.80	0.10–0.50
—	S32520	0.030	1.50	0.035	0.020	0.80	24.0–26.0	5.5–8.0	3.0–4.0	0.20–0.35	0.50–2.00	—
CD4MCu	—	0.04	1.00	0.04	0.04	1.00	24.5–26.5	4.75–6.00	1.75–2.25	—	2.75–3.25	—
CD4MCuN	—	0.04	1.00	0.04	0.04	1.00	24.5–26.5	4.7–6.0	1.7–2.3	0.10–0.25	2.7–3.3	—
255	S32550	0.04	1.50	0.040	0.030	1.00	24.0–27.0	4.5–6.5	2.9–3.9	0.10–0.25	1.50–2.50	—
2507	S32750	0.030	1.20	0.035	0.020	0.80	24.0–26.0	6.0–8.0	3.0–5.0	0.24–0.32	0.50	—
—	S32760	0.030	1.00	0.030	0.010	1.00	24.0–26.0	6.0–8.0	3.0–4.0	0.20–0.30	0.50–1.00	0.50–1.00
CD3M-WCuN	—	0.03	1.00	0.030	0.025	1.00	24.0–26.0	6.5–8.5	3.0–4.0	0.20–0.30	0.5–1.0	0.5–1.0

Zeron 100

[a]A single value is a maximum.
[b]The compositions are grouped according to similar Cr first, then similar N, then Mo.
[c]Originally, Type 2205 was commonly associated with the S31803 composition range. However, when it became well known that low nitrogen produced an unfavorable phase balance in the weld HAZ, Type 2205 was defined as the UNS S32205 composition ranges in ASTM A240/A240M–99a in Volume 1.03 of the 2000 edition of the ASTM standards.

TABLE 7.2 **Composition of Duplex Ferritic–Austenitic Stainless Steel Welding Filler Metals**

		Composition (wt%)[a]										
Class	Source[b]	C	Mn	P	S	Si	Cr	Ni	Mo	N	Cu	W
E2209-XX	A5.4	0.04	0.5–2.0	0.04	0.03	1.00	21.5–23.5	8.5–10.5	2.5–3.5	0.08–0.20	0.75	—
ER2209	A5.9	0.03	0.50–2.0	0.03	0.03	0.90	21.5–23.5	7.5–9.5	2.5–3.5	0.08–0.20	0.75	—
E2209TX-X	A5.22	0.04	0.5–2.0	0.04	0.03	1.0	21.0–24.0	7.5–10.0	2.5–4.0	0.08–0.20	0.5	—
E2552-XX	A5.4	0.04	1.0	0.04	0.03	1.00	24.0–27.0	4.0–6.0	1.5–2.5	0.08–0.22	2.5–3.5	—
E2553-XX	A5.4	0.06	0.5–1.5	0.04	0.03	1.00	24.0–27.0	6.5–8.5	2.9–3.9	0.10–0.25	1.5–2.5	—
E2553TX-X	A5.22	0.04	0.5–1.5	0.04	0.03	0.75	24.0–27.0	8.5–10.5	2.9–3.9	0.10–0.20	1.5–2.5	—
ER2553	A5.9	0.04	1.5	0.04	0.03	1.0	24.0–27.0	4.5–6.5	2.9–3.9	0.10–0.25	1.5–2.5	—
E2593-XX	A5.4	0.04	0.5–2.5	0.04	0.03	1.00	24.7–27.0	8.5–11.0	2.9–3.9	0.08–0.25	1.5–3.0	—
E2594-XX	A5.4	0.04	0.5–2.0	0.04	0.03	1.00	24.0–27.0	8.0–10.5	3.5–4.5	0.20–0.30	0.75	—

[a] A single value is a maximum.
[b] AWS standard.

Filler metals with boosted nickel contents, not yet addressed in filler metal standards, are available under various trade names for joining newer high-alloy base metals (called *superduplex alloys*, such as Alloy 2507). High-alloy austenitic filler metals such as 309L may be used when making dissimilar joints with austenitic stainless steels. Ni-base alloys can be selected for dissimilar combinations or applications requiring extreme corrosion resistance.

7.2 PHYSICAL METALLURGY

7.2.1 Austenite–Ferrite Phase Balance

Duplex stainless steels are based on the Fe–Cr–Ni–N alloy system. The chemical compositions of these steels have been adjusted such that the base metal microstructure consists of nominally 50% ferrite and 50% austenite, hence the term *duplex*. However, all duplex stainless steels solidify as virtually 100% ferrite and depend on partial solid-state transformation to austenite for this balanced microstructure. Nitrogen is usually added as an alloying element to accelerate formation of the austenite phase and stabilize it, and also to improve pitting corrosion resistance. Molybdenum, tungsten, and/or copper are added to some alloys to improve corrosion resistance. Because of both the higher alloy contents, and, especially, the meticulous and difficult thermo-mechanical processing of wrought products, the duplex stainless steels are more costly to produce than the austenitic stainless steels.

Duplex stainless steel base metals in general have a higher ratio of ferrite-promoting elements to austenite-promoting elements than do nominally austenitic stainless steels. This accounts for the fact that duplex stainless steels solidify as virtually 100% ferrite. Figure 7.1 indicates that alloys whose WRC-1992 chromium equivalent is about 1.85 times their nickel equivalent, or more, will solidify as 100% ferrite. The duplex stainless steel base metals typically have ratios between about 2.25 and 3.5. At high temperatures (above the ferrite solvus), the alloys remain 100% ferrite. Austenite can only nucleate and grow below the ferrite solvus. Annealing and hot-working operations of these steels are generally performed at temperatures below the ferrite solvus, where the austenite and ferrite can coexist in equilibrium. By controlling the processing temperature and cooling rate from that temperature, the proportion and distribution of ferrite and austenite can be controlled in the wrought product.

Although constitution tie-lines cannot be drawn with any reasonable accuracy in a pseudobinary diagram, it should be evident from Figure 7.1 that alloys annealed or hot worked at temperatures near the ferrite solvus, followed by quenching, can be expected to contain mostly ferrite, with only small amounts of grain boundary austenite. As the temperature of annealing or hot working is reduced, the equilibrium microstructure will contain progressively more austenite and less ferrite. Under equilibrium conditions, there is a partitioning of the alloy elements, so that the ferrite-promoting elements (chromium, molybdenum, tungsten) will concentrate, by diffusion, into the ferrite. At the same time, the austenite-promoting elements (nickel, carbon, nitrogen, and copper) will concentrate, by diffusion, into the austenite. This effect has been illustrated by Ogawa and Koseki [11] (Figure 7.2). With falling temperature, the equilibrium compositions of the ferrite and austenite are continually changing, according to the austenite

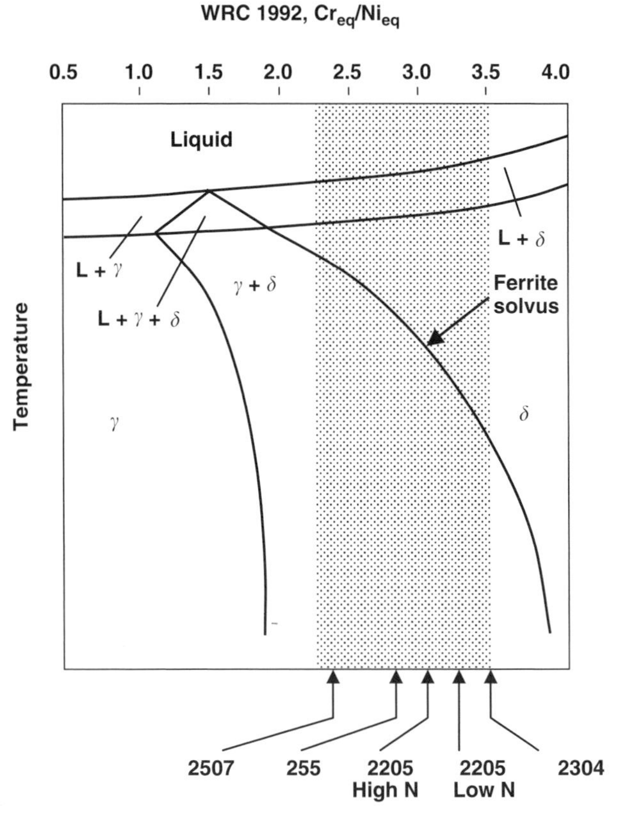

Figure 7.1 Elevated temperature region of a pseudobinary phase diagram for duplex stainless steel compositions. Shaded region represents the range for commercial alloys.

and ferrite solvus line compositions. At the same time, diffusion is slowing with falling temperature. At some temperature during cooling, diffusion can no longer keep pace with the equilibrium compositons of the two phases, and whatever compositions and phase balance are present at that time and temperature are then frozen into the alloy, as if it had been quenched from that temperature.

This concept of an *effective quench temperature* has been proposed by Vitek and David [12]. In a duplex stainless steel, if the effective quench temperature is high (close to the ferrite solvus temperature), the phase balance will be very high in ferrite. If the effective quench temperature is well below the ferrite solvus temperature, there will be more austenite. Hot working speeds diffusion, so that a lower effective quench temperature is more easily obtained for wrought duplex stainless steels than for cast duplex stainless steels or for weld metals. Cast duplex stainless steels obtain their ambient temperature microstructures by annealing at temperatures at or above 1040°C (1900°F), followed by quenching [13]. In practice, the annealing temperature, or hot-working temperature, is chosen as low as possible but sufficiently high to take into solution any precipitate phases [1]. This practice minimizes the amount of nitrogen remaining in the ferrite phase.

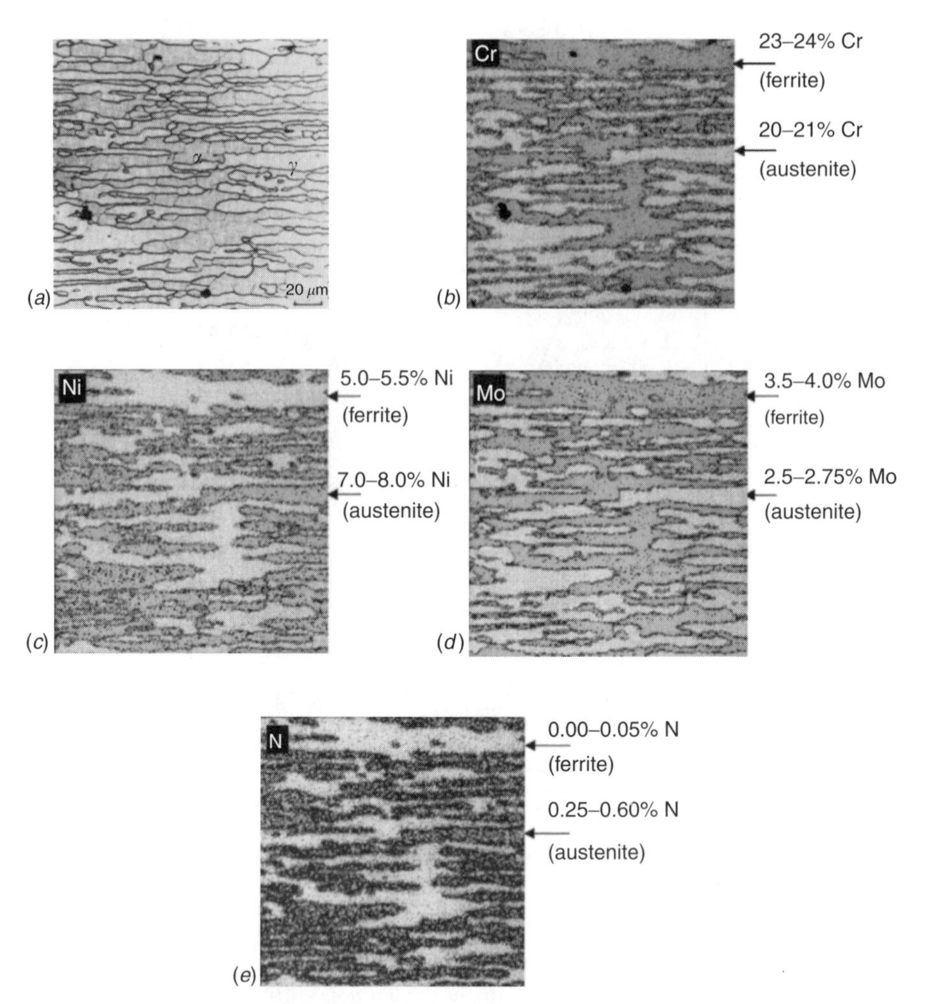

Figure 7.2 Partitioning of alloy elements in 2205. Composition: (22% Cr, 6% Ni, 3% Mo, 0.12% N). (*a*) Microstructure of low-nitrogen duplex stainless steel base metal; white, austenite; gray, ferrite. (*b*) Cr distribution; near-white, 20 to 21% Cr; black and dark gray, 21 to 23% Cr; light gray, 23 to 24% Cr. (*c*) Ni distribution; near-white, 5.0 to 5.5% Ni; black and dark gray, 5.5 to 7.0% Ni; light gray, 7.0 to 8.0% Ni. (*d*) Mo distribution; near-white, 2.50 to 2.75% Mo; black and dark gray, 2.75 to 3.50% Mo; light gray, 3.5 to 4.0% Mo. (*e*) N distribution; near-white, 0.00 to 0.05% N; black and dark gray, 0.05 to 0.25% N; light gray, 0.25 to 0.60% N. (From Ogawa and Koseki [11].)

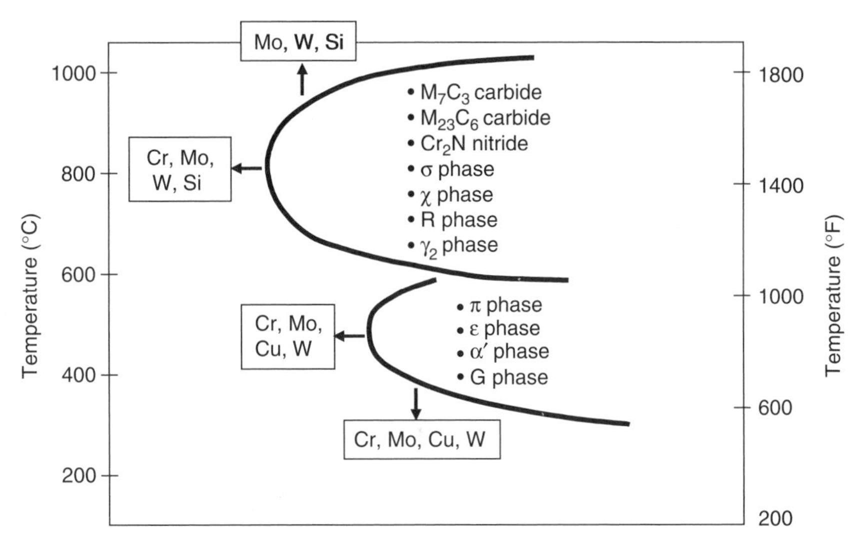

Figure 7.3 Precipitate formation in duplex stainless steels. (Redrawn from Charles [1].)

7.2.2 Precipitation Reactions

The curvature of the two-phase ferrite + austenite region of Figure 7.1 indicates that more austenite will be formed, at the expense of ferrite, at lower hot-working or annealing temperatures (effective quench temperatures). However, a lower limit on these temperatures is imposed by the appearance of undesirable precipitates within the ferrite phase. Because of the complex alloying of the duplex grades, a number of precipitation reactions can occur over a range of temperatures from below approximately 1000°C (1830°F). All of these precipitation reactions are time and temperature dependent, as indicated schematically in Figure 7.3. Many of these precipitates embrittle the duplex alloys and are to be avoided. These include sigma (σ), chi (χ), and alpha prime (α′), as well as chromium nitride. Also, it should be noted that the addition, or increased levels, of Cr, Mo, and W tend to accelerate the formation of these precipitates, particularly the sigma and chi phases. This acceleration has potential implications with respect to embrittlement during postweld heat treatment or multipass welding. The lower-temperature precipitation reactions effectively limit the use of the duplex alloys to temperatures below 280°C (535°F).

7.3 MECHANICAL PROPERTIES

Tensile property requirements are included in ASTM A240 for wrought duplex stainless steels and in ASTM A890 for cast duplex stainless steels. These minimum requirements are listed in Table 7.3. Note that the alloys containing higher levels of chromium and nitrogen, the superduplex alloys, have higher strength than the standard alloys, such as Alloy 2205.

TABLE 7.3 Mechanical Properties of Duplex Ferritic–Austenitic Stainless Steels

Type[a]	UNS No.[a]	Tensile Strength[b]		Yield Strength[b]		Elongation[b] (%)
		MPa	ksi	MPa	ksi	
—	S32201	620	90	450	65	25.0
2304	S32304	600	87	400	58	25.0
2205[c]	S31803	620	90	450	65	25.0
2205[c]	S32205	620	90	450	65	25.0
329	S32900	620	90	485	70	15.0
—	S32950	690	100	485	70	15.0
—	S31260	690	100	485	70	20.0
—	S32520	770	112	550	80	25.0
CD4MCu	—	690	100	485	70	16
CD4MCuN	—	690	100	485	70	16
255	S32550	760	110	550	80	15.0
2507	S32750	795	116	550	80	15.0
—	S32760	750	108	550	80	25.0
CD3M-WCuN	—	690	100	450	65	25

[a]The compositions are grouped according to similar Cr first, then similar N, then Mo.
[b]A single value is a minimum.
[c]Originally, Type 2205 was commonly associated with the S31803 composition range. However, when it became well known that low nitrogen produced an unfavorable phase balance in the weld HAZ, Type 2205 was defined as the UNS S32205 composition ranges in ASTM A240/A240M-99a in Volume 1.03 of the 2000 edition of the ASTM standards.

7.4 WELDING METALLURGY

7.4.1 Solidification Behavior

All the duplex stainless steels solidify as ferrite and are fully ferritic at the end of solidification. Depending on composition, the ferrite phase is stable over some range of elevated temperature before it falls below the ferrite solvus temperature and transformation to austenite begins. The nature of the ferrite-to-austenite transformation is dependent on both composition and cooling rate. It is this transformation that determines the ultimate ferrite–austenite balance and austenite distribution in the weld metal. The transformation sequence for duplex stainless steels is as follows:

$$L \rightarrow L + F \rightarrow F \rightarrow F + A$$

Duplex stainless steel weld metals contain a mixture of austenite and ferrite. Since solidification occurs as ferrite and no austenite forms at the end of solidification (see Figure 6.13, Type F solidification), ferrite is stable in the solid state at elevated temperature. When the transformation to austenite begins below the ferrite solvus, austenite first forms along the ferrite grain boundaries. This occurs by a nucleation and growth process and usually results in complete coverage of the ferrite grain boundaries by austenite. Additional austenite may form as Widmanstätten side plates off the grain boundary austenite, or intragranularly within the ferrite grains. Examples of duplex stainless steel weld metal microstructures with high and moderate ferrite content are shown in Figure 7.4.

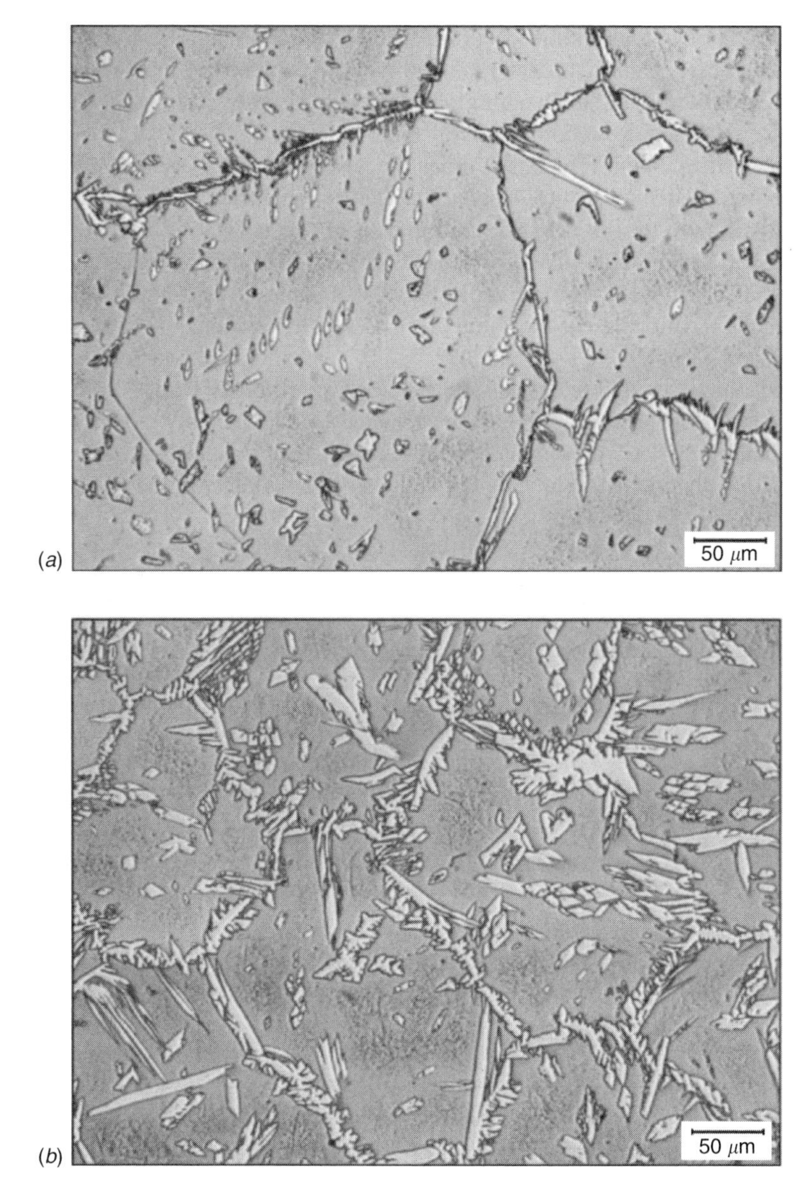

Figure 7.4 Weld metal microstructure of duplex stainless steels: (*a*) high ferrite content (FN 100); (*b*) medium ferrite content (FN 70).

7.4.2 Role of Nitrogen

All the more modern duplex stainless steels have intentional additions of nitrogen to improve strength and pitting corrosion resistance. The range of nitrogen content is between 0.08 and 0.35 wt% (Table 7.1). As can be seen from the solubility versus temperature relationship (Figure 7.5), these levels of nitrogen are well above the solubility limit in ferrite at temperatures below about 1000°C (1830°F). By contrast, solubility in austenite is much higher. This differential in nitrogen solubility has important implications with respect to nitride precipitation in these alloys. When the microstructure is balanced (nearly a 50/50 mixture achieved by an elevated temperature hold followed by rapid cooling), the nitrogen partitions between the ferrite and austenite and remains predominantly in solid solution.

If the ferrite content is high, such as in the weld metal and HAZ under rapid cooling conditions, an intense nitride precipitation reaction occurs upon cooling since the solubility limit of the ferrite is exceeded and the nitrogen has insufficient time to partition to the austenite. In most cases, these nitrides are Cr-rich and are thought to be primarily Cr_2N [14,15]. The consequences of this intense precipitation are similar to those in the ferritic alloys—most notably, loss of ductility, toughness, and corrosion resistance.

Welds, however, and their heat-affected zones, are rapidly cooled from temperatures near the ferrite solvus, so there is a tendency for appreciably more ferrite in the weld metal and HAZ of a duplex stainless steel than there is in the base metal. Because the phase balance in a duplex stainless steel weldment is heavily dependent

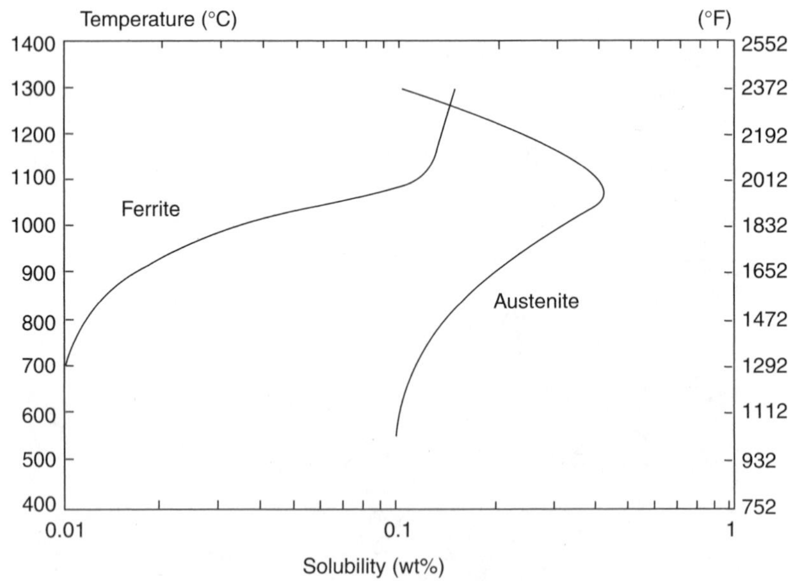

Figure 7.5 Nitrogen solubility in ferrite and in austenite. (Constructed by J. C. Lippold, based on several literature sources.)

on diffusion, nitrogen is the key element for lowering the effective quench temperature so that the HAZ phase balance can approach that of the wrought or cast duplex stainless steel base metal. This is so because all of the alloying elements in duplex stainless steels, except carbon and nitrogen, are large substitutional atoms. As such, their diffusion rates are rather low. In contrast, carbon and nitrogen are small interstitial atoms, with much more rapid diffusion rates at temperatures at and above the normal annealing temperatures for duplex stainless steels [1040°C (1900°F) and higher] up to the ferrite solvus temperature. Carbon, of course, is not a desirable alloying element, due to adverse effects on corrosion resistance, and its concentration is generally held as low as possible. So nitrogen content becomes the key to manipulating the phase balance under weld cooling conditions.

Nitrogen is much more soluble in austenite than in ferrite, in the temperature range above 1040°C (1900°F) to the ferrite solvus temperature. Partitioning of nitrogen in a nominally 0.127% nitrogen 2205 base metal was found by Ogawa and Koseki [11] to be such that the ferrite contained less than 0.05 wt%, while the austenite contained up to 0.30 wt%. When the alloy is at a temperature below its ferrite solvus temperature, nitrogen diffuses from ferrite toward austenite. If the cooling is too rapid for the nitrogen to reach austenite, some of the nitrogen will become trapped within the ferrite and subsequently precipitates as chromium nitrides.

Figure 7.2 shows the partitioning of the alloy elements in wrought Alloy 2205, while Figure 7.6 shows the fusion zone of the same alloy melted by gas tungsten arc welding without filler metal. There is a very large increase in ferrite content of the fused metal versus the wrought metal—Ogawa and Koseki [11] claim 74% ferrite in the fused metal versus 49% ferrite in the wrought metal. It can be noted from Figure 7.6*a* that there is austenite formation mainly along the ferrite grain boundaries, with only a few small austenite platelets within the ferrite grain. The fused metal also exhibits extensive chromium nitride precipitation (fine dark specks in Figure 7.6*a* in the interior of the ferrite grain, some distance from the austenite). The areas close to the austenite are free of nitrides, because the nitrogen originally in those regions had sufficient time to diffuse to the austenite.

In contrast to the wrought metal of Figure 7.2, the fusion zone shows very little partitioning of Cr, Ni and Mo, as can be seen in Figure 7.6*b*–*d*. Due to the relatively rapid cooling rate through the transformation temperature range, these substitutional atoms have little opportunity to partition. The effective quench temperature for Cr, Ni, and Mo is approximately the ferrite solvus temperature. On the other hand, the interstitial nitrogen partitions strongly to austenite where possible, as shown in Figure 7.6*e*. The nitrogen level in the austenite of the fusion zone approaches that of the austenite in the wrought material of Figure 7.2. Nitrogen, therefore, has an effective quench temperature significantly below the ferrite solvus. Thus, it is quite mobile in the microstructure at temperatures where substitutional elements have extremely low diffusivity.

Ogawa and Koseki [11] further show that by increasing the nitrogen from 0.12% to 0.18% in Alloy 2205, austenite nucleated at a higher temperature and intragranularly, so that nitrogen diffusion was more rapid and the distance over which nitrogen had to diffuse below the ferrite solvus temperature was reduced. The end result is that nitride formation was eliminated in the fusion zone, as shown in Figure 7.7*a*. The partitioning

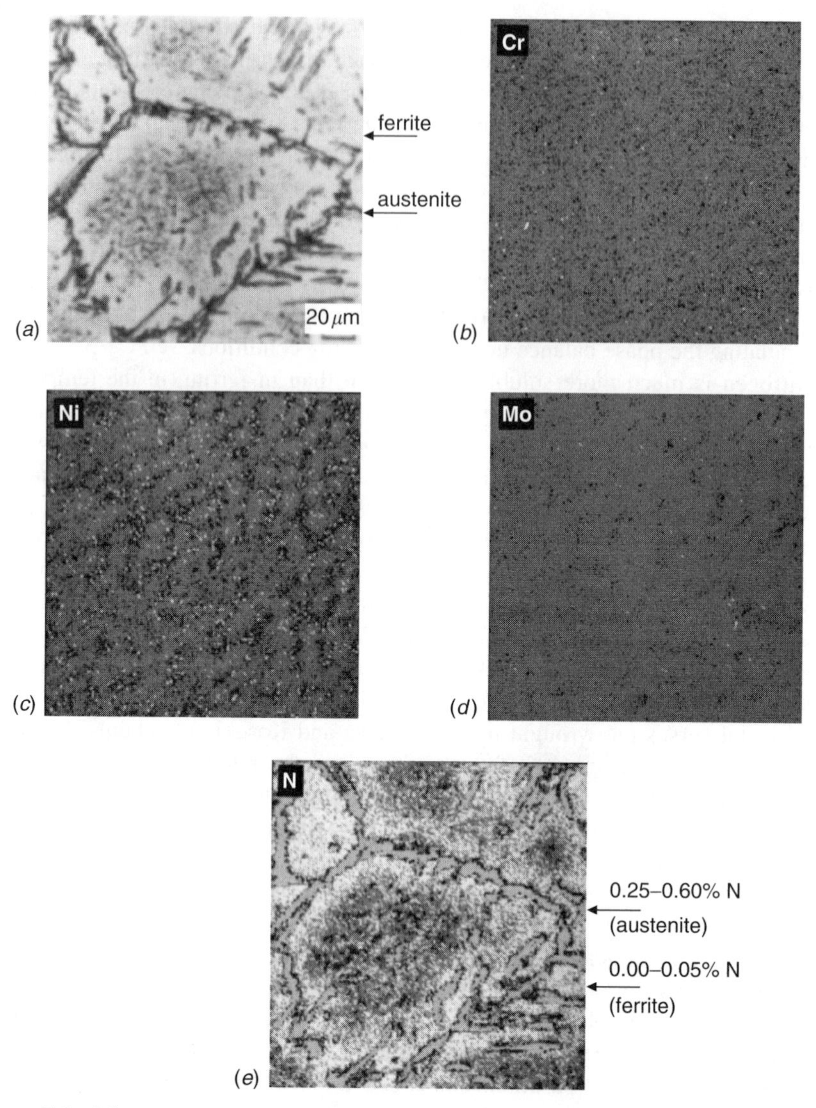

Figure 7.6 Microstructure and alloy element partitioning in low-N 2205 weld fusion zone. Composition: 22% Cr, 6% Ni, 3% Mo, 0.12% N. (*a*) Microstructure of low-nitrogen duplex stainless steel fusion zone. (*b*) Cr distribution; near-white, 20 to 21% Cr; black and dark gray, 21 to 23% Cr (no significant partitioning of Cr between ferrite and austenite is evident; the composition is uniformly 21 to 23% Cr); light gray, 23 to 24% Cr. (*c*) Ni distribution; white, 5.0 to 5.5% Ni; black and dark gray, 5.5 to 7.0% Ni (no significant partitioning of Ni between austenite and ferrite is evident, although some cellular segregation is evident; the composition is entirely between 5.5 and 7.0% Ni); light gray, 7.0 to 8.0% Ni. (*d*) Mo distribution; white, 2.50 to 2.75% Mo; black and dark gray, 2.75 to 3.50% Mo (no significant partitioning of Mo between austenite and ferrite is evident; the composition is entirely between 2.75 and 3.50% Mo); light gray, 3.5 to 4.0% Mo. (*e*) N distribution; near-white, 0.00 to 0.05% N; black and dark gray, 0.05 to 0.25% N; light gray, 0.25 to 0.60% N. (From Ogawa and Koseki [11].)

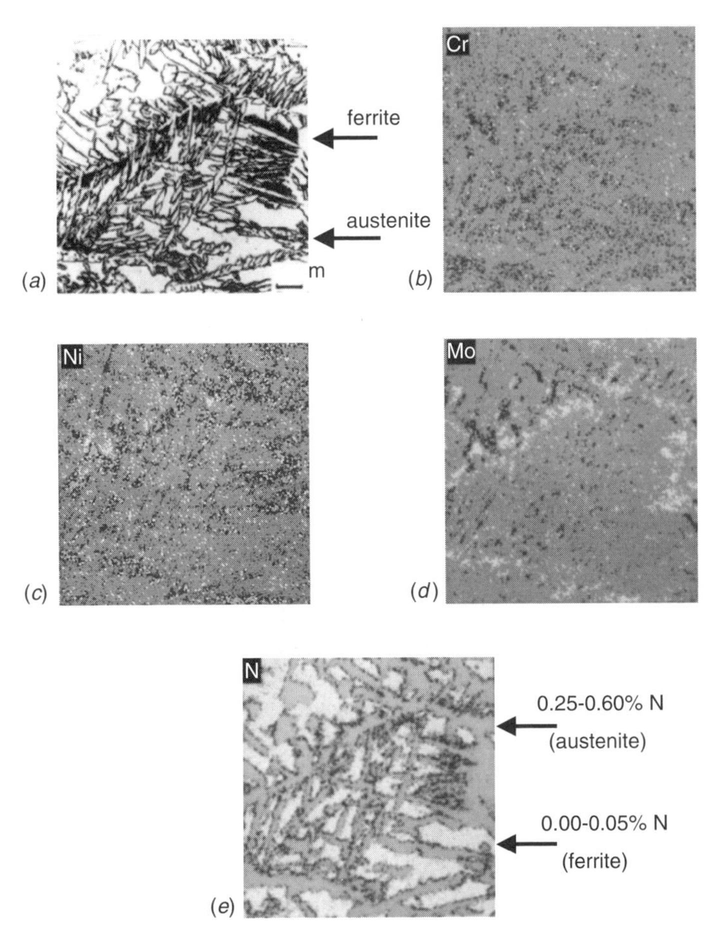

Figure 7.7 Microstructure and alloy element partitioning in high-N 2205 weld fusion zone. Composition: 22% Cr, 6% Ni, 3% Mo, 0.18% N. (*a*) Microstructure. (*b*) Cr distribution; near-white, 20 to 21% Cr; black and dark gray, 21 to 23% Cr (no significant partitioning of Cr between austenite and ferrite is evident; the composition is almost entirely 21 to 23% Cr); light gray, 23 to 24% Cr. (*c*) Ni distribution; near-white, 5.0 to 5.5% Ni; black and dark gray, 5.5 to 7.0% Ni (no significant partitioning of Ni between austenite and ferrite is evident; the composition is almost entirely between 5.5 and 7.0% Ni); light gray, 7.0 to 8.0% Ni. (*d*) Mo distribution; near-white, 2.50 to 2.75% Mo (some partitioning of Mo between austenite and ferrite is evident; 2.50–2.75% Mo appears in the austenite along prior ferrite grain boundaries only); black and dark gray, 2.75 to 3.50% Mo (no significant partitioning of Mo is evident between austenite plates that are not on the prior ferrite grain boundaries and ferrite; except for the austenite at the prior ferrite grain boundaries, the composition is almost entirely between 2.75 and 3.50% Mo); light gray, 3.5 to 4.0% Mo. (*e*) N distribution; near-white, 0.00 to 0.05% N; black and dark gray, 0.05 to 0.25% N; light gray, 0.25 to 0.60% N. (From Ogawa and Koseki [11].)

of Cr, Ni, and Mo progressed very little, as can be seen in Figure 7.7*b–d*. But the partitioning of nitrogen is essentially completed, as can be seen in Figure 7.7*e*. The amount of austenite was greatly increased and distributed intragranularly, so that nitrogen had to diffuse a shorter distance to reach austenite than in the lower N alloy of Figure 7.6.

The weld metal is not the only zone in which nitrogen plays a critical role in development of a proper phase balance. Lippold et al. [16] examined simulated HAZ microstructures and properties of Alloy 2205 (0.13% N) and Alloy 255 (0.17% N) heated to 1300°C (2370°F) (above the ferrite solvus temperature), then cooled at rates varying from 75°C (135°F) per second to 2°C (3.6°F) per second. For both alloys, nitride precipitation was reported to be extensive at a 75°C (135°F)/s cooling rate, moderate at a 50°C (90°F)/s cooling rate, low at a 20°C (36°F)/s cooling rate, and very low at 2°C (3.6°F)/s cooling rate. Ferrite grain growth was noted when the holding time at 1300°C (2370°F) was increased from 1 second to 10 seconds, but this did not affect the extent of nitride precipitation. The ferrite content of these simulated heat-affected zones was reported to decrease from very high values at high cooling rates to values approaching, but not reaching, the base metal ferrite content at low cooling rates.

The effect of higher nitrogen on increasing the rate of austenite formation in the weld metal, as noted previously from the work of Ogawa and Koseki [11], appears to have had some influence on the base metal specifications as well as the design of filler metals. Countless authors in the proceedings of the international conferences on duplex stainless steels [5–10] have referred to Alloy 2205 by the UNS S31803 number. ASTM A240 specifies 0.08 to 0.20% N for UNS S31803. But beginning with ASTM A240/A240M-99a, first published in the 2000 *Annual Book of ASTM Standards*, Volume 01.03, ASTM attached the Alloy 2205 designation to UNS S32205 (0.14 to 0.20% N). This effective elimination of the lower half of what had been considered the nitrogen range for Alloy 2205 has the net effect of improving the weld heat-affected zone properties for this alloy.

7.4.3 Secondary Austenite

Under rapid cooling conditions, such as those experienced in duplex stainless steel weld metals and HAZs, the ferrite–austenite phase balance tends to be higher than equilibrium for a given composition (due to a high effective quench temperature). As a result, reheating of the weldment allows for additional diffusion to take place (a lowering of the effective quench temperature), that can result in further growth of existing austenite or nucleation of new austenite. This second nucleation results in what is termed *secondary austenite* (γ_2). Secondary austenite formation is most prevalent in the weld metal and HAZ during multipass welding and can significantly alter the ferrite–austenite balance of the microstructure. Figure 7.8 shows secondary austenite in an Alloy 2205 simulated HAZ.

Secondary austenite precipitation can markedly improve the toughness of deposits that would otherwise have high ferrite contents. During multipass welding, weld heat inputs and thermal cycles can be controlled to promote extensive secondary austenite formation by optimizing the reheating of previously deposited weld metal. There is

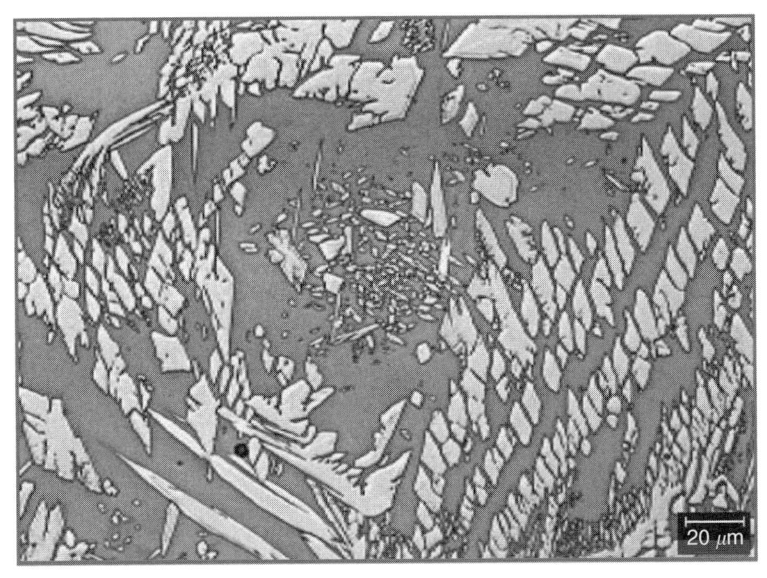

Figure 7.8 Secondary austenite in a simulated HAZ in Alloy 2205. Secondary austenite is the fine light-etching phase in the center of a prior ferrite grain. The thermal treatment was 1350°C (2460°F) for 10 seconds, cooling, then reheating to 1000°C (1830°F) for 10 seconds. (From Ramirez [17].)

some evidence to suggest that the presence of secondary austenite may reduce pitting corrosion resistance, since pit nucleation seems to prefer the secondary austenite–ferrite interface [18]. Nilsson et al. [19] reported about half as much nitrogen in secondary austenite (0.19 to 0.26% N) as in primary austenite (0.43 to 0.54% N). Since nitrogen is a very important alloy element as regards pitting resistance, reduced nitrogen in secondary austenite would seem to explain the reduced pitting corrosion resistance. Nilsson et al. [19] also note that practical pitting problems associated with secondary austenite formation have not been encountered because the reheated zones where secondary austenite tends to form are seldom, if ever, exposed to the surface, and observed volumes of secondary austenite are low (5% or less).

The mechanism of secondary austenite formation has been studied in great detail by Ramirez et al. [17,20]. This work has shown that there are two distinct forms of secondary austenite. One form simply grows off the existing austenite, as shown in Figure 7.9. The other form nucleated within the ferrite phase and is associated with chromium nitrides that have previously precipitated. The cooperative precipitation mechanism for γ_2 growth at the α–γ_1 interface proposed by Ramirez et al. [20] is shown in Figure 7.10. According to this mechanism, Cr_2N first nucleates at the interphase interface, resulting in a local depletion of ferrite-promoting elements Cr and Mo. This local depletion then leads to the nucleation of γ_2 at the interface and subsequent growth. The original Cr_2N then dissolves since it is isolated from the ferrite. This results in the form of γ_2 seen in Figure 7.9.

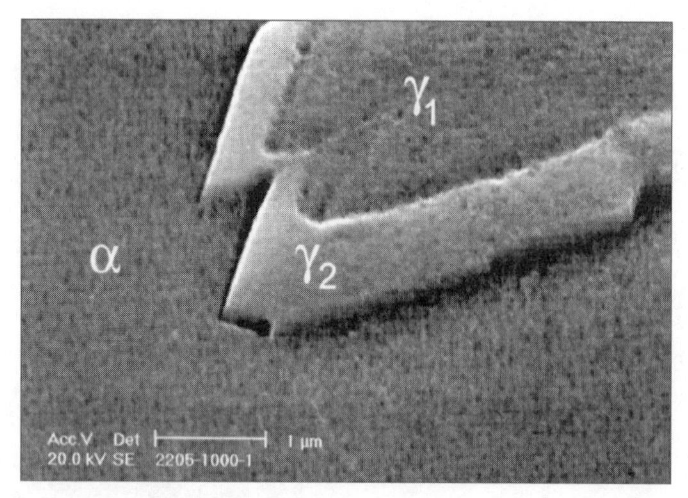

Figure 7.9 Secondary austenite (γ_2) resulting from growth off of primary austenite (γ_1) in Alloy 2205. (From Ramirez et al. [20].)

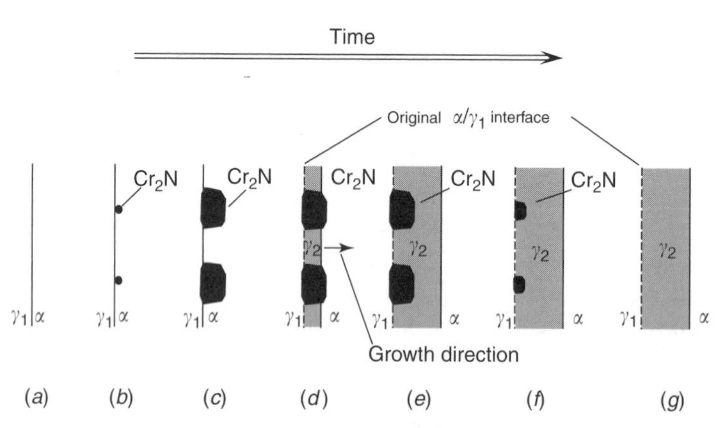

Figure 7.10 Cooperative growth mechanism for formation of secondary austenite. (From Ramirez et al. [20].)

Intragranular Cr_2N is also a favorable nucleation site for γ_2 and significant intragranular precipitation occurs in duplex stainless steels during reheating. Multiple reheating cycles such as those encountered during multipass welding can lead to a very high proportion of austenite in the weld metal, as shown in Figure 7.11.

7.4.4 Heat-Affected Zone

The HAZ thermal cycle in the area just adjacent to the fusion boundary can be divided into three distinct time regions, as regards the temperature cycle. Figure 7.12

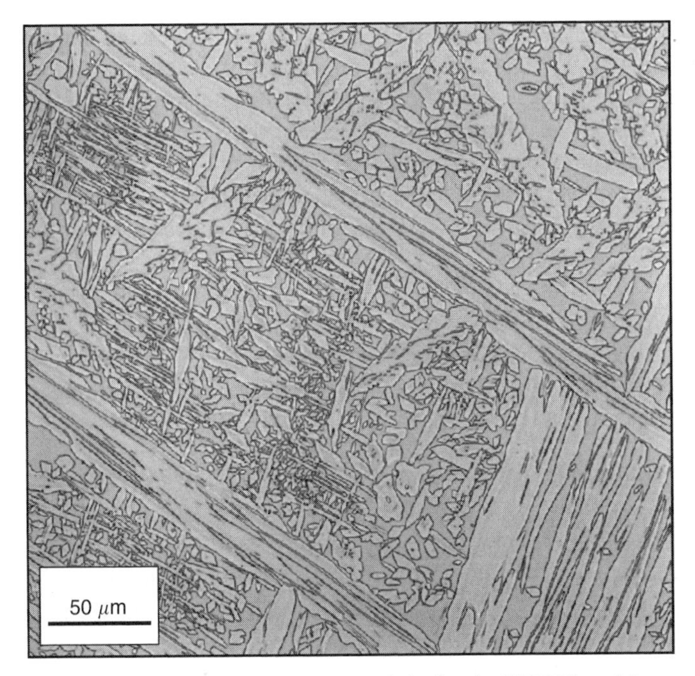

Figure 7.11 Extensive secondary austenite precipitation in ER2209 multipass weld metal deposit.

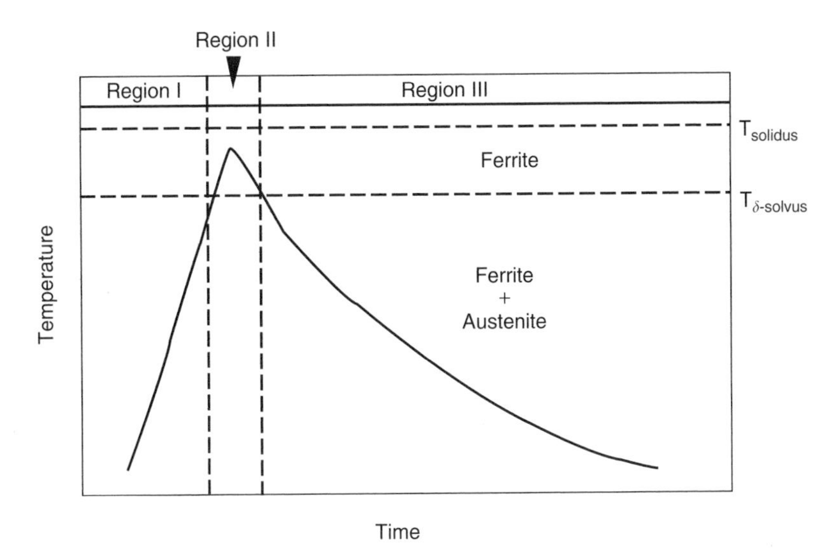

Figure 7.12 HAZ thermal cycle close to the fusion line of a duplex stainless steel with high Cr_{eq}/Ni_{eq}. Region reactions: region 1, austenite transforms to ferrite and precipitate dissolution; region 2, fully ferritic structure and grain growth; region 3, austenite reformation and precipitation reactions (carbides and nitrides). (From Varol et al. [21].)

represents a duplex stainless steel that is fully ferritic at elevated temperature, such as Alloy 2205 with 0.1% N, which was shown previously. In time region 1, the base metal is heated to temperatures approaching the ferrite solvus. In this temperature range austenite begins to transform to ferrite via a diffusion-controlled growth process, until eventually the entire structure is ferritic. Over the same temperature range, most precipitates present in the structure due to prior thermo-mechanical processing will also begin to dissolve. These precipitates consist primarily of carbides and, especially, nitrides.

In time region 2, above the ferrite solvus, ferrite grain growth occurs since there is no second phase (austenite) or precipitates to inhibit growth. This is similar to the rapid grain growth observed in ferritic stainless steels. The lower the ferrite solvus, the more pronounced this grain growth will be. The time above the ferrite solvus, where the microstructure is fully ferritic, is proportional to the amount of grain growth.

Upon cooling below the ferrite solvus, in time region 3, austenite will nucleate and grow and precipitates will reform. The ferrite-to-austenite transformation for a given alloy is controlled by the cooling rate, with higher cooling rates retarding the transformation and resulting in higher HAZ ferrite contents. The cooling rate between 1200 and 800°C (2190 and 1470°F) ($\Delta T_{12\text{-}8}$) is often used to quantify the effect of cooling rate on ferrite content. The degree of precipitation is also a function of cooling rate. At higher cooling rates, which promote ferrite retention, the precipitation of carbides and nitrides in the ferrite phase is much more pronounced.

As mentioned previously, time above the ferrite solvus temperature has a dramatic effect on ferrite grain growth. Above this temperature, there is effectively no impediment to grain growth and grain size increases dramatically. Since ferrite grain size has a strong effect on toughness and ductility, it is generally advisable to minimize the time in the fully ferritic region. This can be done by controlling the composition (alloy selection) or weld heat input and thermal conditions. As the Cr_{eq}/Ni_{eq} decreases, the ferrite solvus increases and the time above this temperature will decrease for a given HAZ thermal cycle (see Figure 7.1). For a fixed Cr_{eq}/Ni_{eq}, decreasing the heat input promotes steeper thermal gradients and minimizes the time in the fully ferritic region.

The ferrite solvus temperature ranges from about 1250 to 1350°C (2280 to 2460°F) for duplex stainless steels, depending on composition. Thus, the width of the fully ferritic region in the HAZ can vary significantly. Alloys such as 2205 (with low nitrogen) and 2304 tend to have a relatively low ferrite solvus, while 2205 with higher nitrogen and the superduplex alloys have ferrite solvus temperatures at or above 1350°C (2460°F), or not at all, as shown in the pseudobinary diagram for Alloy 2205 in Figure 2.8. Recent work by Ramirez [17,20] has shown that achieving fully ferritic simulated HAZ microstructures at 1350°C (2460°F) in all but Alloy 2304 was not possible. Thus, it is expected that the grain coarsened region in the high-nitrogen 2205 and superduplex alloys will be extremely narrow.

An example of the HAZ in Alloy 2205 (low nitrogen) and Alloy 2507 is shown in Figure 7.13. Note that the grained coarsened region in the 2205 HAZ is much

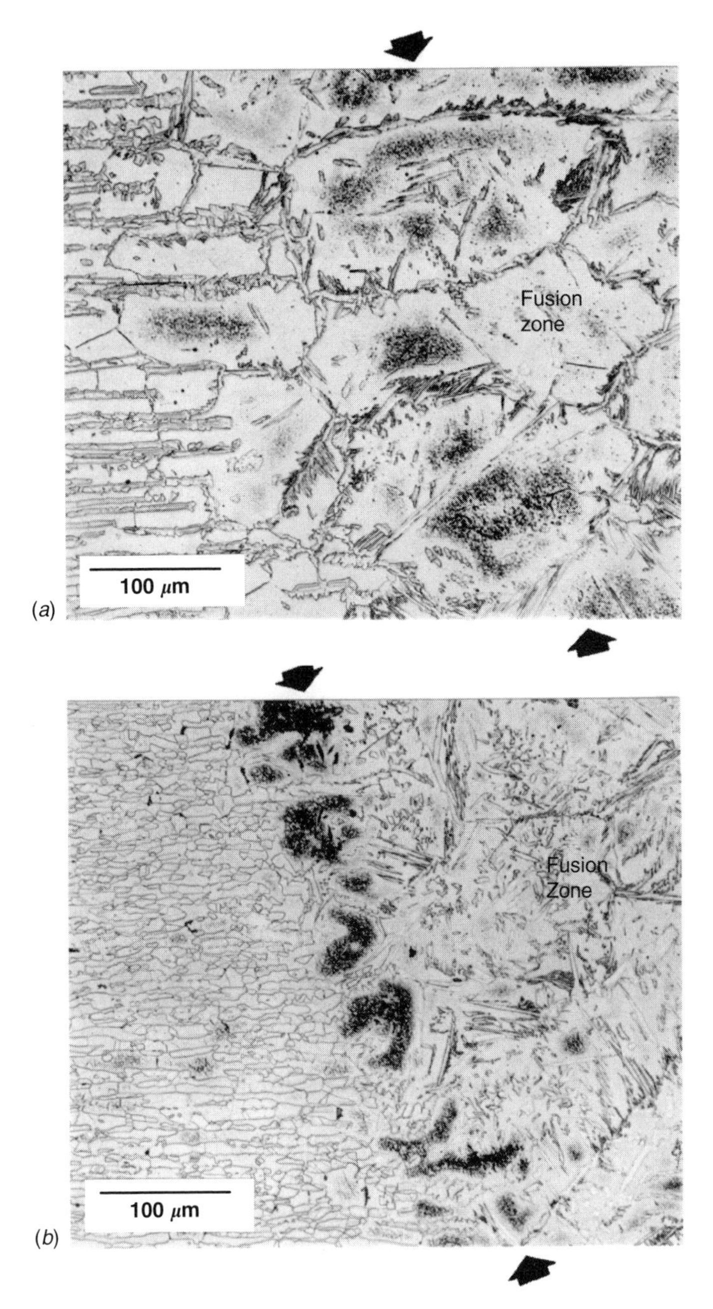

Figure 7.13 HAZ in autogenous GTA welds in duplex stainless steel: (*a*) 2205 (0.12 N); (*b*) 2507.

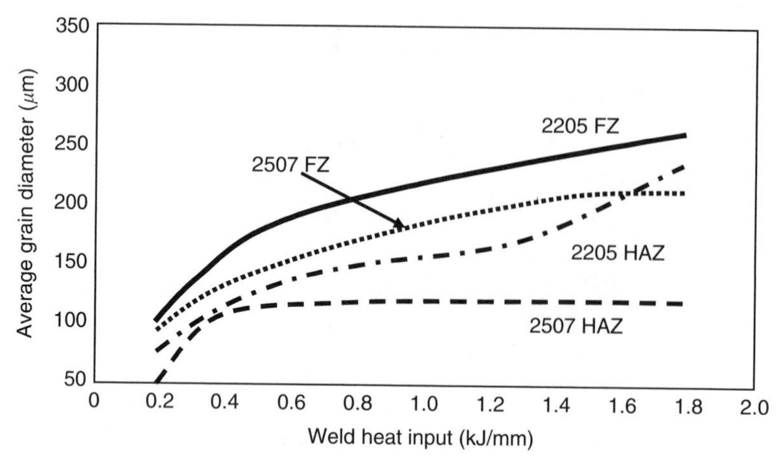

Figure 7.14 Effect of heat input (time above the ferrite solvus) on ferrite grain growth. (Data from Lippold et al. [16].)

wider than that of 2507. This is due to complete transformation to ferrite, as depicted in Figure 7.12. The 2507 HAZ only shows grain coarsening immediately adjacent to the fusion boundary. The grain-coarsened region in this case is essentially one grain diameter wide. The peak temperature in the HAZ may have only just exceeded the ferrite solvus, allowing virtually no grain coarsening. It should also be noted that there is a much more intense precipitation of Cr-rich nitrides in the 2507 HAZ.

Both weld metal and HAZ grain size increase as a function of heat input. Figure 7.14 shows that for autogenous GTA welds in Alloy 2205, both fusion zone and HAZ grain size increase by nearly a factor of 5 over the heat input range from 0.25 to 1.7 kJ/mm. In the superduplex alloy 2507, the fusion zone also exhibits fairly dramatic grain growth as a function of heat input. HAZ grain growth in this alloy is retarded, however, due to the high ferrite solvus that restricts the fully ferritic region to a very narrow band along the fusion boundary.

7.5 CONTROLLING THE FERRITE–AUSTENITE BALANCE

Weld metal ferrite content is controlled through a combination of composition and weld thermal conditions. To overcome the rapid cooling rate effects that can promote higher than optimum ferrite contents, many filler metals contain higher levels of nickel than the base metal. Higher nitrogen in the filler metal and base metal is also useful, although an upper limit for filler metal of approximately 0.30 to 0.35 wt% is imposed by tendencies toward porosity formation. Higher nickel and/or higher nitrogen effectively reduces the Cr_{eq}/Ni_{eq} ratio and allows austenite to form at higher temperature due to a higher ferrite solvus temperature. It also promotes more rapid austenite formation during cooling. Preheat and interpass control can be used, to a limited extent, to slow down the weld cooling rate and allow a more complete ferrite-to-austenite transformation. Postweld heat treatment may also be employed, but precautions against embrittlement, which are discussed more fully later, must be taken.

Once the base metal is chosen, ferrite–austenite balance in the HAZ can only be controlled by adjusting the thermal cycle. It is advisable, early in the design phase of a duplex stainless steel weldment, to select higher-nitrogen base metal where available. For example, in specifying 2205 base metal, it is useful to reference the UNS S32205 composition, rather than the now outdated UNS S31803 composition. Rapid cooling rates as a consequence of low weld heat inputs or in thick sections can result in a highly ferritic HAZ microstructure adjacent to the fusion boundary (Figure 7.13*a*). Preheat, interpass, and weld heat input control measures can all be used to control HAZ microstructure in a given alloy.

7.5.1 Heat Input

As discussed previously, composition is the most effective way to control ferrite content in duplex stainless steel weld metals. Over a reasonable range of heat inputs, weld ferrite contents vary only slightly as shown for autogenous GTA welds in Alloys 2205 and 2507 (Figure 7.15). Note that the difference in Cr_{eq}/Ni_{eq} between the two alloys is a much more dominant effect than heat input. Only at relatively high heat inputs do the resultant cooling rates reduce weld ferrite contents.

7.5.2 Cooling Rate Effects

At the extremes of cooling rate from above the ferrite solvus, the ferrite–austenite balance can be influenced dramatically. Figure 7.16 shows an Alloy 255 microstructure

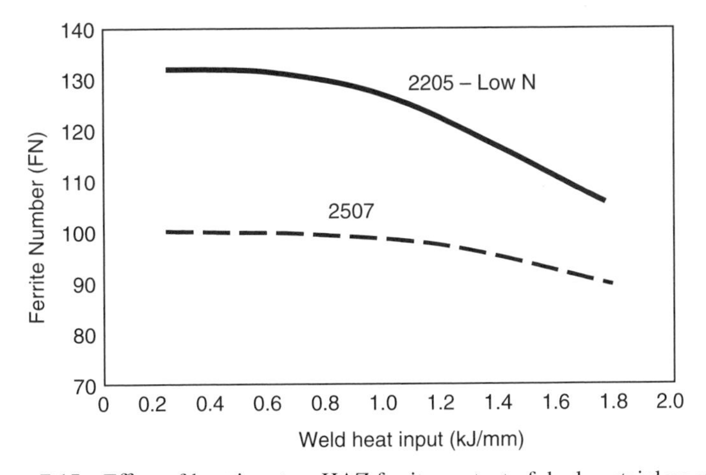

Figure 7.15 Effect of heat input on HAZ ferrite content of duplex stainless steel.

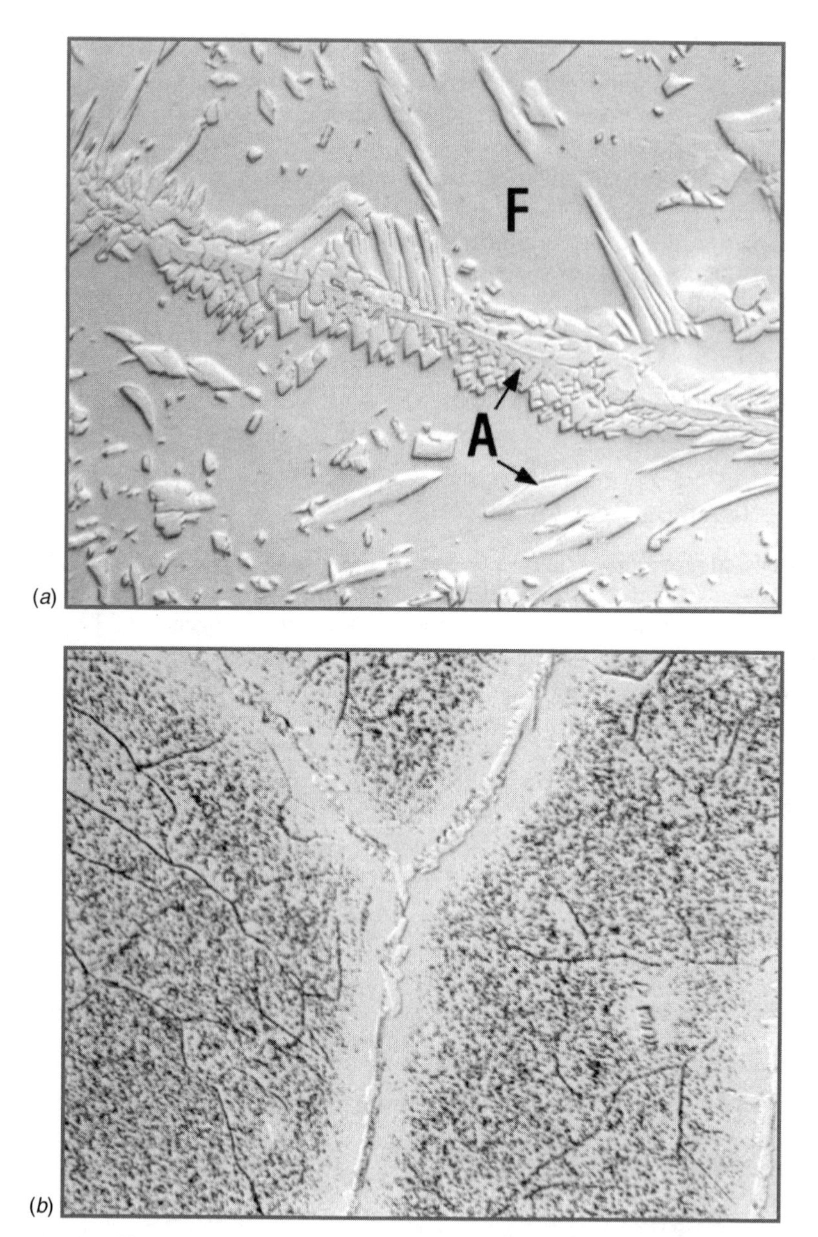

Figure 7.16 Effect of cooling rate from above the ferrite solvus on microstructure of duplex stainless steel: (*a*) normal weld cooling rate; (*b*) quenched from above ferrite solvus.

resulting from normal weld cooling rates with that of the same alloy quenched from 1350°C (2460°F) (above the ferrite solvus). In the quenched microstructure, only a small amount of austenite is observed along the ferrite grain boundaries. There is no austenite within the large ferrite grains that existed above the ferrite solvus temperature. In addition, there is a strong precipitation reaction within the ferrite grains. These are Cr_2N precipitates that result from the rapid supersaturation of nitrogen in the ferrite upon quenching. This situation can exist in resistance welds, laser beam welds, electron-beam welds, and stud welds involving duplex stainless steels. Mechanical properties (ductility and toughness) are severely reduced by the extensive ferrite, and corrosion resistance is diminished by the chromium nitride formation. Precautions should be taken when considering the use of high-energy density processes for welding of the duplex stainless steels.

7.5.3 Ferrite Prediction and Measurement

The same tools for ferrite prediction and measurement used for the austenitic alloys are applicable with the duplex stainless steels. The WRC-1992 diagram allows ferrite prediction based on composition up to 100 FN. Another version of this diagram, plotted on extended axes, is useful when using dissimilar filler metals or when welding duplex alloys to dissimilar base metals. Since ferrite is ferromagnetic, magnetic instruments such as the MagneGage®, Feritscope®, or Inspector Gauge® are often used to measure the ferrite content in terms of Ferrite Number. Metallographic techniques are sometimes used, particularly to determine ferrite content in the HAZ, where the narrowness of this region makes the use of magnetic probes difficult. These techniques are cumbersome and time consuming and provide a value in terms of volume percent ferrite rather than FN. A rough conversion from FN to volume percent for the duplex alloys is 70% [22]. For example, 100 FN is considered to be approximately 70 vol% ferrite. The WRC-1992 diagram can be used to estimate the ferrite content (in terms of FN) and solidification behavior based on composition. This diagram was developed through the efforts of the Welding Research Council (WRC), Subcommittee on Welding Stainless Steels, and the International Institute of Welding (Commission II). Hundreds of welds were analyzed at laboratories around the world to determine Ferrite Number as a function of composition. The WRC-1992 diagram can be used for both austenitic and duplex alloys since the FN range extends to 100 FN. An extended version of this diagram, plotted on axes that start from a value of zero Cr_{eq} and Ni_{eq}, is also available.

The FN range of conventional duplex alloys is superimposed on the diagram in Figure 7.17. Note that ferrite contents can range from 30 to over 100 FN and that this entire range lies within the F-type (fully ferritic) solidification field. Weld metal from flux shielded processes (SMAW, FCAW, and SAW) has been found [23] to be adversely affected with regard to ductility and toughness above about 60 to 70 FN.

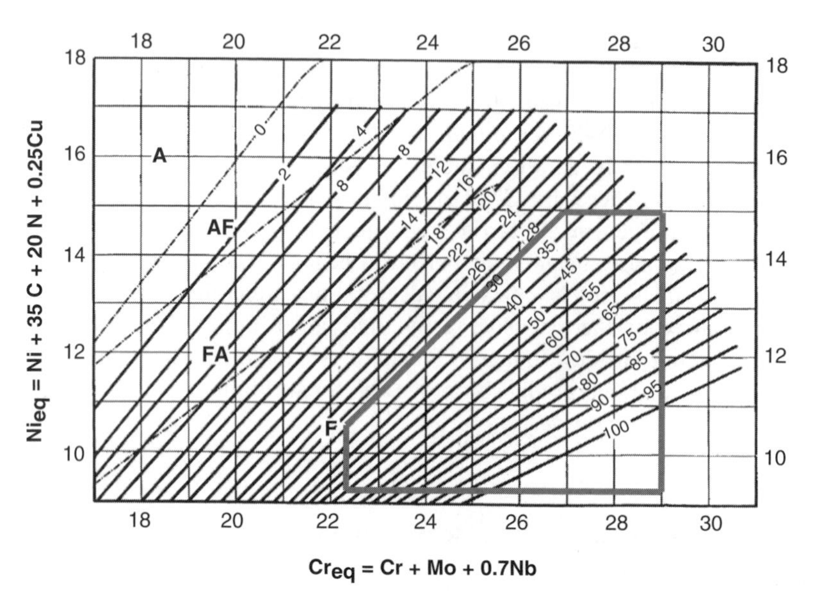

Figure 7.17 WRC-1992 diagram with approximate range of duplex stainless steel compositions indicated.

7.6 WELDABILITY

7.6.1 Weld Solidification Cracking

Weld solidification cracking is a strong function of composition, as shown schematically in Figure 7.18 as a plot of susceptibility versus Cr_{eq}/Ni_{eq} ratio. Essentially all of the duplex alloys solidify in the F mode and have inherently higher solidification cracking susceptibility than the austenitic alloys that solidify in the FA mode. As discussed in Chapter 6, this increase is due to the presence of ferrite–ferrite boundaries at the end of solidification that are more easily wet by liquid films than ferrite–austenite boundaries. The susceptibility to solidification cracking is much lower than austenitic alloys that solidify in the A mode.

In practice, the duplex alloys are generally quite resistant to weld solidification cracking. This is probably due to the fact that they have relatively low impurity levels and the likelihood of continuous grain boundary liquid films is small. Some instances of cracking have been reported under high restraint conditions and some precautions may be warranted in those cases. The use of austenitic filler materials (such as Type 309) will improve cracking resistance but will reduce the corrosion resistance of the weld metal relative to that of the base metal.

7.6.2 Hydrogen-Induced Cracking

Although duplex stainless steels are considered to be resistant to hydrogen-induced cracking, a number of costly failures have occurred in duplex stainless steel weldments

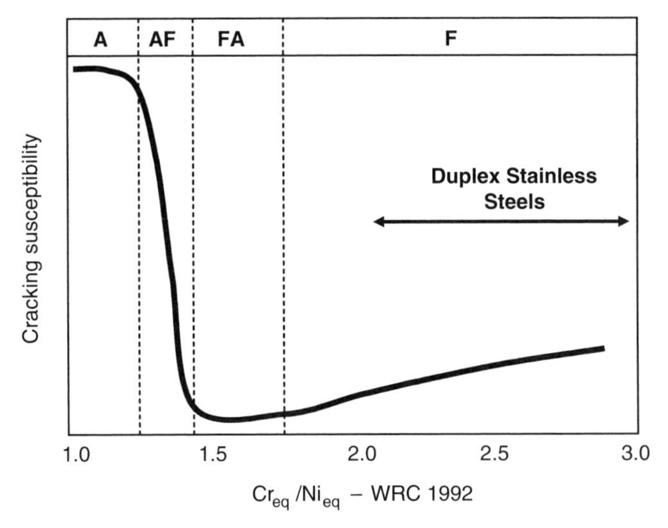

Figure 7.18 Weld solidification cracking susceptibility versus Cr_{eq}/Ni_{eq}. (From Lippold [24].)

due to a combination of high hydrogen contents and poor microstructure control. High FN deposits are susceptible to hydrogen cracking if sufficient hydrogen and stress are present [23,25–27]. Low-hydrogen welding practices are usually recommended for the SMAW and SAW processes, particularly in moist environments. Use of hydrogen additions to argon gas shielding for GMAW or GTAW, sometimes used for welding austenitic stainless steels, is generally inadvisable for duplex stainless steels due to cracking possibilities. The use of 2 to 5% hydrogen in argon shielding gas may be used for GTAW if the weld metal ferrite content can be maintained below 70 FN or solution annealing can be used after welding [28].

The best insurance against hydrogen cracking in duplex alloys is control of the deposit FN. With sufficient austenite in the structure, there are continuous networks of austenite, not only along grain boundaries but also within ferrite grains, effectively limiting hydrogen diffusion. This encapsulation of the ferrite phase effectively limits hydrogen diffusion in the microstructure. The austenite also provides a "sink" for the hydrogen and impedes crack growth.

7.6.3 Intermediate-Temperature Embrittlement

Formation of intermetallic phases is detrimental to ductility, toughness, and corrosion resistance. Due to the high Cr and Mo contents of duplex stainless, precipitation of these phases is likely if they are subjected to the appropriate temperature range (see Figure 7.3). Duplex stainless steels are limited to service temperatures of approximately 280°C (535°F) maximum as a result of this. Welding procedures with proper heat inputs should not result in embrittlement, although the weld metals and HAZs will be more prone to intermetallic formation when exposed to elevated temperature. Karlsson [29] has prepared an extensive review of intermetallic phases in duplex stainless steel weldments. The reader is referred to that source for more detailed consideration than found in this book.

7.6.3.1 Alpha-Prime Embrittlement Figure 7.19 demonstrates the strong influence of microstructure on loss of toughness after exposure of Alloy 2205 base metal and HAZ to the 475°C (885°F) embrittlement range. In this case, the simulated HAZ microstructure contained a higher level of ferrite than the base metal (100 FN versus 70 FN) because it was an older, low-nitrogen heat. This dramatic loss of HAZ toughness within a few minutes of exposure demonstrates why the use of multipass welding and postweld heat treatment can create potential embrittlement problems if weld microstructure (i.e., phase balance) is not properly controlled.

In contrast, the 2507-simulated HAZ of Figure 7.19*c* is much slower to lose toughness due to its higher nitrogen content and resulting lower FN (80 FN versus 100 FN for Alloy 2205). Since the 2507 contains higher Cr, alpha-prime embrittlement of the ferrite should be more rapid. However, the higher austenite content of the 2507-simulated HAZ mitigates the effect of the ferrite embrittlement, resulting in much longer aging times for comparable loss of toughness. Nevertheless, after 100 hours, both of the alloys have severely reduced toughness. Even a balanced HAZ microstructure cannot prevent eventual embrittlement, which demonstrates why temperatures above 280°C (535°F) are not suitable for duplex stainless steel service.

7.6.3.2 Sigma Phase Embrittlement As indicated schematically in Figure 7.3, higher temperatures than those which produce alpha-prime precipitation produce intermetallic compound precipitation. The most important of the intermetallic compounds is sigma phase (approximately FeCr), although chi phase ($Fe_{36}Cr_{12}Mo_{10}$ or approximately Fe_3CrMo) can also be formed. Intermetallic compound formation begins at about 570°C (1000°F) and is most rapid at about 800 to 850°C (1470 to 1560°F). Above about 1000°C (1830°F), depending on alloy composition, these intermetallic compounds dissolve again.

The weld HAZ and reheated weld metal invariably have areas that experience single or multiple exposures to the temperature range 570 to 1000°C (1000 to 1830°F), where sigma phase and other intermetallic phases form. In a 22% Cr duplex stainless, formation of intermetallic phases in the as-welded condition is normally not significant. However, in the 25% Cr duplex stainless steels, the formation of intermetallic phases is more rapid, and total avoidance of such phases in the as-welded condition is unlikely. But, when these phases are confined to small discontinuous zones, they have little effect on weldment properties [29].

Duplex stainless steel weldments, especially those involving castings, may require annealing to optimize their microstructures. The ASTM A240 standard for wrought stainless steels and the ASTM 890 standard for cast duplex stainless steels both require annealing at 1040°C (1900°F) minimum temperature, followed by water quench. However, in selecting an annealing temperature for a weldment, it is necessary to take into account the use of weld filler metals with boosted nickel content. Grobner [30] found that increased nickel in duplex stainless steel alloys increased the

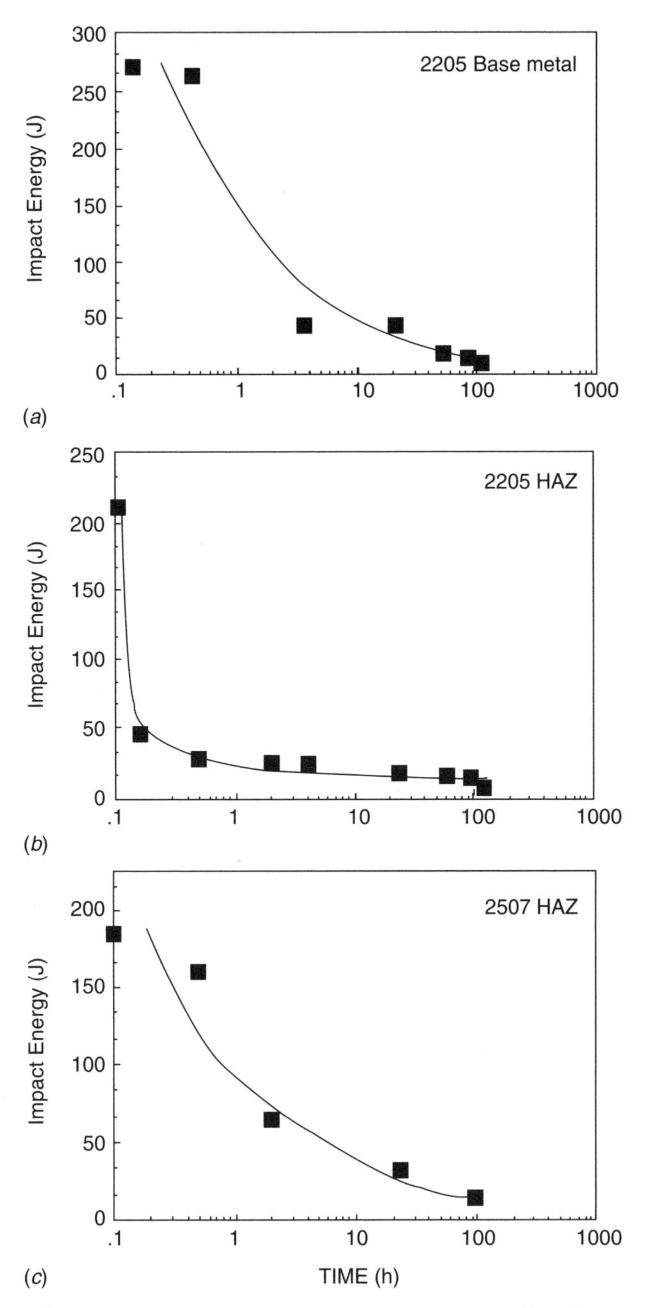

Figure 7.19 Effect of 475°C (885°F) exposure on toughness of duplex stainless steels: (*a*) Alloy 2205 base metal; (*b*) Alloy 2205 HAZ; (*c*) Alloy 2507 HAZ. (From Lippold et al. [16].)

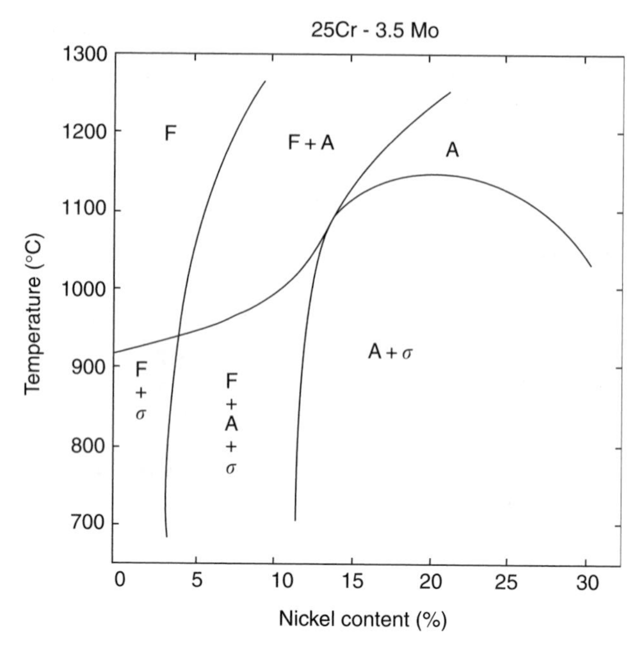

Figure 7.20 Grobner diagram showing sigma stability at increasing temperature with increasing nickel content. (From Grobner [30]. Courtesy of AMAX Metals.)

maximum temperature at which sigma phase is stable. Grobner's diagram is shown in Figure 7.20.

Extensive sigma phase was found [31] in 2205 and 255 duplex stainless steel weld metals containing 8 to 10% Ni after annealing at 1040°C (1900°F) with tensile elongation of 4% or less. Figure 7.21 shows sigma mixed with ferrite and austenite in 8.3% Ni 2205 weld metal after annealing for 4 hours at 1065°C (1950°F). Some sigma was found to persist after annealing at 1095°C (2000°F). Annealing at 1120 to 1150°C (2050 to 2100°F) eliminated sigma.

The higher annealing temperatures necessary to eliminate sigma in weld metal with boosted nickel content are not always desirable from a corrosion resistance point of view because the ferrite retains too much nitrogen in solution (see Figure 7.5). Quenching from 1120 to 1150°C (2050 to 2100°F) can result in some nitride precipitation. To prevent this, a step-annealing procedure was devised, consisting of holding at 1150°C (2100°F) to dissolve all sigma formed during heating, furnace cooling to 1040°C (1900°F), and quenching after 2 hours at 1040°C. Once the sigma formed on heating was dissolved at 1150°C, sigma has to be nucleated at 1040°C, at which temperature sigma nucleation is slow (see Figure 7.3). The quench is then applied before nucleation of sigma can occur, and the resulting weld metal has high ductility and toughness [31]. It should be pointed out that the mechanical strength of duplex stainless steels is very low at the annealing temperature of 1040°C and that precautions should be taken to support the component properly during heat treatment.

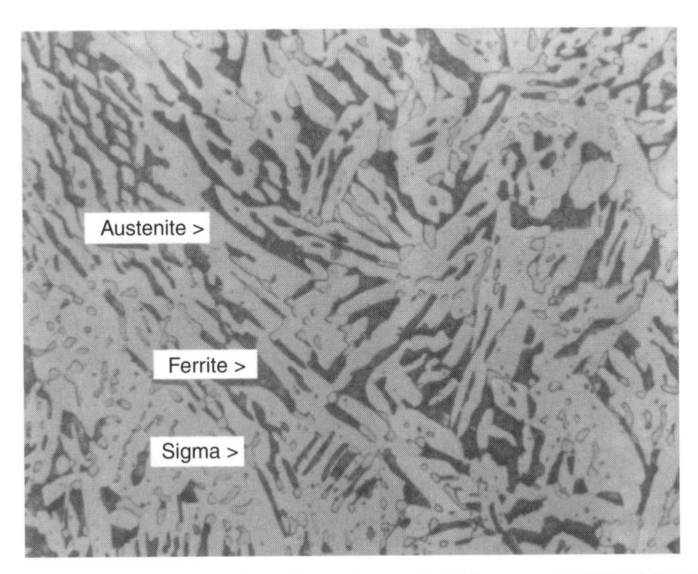

Figure 7.21 8.3% Ni 2205 weld metal annealed 4 hours at 1065°C (1950°F).

TABLE 7.4 Mechanical Properties of Duplex Ferritic–Austenitic Stainless Steel Welding Filler Metals

		Tensile Strength[b]		Yield Strength[b]		Elongation[b]
Class	Source[a]	MPa	ksi	MPa	ksi	(%)
E2209-XX	A5.4	690	100	Not specified		20
ER2209	A5.9	Not specified		Not specified		Not specified
E2209TX-X	A5.22	690	100	Not specified		20
E2552-XX	A5.4	760	110	Not specified		10
E2553-XX	A5.4	760	110	Not specified		15
E2553TX-X	A5.22	760	110	Not specified		15
ER2553	A5.9	Not specified		Not specified		Not specified
E2593-XX	A5.4	760	110	Not specified		15
E2594-XX	A5.4	760	110	Not specified		15

[a]AWS standard.
[b]A single value is a minimum.

7.7 WELD MECHANICAL PROPERTIES

Minimum tensile property requirements for duplex stainless steel weld metal, in the as-welded condition, are given in the AWS A5.4 and A5.22 specifications, but not in the AWS A5.9 specification. These are collected in Table 7.4. In general, matching composition weld metal (albeit with boosted nickel in most cases) provides matching tensile properties. Annealing of weldments in general reduces the yield strength of the joint. In certain cases, annealed weld metal may not meet minimum yield strength requirements for base metals [31].

In general, base metal and welding filler metal specifications do not include requirements for toughness. Some European standards (EN 14532-1) and industrial application standards require 40 J (about 30 ft-lb) of CVN absorbed energy at the minimum use temperature. Therefore, control of toughness can be a key element in the use duplex alloys in certain applications. Because they may have high ferrite contents, the weld metal and HAZ of these alloys may behave similarly to the ferritic stainless steels (i.e., grain size and composition of the ferrite can influence toughness). Both base metals and weldments exhibit a ductile-to-brittle transition temperature because of the high proportion of ferrite. When a toughness requirement is imposed for the HAZ, it is often essential to use base metal of high nitrogen content to meet the requirement.

Weld metal toughness is affected by ferrite content and by oxygen content [32,33]. As shown in Figure 7.22, welding process and filler metal selection can have a dramatic effect on impact toughness. Welding processes that produce weld metal of lower oxygen content, in general, produce weld metal of higher toughness. Low-heat-input SMAW or GTAW with matching filler metals will exhibit lower as-welded toughness due to high ferrite contents. The use of "Ni-boosted" filler metals which lower the deposit FN can result in marked improvements, almost approaching base metal toughness levels, with the low-oxygen processes GMAW and GTAW.

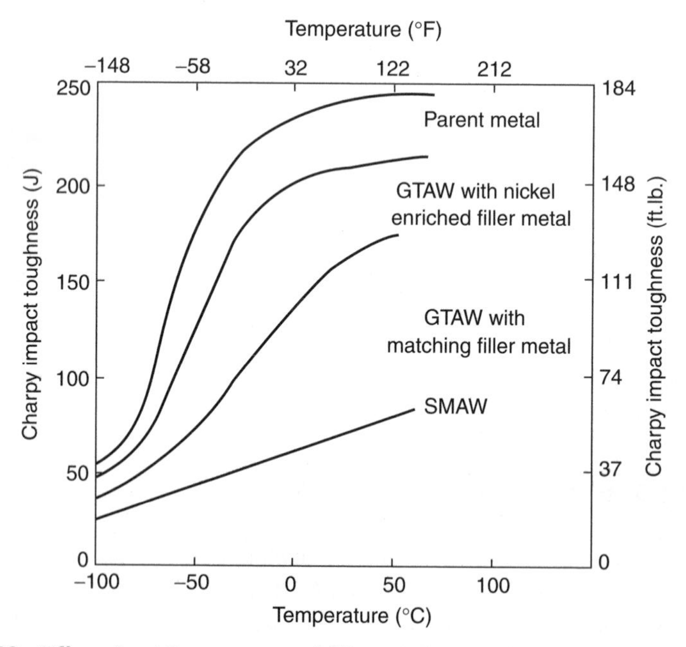

Figure 7.22 Effect of welding process and filler metal composition on impact toughness of 2205 weld metal. (From Larson and Lundqvist [33]. Courtesy of Sandvik Steel.)

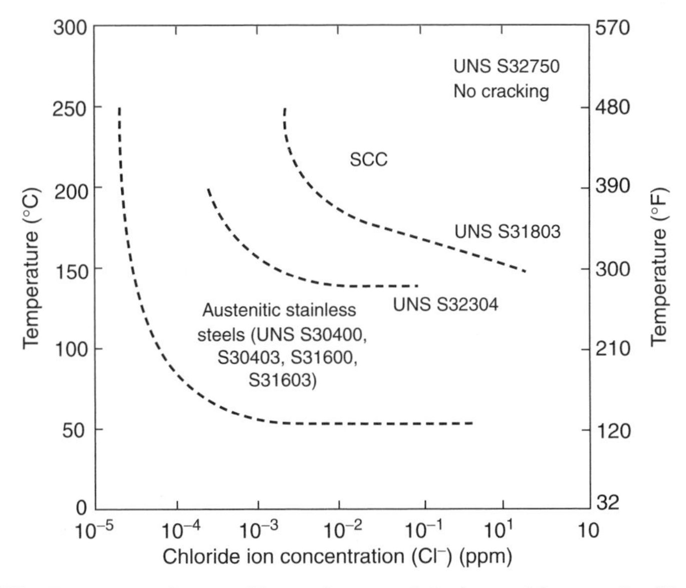

Figure 7.23 Stress corrosion cracking resistance of duplex stainless steels. (From Larson and Lundqvist [33]. Courtesy of Sandvik Steel.)

7.8 CORROSION RESISTANCE

7.8.1 Stress Corrosion Cracking

Duplex stainless steels are most often chosen for their superior corrosion resistance relative to austenitic stainless steels. They are especially resistant to SCC in chloride environments because of their low-Ni and high-Cr contents. Figure 7.23 shows resistance to SCC as a function of temperature and Cl^- concentration. Note that Alloy 2205 (UNS S31803) is superior to Alloy 2304. The superduplex Alloy 2507 (UNS S32750) shows no susceptibility to SCC under the temperature and Cl^- ion concentrations shown in Figure 7.23.

7.8.2 Pitting Corrosion

Pitting corrosion is characterized by localized surface attack that rapidly grows into deep pits that can severely compromise structural integrity. In the stainless steels, resistance to pitting is primarily a function of composition, and several pitting resistance equivalent (PRE) formulas have been developed to represent this effect. The most popular of these is shown below. Note the potent effect of nitrogen on pitting resistance. PRE_N values greater than 40 are loosely taken to identify an alloy as *superduplex*.

$$PRE_N = Cr + 3.3(Mo + 0.5W) + 16N \tag{7.1}$$

Although a secondary effect, microstructure can also influence pitting corrosion resistance, particularly when there are high ferrite contents and extensive nitride precipitation. It is not clear how effectively PRE values can be used with weld metals and HAZs that may have unbalanced microstructures (high ferrite) and/or significant precipitation.

REFERENCES

1. Charles, J. 1991. Super duplex stainless steels: structure and properties, in *Duplex Stainless Steels '91*, Vol. 1, Les Éditions de Physique, Les Ulis, France, pp. 3–48.
2. Roscoe, C. V., and Gradwell, K. J. 1986. The history and development of duplex stainless steels: all that glistens is not gold, in *Duplex Stainless Steels '86*, Nederlands Instituut voor Lastechniek, The Hague, The Netherlands, pp. 126–135.
3. Kearns, W. H., ed. 1982. *Welding Handbook*, 7th ed., Vol. 4, *Metals and Their Weldability*, American Welding Society, Miami, FL, p. 99.
4. CTC. 1983. *Selecting Carpenter Stainless Steels*, Carpenter Technology Corporation, Reading, PA.
5. Lula, R. A., ed. 1983. *Duplex Stainless Steels*, ASM International, Materials Park, OH.
6. NIL. 1986. *Duplex Stainless Steels '86*, Nederlands Instituut voor Lastechniek, The Hague, The Netherlands.
7. Charles, J., and Bernhardsson, S., eds. 1991. *Duplex Stainless Steels '91* (2 volumes), Les Éditions de Physique, Les Ulis, France.
8. *Duplex Stainless Steels '94* (3 volumes), Abington Publishing, Cambridge.
9. *Duplex Stainless Steels '97* (2 volumes), KCI Publishing, Zutphen, The Netherlands.
10. *Duplex America 2000*, KCI Publishing, Zutphen, The Netherlands.
11. Ogawa, T., and Koseki, T. 1989. Effect of composition profiles on metallurgy and corrosion behavior of duplex stainless steel weld metals, *Welding Journal*, 68(5):181s–191s.
12. Vitek, J. M., and David, S. A. 1985. The concept of an effective quench temperature and its use in studying elevated-temperature microstructures, *Metallurgical Transactions A*, 16A(8):1521–1523.
13. ASTM. 1999. *Standard Specification for Castings, Iron–Chromium–Nickel–Molybdenum Corrosion Resistant, Duplex (Austenitic/Ferritic) for General Application*, ASTM A890/A890M-99, American Society for Testing and Materials, West Conshohocken, PA.
14. Brandi, S. D., and Ramirez, A. J. 1997. In *Duplex Stainless Steels '97*, KCI Publishing, Zutphen, The Netherlands, pp. 405–411.
15. Brandi, S. D., and Lippold, J. C. 1997. The corrosion resistance of simulated multipass welds of duplex and superduplex stainless steels, in *Duplex Stainless Steels '97*, KCI Publishing, Zutphen, The Netherlands, pp. 411–418.
16. Lippold, J. C., Varol, I., and Baeslack, W. A., III. 1994. The influence of composition and microstructure on the HAZ toughness of duplex stainless steels at −20°C, *Welding Journal*, 73(4):75s–79s.
17. Ramirez, A. 2001. Ph.D. dissertation, University of São Paulo, São Paulo, Brazil.
18. Serna, C. P., Ramirez, A. J., Alonso-Falleros, N., and Brandi, S. D. 2003. Pitting corrosion resistance of duplex stainless steel multipass welds, in *Proceedings of the 6th*

International Conference on Trends in Welding Research, ASM International, Materials Park, OH, pp. 17–22.

19. Nilsson, J.-O., Jonsson, P., and Wilson, A. 1994. Formation of secondary austenite in super duplex stainless steel weld metal and its dependence on chemical composition, Paper 39 in *Duplex Stainless Steels '94*, Vol. 1, Abington Publishing, Cambridge.
20. Ramirez, A. J., Brandi, S., and Lippold, J. C. 2003. The relationship between chromium nitride and secondary austenite precipitation in duplex stainless steels, *Metallurgical Transactions A*, 34A(8):1575–1597.
21. Varol, J. C., Lippold, J. C., and Baeslack, W. A., III. 1990. Microstructure/property relationships in simulated heat-affected zones in duplex stainless steels, in *Recent Trends in Welding Science and Technology*, S. A. David and J. M. Vitek, eds., ASM International, Materials Park, OH, pp. 757–762.
22. Kotecki, D. J. 1997. Ferrite determination in stainless steel welds: advances since 1974, *Welding Journal*, 76(1):24s–37s.
23. Kotecki, D. J. 1986. Ferrite control in duplex stainless steel weld metal, *Welding Journal*, 65(10):273s–278s.
24. Lippold, J. C. Unpublished Varestraint test data from a variety of austenitic and duplex stainless steels.
25. Fekken, U., van Nassau, L., and Verwey, M. 1986. Hydrogen induced cracking in austenitic/ferritic duplex stainless steel, in *Duplex Stainless Steel '86*, Nederlands Instituut voor Lastechniek, The Hague, The Netherlands, pp. 268–279.
26. Van der Mee, V., Meelker, H., and van der Schelde, R. 1997. How to control hydrogen level in (super) duplex stainless steel weldments using the GTAW or GMAW process, in *Duplex Stainless Steels '97*, Vol. 1, KCI Publishing, Zutphen, The Netherlands, pp. 419–432.
27. Shinozaki, K., Ke, L., and North, T. H. 1992. Hydrogen cracking in duplex stainless steel weld metal, *Welding Journal*, 71(11):387s–396s.
28. Lincoln-Smitweld Laboratory. Private communication with Leo van Nassau.
29. Karlsson, L. 1999. Intermetallic phase precipitation in duplex stainless steels and weld metals: metallurgy, influence on properties and welding aspects, *Welding in the World*, 43(5):20–41. Also available as WRC Bulletin 438, Welding Research Council, formerly of New York, currently of Shaker Heights, OH.
30. Grobner, P. J. 1985. *Phase Relations in High Molybdenum Duplex Stainless Steels and Austenitic Corrosion Resistant Alloys*, Report RP-33-84-01/82-12, AMAX Metals Group, Ann Arbor, MI.
31. Kotecki, D. J. 1989. Heat treatment of duplex stainless steel weld metals, *Welding Journal*, 68(11):431s–441s.
32. Perteneder, E., Tösch, J., Zieerhofer, J., and Rabensteiner, G. 1997. Characteristic profiles of modern filler metals for duplex stainless steel welding, in *Duplex Stainless Steels '97*, Vol. 1, KCI Publishing, Zutphen, The Netherlands, pp. 321–327.
33. Larson, B., and Lundqvist, B. 1987. *Fabricating Ferritic–Austenitic Stainless Steels*, Sandvik Steel Trade Literature, Pamphlet s-51-33-ENG, October; also in *ASM Metals Handbook*, 10th ed., Vol. 6, ASM International, Materials Park, OH.

CHAPTER 8

PRECIPITATION-HARDENING STAINLESS STEELS

The first commercial precipitation-hardening stainless steel is attributed by Funk and Granger [1] to Smith et al. [2]. It was marketed by U.S. Steel Corporation as Stainless W. Its composition (wt%) was about 17% Cr, 7% Ni, and 0.7% Ti, making it most similar to Type 635 (UNS S17600) stainless as we know it today. Titanium was the alloying element responsible for precipitation hardening in this steel. Before long, aluminum, copper, and beryllium were found to produce similar effects [1]. The beryllium-containing alloy (nominally 19% Cr, 10% Ni, 3% Si, 3% Mo, 2% Cu, 0.15% Be), known as V2B, was produced only as castings and appears to be entirely obsolete today, probably due to health concerns regarding beryllium.

For some years it was thought that precipitation could only be obtained in martensitic or ferritic microstructures in stainless steels [1], but 10 years after the first precipitation-hardening stainless steel was introduced, Linnert [3] noted the development of two precipitation-hardening austenitic stainless steels. One, known as 17-10 P, contained nominally 17% Cr, 10% Ni, and 0.25% P. Precipitation hardening was obtained by formation of phosphides, but solidification cracking problems led to the demise of this composition. The other, known as 3311, contained nominally 22% Cr, 23% Ni, and 3.25% Al and is also obsolete today.

Precipitation-hardening (PH) stainless steels are designated as such because they derive a significant part of their strength from precipitation reactions. They are grouped together because they contain elements that will form fine precipitates when heat treated. Hardening of most of these steels is also accomplished by the formation

Welding Metallurgy and Weldability of Stainless Steels, by John C. Lippold and Damian J. Kotecki
ISBN 0-471-47379-0

of martensite. Precipitation-hardening stainless steels, often designated with a suffix PH, can be subclassified by the predominant microstructure that constitutes the alloys. These subgroupings are martensitic, semiaustenitic, and austenitic. Linnert [3] noted that the now-obsolete V2B casting alloy had a duplex ferritic–austenitic mixed microstructure, but this microstructure seems to be unused today in commercially available precipitation-hardening stainless steels.

Precipitation-hardening stainless steels are capable of achieving high tensile strengths, over 1520 MPa (220 ksi) in some grades. In addition to their strength, they have good ductility and toughness if properly heat treated. Service temperatures are typically limited to about 315°C (600°F) for continuous service, although austenitic PH grades may be used at up to 650°C (1200°F) or even higher. Corrosion resistance of most PH stainless steels approaches that of the austenitic Type 304 alloy. Heat treatment conditions affect the corrosion resistance and should be carefully controlled when corrosion is an issue.

Common applications for the martensitic types include valves, gears, splines, and shafts. Semiaustenitic PH alloys find applications in items such as pressure vessels, aircraft frames, and surgical instruments. Jet engine frames, hardware, and turbine blades are constructed from the austenitic types. For example, Space Shuttle engines use large quantities of the austenitic alloy A-286, also known as Type 660.

Despite the combination of high strength and corrosion resistance, the PH stainless steels are not as widely used as the other grades. Many of the martensitic and semi-austenitic alloys are used in demanding aerospace and defense applications. For example, high-pressure gas bottles (or "work" bottles) used to actuate wings, rudders, and other guidance systems on self-guided missiles are built from PH stainless steel. The missile launch tubes on nuclear submarines are also constructed from PH alloys (usually, 17-4PH). The combination of high specific strength and corrosion resistance meets the needs of these applications.

Because many of these alloys are martensitic and precipitation strengthened, they are more difficult to fabricate than other stainless steels and usually require special heat treatments. For example, some of the alloys must be refrigerated following welding and before PWHT to ensure transformation of any retained austenite. Because of their processing requirements and low-volume use, these alloys are generally more expensive to apply than other grades of stainless steels. They are most often selected in applications where high-strength structural steels do not provide adequate corrosion resistance.

8.1 STANDARD ALLOYS AND CONSUMABLES

The PH stainless steels do not use the standard 200, 300, 400 series numbering system of the other stainless steels. Most of the alloys are known by their trade names (e.g., many PH grades were originally developed by Armco and Custom 450 was a Carpenter Technology alloy). They can also be generically identified by their UNS numbers or ASTM types. Table 8.1 lists some PH stainless steels. Additional alloys are given in Appendix 1.

TABLE 8.1 Compositions of Some Precipitation-Hardening Stainless Steels

UNS No.	ASTM Type	Common Name	Nominal Composition (%)[a]										
			C	Mn	P	S	Si	Cr	Ni	Mo	Al	Ti	Other
Martensitic Types													
S13800	XM-13	13-8Mo	0.05	0.20	0.010	0.008	0.10	12.25–13.25	7.50–8.50	2.00–2.50	0.90–1.35	—	N: 0.01
S15500	XM-12	15-5PH	0.07	1.00	0.040	0.030	1.00	14.00–15.50	3.50–5.50	—	—	—	Cu: 2.50–4.50 Nb[b]: 0.15–0.45
S17400	630	17-4PH	0.07	1.00	0.040	0.030	1.00	15.00–17.50	3.00–5.00	—	—	—	Cu: 3.00–5.00 Nb[b]: 0.15–0.45
S17600	635	—	0.08	1.00	0.040	0.030	1.00	16.00–17.50	6.00–7.50	—	0.40	0.40–1.20	—
S45000	XM-25	Custom 450	0.05	1.00	0.030	0.030	1.00	14.00–16.00	5.00–7.00	0.50–1.00	—	—	Cu: 1.25–1.75 Nb[b]: 8 × C − 0.75
S45500	XM-16	Custom 455	0.05	0.50	0.040	0.030	0.50	11.00–12.50	7.50–9.50	0.50	—	0.80–1.40	Cu: 1.50–2.50 Nb[b]: 0.10–0.50
Semi-Austenitic Types													
S15700	632	15-7Mo	0.09	1.00	0.040	0.030	1.00	14.00–16.00	6.50–7.75	2.00–3.00	0.75–1.50	—	—
S17700	631	17-7PH	0.09	1.00	0.040	0.030	1.00	16.00–18.00	6.50–7.75	—	0.75–1.50	—	—
S35000	633	AM350	0.07–0.11	0.50–1.25	0.040	0.030	0.50	16.00–17.00	4.00–5.00	2.50–3.25	—	—	N: 0.07–0.13
S35500	634	AM355	0.10–0.15	0.50–1.25	0.040	0.030	0.50	15.00–16.00	4.00–5.00	2.50–3.25	—	—	N: 0.07–0.13
Austenitic Types													
S66220	662	Discaloy	0.08	1.50	0.040	0.030	1.00	12.00–15.00	24.00–28.00	2.50–3.50	0.35	1.55–2.00	B: 0.0010–0.010
S66286	660	A-286	0.08	2.00	0.040	0.030	1.00	13.50–16.00	24.00–27.00	1.00–1.50	0.35	1.00–1.50	V: 0.10–0.50 B: 0.0010–0.010
—	—	JBK-75[c]	0.01–0.03	0.20	0.010	0.006	0.10	13.5–16.0	29.0–31.0	1.0–1.5	0.15–0.35	2.0–2.3	V: 0.1–0.5; B: 0.0020 O: 0.005; N: 0.010

[a]A single value is a maximum unless stated otherwise.
[b]Nb includes incidental Ta.
[c]No UNS number or ASTM specification is known for this alloy, but it is found in a company specification [5,6].

There is some doubt as to whether or not the alloys S35000 and S35500 should be considered as precipitation hardening. Both are listed in ASTM A693, and S35500 is listed in ASTM A564, as precipitation hardening, but neither contains copper, niobium, aluminum, or titanium. They apparently obtain some hardening from nitride precipitation, but their strength does not increase due to precipitation.

Most of the alloys are martensitic, or a mixture of martensite, ferrite, and small amounts of austenite. The austenitic Type 660 (A-286) and Type 662 alloys are strengthened by a gamma-prime precipitate (similar to Ni-base superalloys) and are often called *iron-base superalloys*. Major solidification cracking problems have accompanied welding of the austenitic Types 660 and 662. Research led to the development of a modified version of Type 660 known as JBK-75, which appears to have significantly greater resistance to solidification cracking [4]. Although not included in the ASTM standards, JBK-75 found application in superconducting magnet structures for the fusion energy program, with material procurement according to a company specification [5,6].

Except for the 17-4PH alloy (Type 630), filler metals of matching composition to the base metals of Table 8.1 are not classified by AWS. Most of them are produced only as solid rods or wires for GTA and/or GMA welding and are classified according to the AMS (Aerospace Materials Specifications) published by the Society of Automotive Engineers (SAE). Table 8.2 lists compositions of a number of these and the AMS specification in which each can be found. Note that the filler metal, in several instances, has a slightly different UNS number than the corresponding base metal, indicating, typically, a slightly different, often more restrictive chemical composition for the filler metal.

8.2 PHYSICAL AND MECHANICAL METALLURGY

Figure 8.1 shows a pseudobinary diagram for 16% Cr–1% Ti alloys with varying nickel content, constructed using ThermoCalc®. For alloys of the composition range for martensitic and semi-austenitic precipitation-hardening stainless steels, it can be seen that the sequence of phase appearance during cooling from the liquid state is as follows:

$$L \rightarrow L + F_P \rightarrow F_P \rightarrow F_P + A \rightarrow F_P + A + M \rightarrow F_P + M$$

On the other hand, the austenitic precipitation-hardening stainless steels all solidify as 100% austenite, due to their very high nickel content—25% or more, as can also be seen in Figure 8.1. For the austenitic PH stainless steels, the sequence of phase appearance during cooling from the liquid state is

$$L \rightarrow L + A \rightarrow A$$

If aluminum is substituted for titanium, the pseudobinary diagram, at elevated temperatures, is quite similar, and the sequence of phases present is essentially the same.

TABLE 8.2 Compositions of Some Precipitation-Hardening Stainless Steel Welding Filler Metals

UNS No.	AMS Spec.	Common Name	Nominal Composition (%)[a]										
			C	Mn	P	S	Si	Cr	Ni	Mo	Al	Ti	Other
							Martensitic Types						
S13889	5840	13-8Mo	0.05	0.10	0.008	0.010	0.10	12.25–13.25	7.50–8.50	2.00–2.50	0.90–1.35	—	N: 0.01; H: 0.0025
S15500	5826	15-5PH	0.07	1.00	0.040	0.030	1.00	14.00–15.50	3.50–5.50	—	—	—	Cu: 2.50–4.50 Nb[b]: 0.15–0.45
S17480	5825 (AWS A5.9 ER630)	17-4PH	0.05	0.25–0.75	0.04	0.03	0.75	16.00–16.75	4.50–5.00	—	—	—	Cu: 3.25–4.00 Nb[b]: 0.15–0.30
W37410	AWS A5.4	E630-XX	0.05	0.25–0.75	0.04	0.03	0.75	16.00–16.75	4.5–5.0	0.75	—	—	Cu: 3.25–4.00 Nb[b]: 0.15–0.30
S45000	5763	Custom 450	0.05	1.00	0.030	0.030	1.00	14.00–16.00	5.00–7.00	0.50–1.00	—	—	Cu: 1.25–1.75 Nb[b]: 8 × C−0.75
S45500	5617 Grade 2	Custom 455	0.010	0.50	0.010	0.010	0.20	11.00–12.50	7.50–9.50	0.50	—	1.00–1.35	Cu: 1.50–2.50; N: 0.010
							Semi-Austenitic Types						
S15789	5812	15-7Mo	0.09	1.00	0.010	0.010	0.50	14.00–15.25	6.50–7.75	2.00–2.75	0.75–1.25	—	H: 0.0025; O: 0.005
S17780	5824	17-7PH	0.09	1.00	0.025	0.025	0.50	16.00–17.25	6.50–7.75	—	0.75–1.25	—	—
S35080	5774	AM350	0.08–0.12	0.50–1.25	0.040	0.030	0.50	16.0–17.0	4.00–5.00	2.50–3.25	—	—	N: 0.07–0.13
S35580	5780	AM355	0.10–0.15	0.50–1.25	0.04	0.03	0.50	15.0–16.0	4.00–5.00	2.50–3.25	—	—	N: 0.07–0.13; Cu: 0.50
							Austenitic Types						
S66286	5805	A-286	0.04	0.35	0.01	0.005	0.25	13.50–16.00	24.00–27.00	1.00–1.50	0.35	1.90–2.30	V: 0.10–0.50; B: 0.003–0.005; O: 0.005; N: 0.005; H: 0.0005
—	5811A	JBK-75[c]	0.01–0.03	0.20	0.010	0.006	0.10	13.5–16.0	29.0–31.0	1.0–1.5	0.15–0.35	2.0–2.3	V: 0.1–0.5; B: 0.0020; O: 0.005; N: 0.010

[a]A single value is a maximum unless stated otherwise.
[b]Nb includes incidental Ta.
[c]No UNS number or ASTM specification is known for this alloy, but it is found in a company specification [5,6].

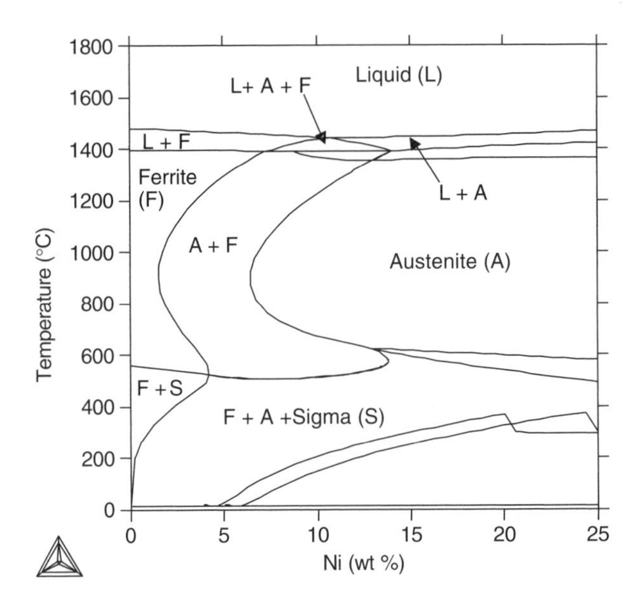

Figure 8.1 Pseudobinary Thermocalc® diagram for Fe–0.05% C–16% Cr–1% Ti–0.3% Mn–0.2% Si, variable 0 to 25% Ni.

At lower temperatures where various precipitation reactions can take place, the equilibrium diagram becomes rather complex, but also becomes irrelevant because equilibrium is not reached. Solution heat treatment is used in the single-phase austenite range for the austenitic PH stainless steels, or in the predominately austenitic range with possibly a little ferrite also present for the martensitic or semi-austenitic PH stainless steels, to dissolve all precipitates. This is followed by cooling at a sufficiently rapid rate to retain the solutes in solution when ambient temperature is reached. Heat treatment is required at an appropriate temperature and time for optimum precipitation hardening to occur.

8.2.1 Martensitic Precipitation-Hardening Stainless Steels

Brooks and Garrison [7] showed that three martensitic PH stainless steels (13-8Mo, 15-5 PH, and Custom 450) solidify as essentially 100% ferrite. During cooling, the ferrite transforms almost entirely to austenite. Whatever ferrite persists will remain to ambient temperatures, with no transformation at all. Pollard [8] states that ferrite stringers are generally found in wrought 17-4PH, sometimes found in 15-5PH and Custom 455, and not found in 13-8Mo and Custom 455. On the other hand, Hochanadel et al. [9] found ferrite in as-cast 13-8Mo. Linnert [3] illustrates ferrite stringers in both annealed and fully hardened 17-4PH (Figure 8.2). It seems clear that ferrite can be present at ambient temperatures in martensitic PH stainless steels. Wrought steel will have the ferrite aligned in directions parallel to the rolling direction (parallel to the

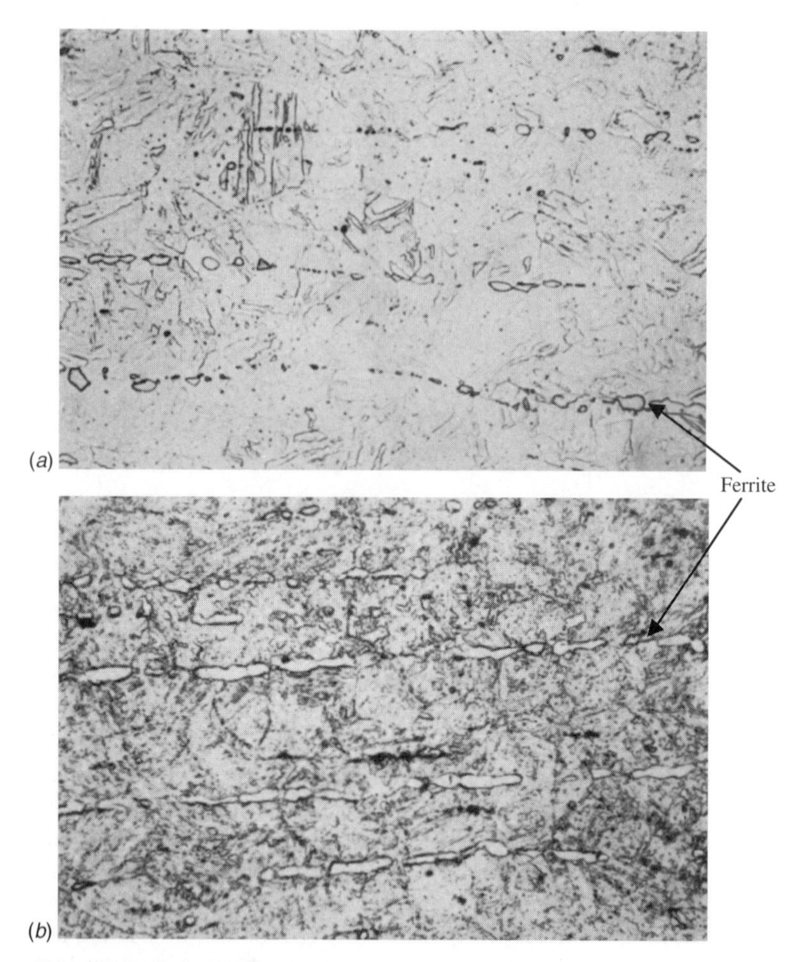

Figure 8.2 Hot-rolled 17-4PH (UNS 17400): (*a*) annealed at 1040°C (1900°F), ferrite in martensite matrix; (*b*) annealed as above, aged at 482°C (900°F), ferrite in precipitation-hardened matrix. (From Linnert [3].)

surface). Funk and Granger [1] propose that this distribution of ferrite, if present, can result in markedly lower strength and ductility in the through-thickness direction compared to properties in directions parallel to the rolled surface.

A homogenization treatment can be used as part of the annealing process when the steel is in the austenite phase. This homogenization, accomplished by holding at temperatures above about 850°C (1560°F), tends to reduce the ferrite content [9]. The austenite subsequently transforms to martensite on cooling to temperatures approaching ambient. The M_S temperature for both 17-4PH and 15-5PH is reported as approximately 132°C (270°F), and the M_F temperature is reported as approximately 32°C (90°F) [10,11]. However, a small amount of austenite may be retained to ambient temperature or even below. Hochanadel et al. [9] used Mössbauer spectroscopy to identify and quantify retained austenite in 13-8Mo. They found on the order of 1%, or less, austenite in solution annealed material, and refrigeration to −78°C (−108°F) essentially removed it by transformation to martensite. So although it is common to think of the martensitic PH stainless steels as single-treatment martensitic steels [3], refrigeration may also be useful.

Once a martensitic PH stainless steel has been transformed to largely martensite, an aging treatment can be performed to precipitate a hardening phase and increase the strength above that of the freshly formed martensite. The aging temperature for maximum strength depends on the specific precipitate being produced. The aged condition is generally indicated by the letter "H" followed by the temperature in degrees Fahrenheit at which the aging took place (e.g., H900 or H1150). Table 8.3 lists the mechanical properties for the PH stainless steels as a function of specific heat treatment.

The precipitates that harden these steels depend on the alloying elements present. There has been some controversy over the specific phases precipitated. A copper-rich precipitate forms in 17-4PH [3] and in 15-5PH. Anthony [14] reported also finding chromium-rich ferrite (alpha prime) and carbides in 17-4PH. Hochanadel et al. [9] identified β-NiAl as the main strengthening precipitate in 13-8Mo, although carbides and Laves phase were also present. Ni_3Ti (also known as gamma prime) appears to be the main strengthening precipitate in the titanium-bearing alloys Custom 455 and Type 635 [8]. A Laves phase containing Fe, Mo, and Nb is considered to be the hardening precipitate in Custom 450 [8].

During the aging treatment, at least with some alloys, austenite reversion can take place. Hochanadel et al. [9] noted austenite reversion during aging of 13-8Mo. An increase of about 1% austenite occurred during aging at 565°C (1050°F) and lower. However, aging at 595°C (1100°F) or 621°C (1150°F) resulted in about 15% austenite after cooling to room temperature. This reverted austenite, distinct from the retained austenite, was found to be enriched in Ni and Mn, and to be very resistant to martensite transformation. Likewise, Anthony [14] noted an increase in austenite content of 17-4PH stainless steel from about 1.5% in the solution annealed condition or aged at 480°C (900°F), to 5.5% after aging at 595°C (1100°F). Subsequent exposure to lower temperatures such as 425 to 480°C (800 to 900°F) for a few thousands of hours resulted in as much as 18% austenite. He also noted the appearance of alpha prime during this prolonged aging, which is consistent with the 475°C (885°F) embrittlement discussed in Chapter 5.

Table 8.3 shows the effects of overaging on the mechanical properties of the martensitic PH stainless steels. The peak strength is achieved at the lowest aging treatment temperature in this table. Higher temperatures still produce strengthening as compared to the solution-annealed condition, but not as much strengthening. Higher aging temperatures, however, are accompanied by higher ductility. Lower strength with increasing aging temperature was thought to be due to precipitate coarsening and loss of precipitate coherency with the matrix. However, Hochanadel et al. [9] dispute this, finding that the β-NiAl precipitates in 13-8Mo remain coherent even after aging at 620°C (1150°F). Instead, the reduction of strength is attributed to formation of

TABLE 8.3 Mechanical Properties of Wrought Precipitation-Hardening Stainless Steels[a]

Type	Precipitation Treatment	Tensile Strength MPa	Tensile Strength ksi	Yield Strength MPa	Yield Strength ksi	Elongation (%)	Hardness (Rockwell C)	Charpy (ft-lbf)	V-Notch (J)
	Martensitic Precipitation-Hardening Stainless Steels								
13-8Mo	Solution at 927°C (1700°F)	—	—	—	—	—	38 max.	—	—
	510°C (950°F), 4 h	1515	220	1410	205	10	45	—	—
	538°C (1000°F), 4 h	1380	200	1310	190	10	43	—	—
15-5PH and 17-4PH	Solution at 1038°C (1900°F)	—	—	—	—	—	38 max.	—	—
	482°C (900°F), 1 h	1310	190	1170	170	8	40–48	—	—
	496°C (925°F), 4 h	1170	170	1070	155	8	38–46	—	—
	552°C (1025°F), 4 h	1070	155	1000	145	8	35–43	10	14
	579°C (1075°F), 4 h	1000	145	860	125	9	29–38	15	20
	593°C (1100°F), 4 h	965	140	790	115	10	29–38	15	20
	621°C (1150°F), 4 h	930	135	725	105	10	26–36	25	34
	760°C (1400°F), 2 h, air cool then 621°C (1150°F), 4 h	790	115	515	75	11	24–34	55	75
635	Solution at 1038°C (1900°F)	825	120	515	75	5	32 max.	—	—
	510°C (950°F), 30 min	1310	190	1170	170	8	39	—	—
	540°C (1000°F), 30 min	1240	180	1105	160	8	38	—	—
	565°C (1050°F), 30 min	1170	170	1035	150	8	36	—	—
Custom 450	Solution at 1038°C (1900°F)	1140	165	1035	150	4	33 max.	—	—
	482°C (900°F), 4 h	1240	180	1170	170	5	40	—	—
	538°C (1000°F), 4 h	1105	160	1035	150	7	36	—	—
	621°C (1150°F), 4 h	860	125	515	75	10	26	—	—
Custom 455	Solution at 829°C (1525°F)	1205	175	1105	160	3	36 max.S	—	—
	510°C (950°F), 4 h	1525	222	1410	205	3	44	—	—
	Semi-Austenitic Precipitation-Hardening Stainless Steels								
15-7Mo	Solution at 1065°C (1950°F)	1035	150	450	65	25	B100 max.	—	—
	760°C (1400°F), 90 min; cool to 15°C (55°F) and hold 30 min; 566°C (1050°F), 90 min; air cool	1310	190	1170	170	4	40	—	—
	954°C (1750°F), 10 min; cool rapidly to ambient; cool to −73°C (−100°F) within 24 h, hold 8 h min. warm to ambient; 510°C (950°F), 1h	1550	225	1380	200	4	45	—	—

17-7PH	Solution at 1065°C (1950°F)	1035	150	450	65	20	B92 max.	—	—
	760°C (1400°F), 90 min; cool to 15°C (55°F) and hold 30 min; 566°C (1050°F), 90 min; air cool	1170	170	965	140	7	38	—	—
	954°C (1750°F), 10 min; cool rapidly to ambient; cool to −73°C (−100°F) within 24 h, hold 8h min.; warm to ambient; 510°C (950°F), 1h	1380	200	1240	180	6	43	—	—
633 (AM350)	Solution at 930°C (1710°F), quench; −73°C (−100°F), 3 h	1380	200	585	85	12	30 max.	—	—
	455°C (850°F), 3 h	1275	185	1035	150	8	42	—	—
	540°C (1000°F), 3 h	1140	165	1000	145	8	36	—	—
634 (AM355)	1038°C (1900°F), quench; −73°C (−100°F), 3 h	—	—	—	—	—	40 max.	—	—
	954°C (1750°F), 10 to 60 min; water quench; cool to −73°C (−100°F), 3 h; 455°C (850°F), 3 h	1310	190	1140	165	10	—	—	—
	954°C (1750°F), 10 to 60 min; water quench; cool to −73°C (−100°F), 3 h; 538°C (1000°F), 3 h	1170	170	1035	150	12	37	—	—
	Austenitic Precipitation-Hardening Stainless Steels								
662	Solution treat at 955 to 1040°C (1750 to 1900°F), 1 h; quench; age at 675 to 760°C (1250 to 1400°F), 5 h min.; slow cool to 650°C (1200°F), hold 20 h; air cool	895	130	585	85	15	—	—	—
A-286	Solution treat at 900°C (1650°F) for 2 h; quench; age at 705 to 760°C (1300 to 1400°F) for 16 h; air cool	895	130	585	85	15	—	—	—

Source: Data from ASTM [12,13].

[a]All values are minimum unless stated otherwise. Requirements are for 12.7mm ($\frac{1}{2}$ in.) thickness, or greatest thickness if that thickness is not made. Requirements for other thicknesses may vary slightly.

reverted austenite in this alloy. The austenite has much lower strength than that of martensite.

8.2.2 Semi-Austenitic Precipitation-Hardening Stainless Steels

As with the martensitic PH stainless steels, solidification of the semi-austenitic grades is as primary ferrite [3,8]. Transformation to austenite occurs at high temperatures, but some ferrite (5 to 15%) is retained to ambient temperatures in 17-7PH, 15-7Mo, AM350, and AM355 [8]. Figure 8.3 shows the microstructure of hot-rolled 17-7PH stainless steel in various conditions of heat treatment, and ferrite stringers are visible in all cases. Solution annealing at 1040 to 1065°C (1900 to 1950°F) causes some of the ferrite to transform to austenite and homogenizes the austenite. Cooling to ambient temperature from the solution anneal leaves the austenite largely untransformed. This is commonly referred to as *condition A*. In condition A, which is the condition in which the alloys are commonly supplied by the steel mill, the steel is relatively soft and ductile (see Table 8.3). The steel can then be machined, cold-formed, or welded with ease. Cold forming may induce some martensite transformation.

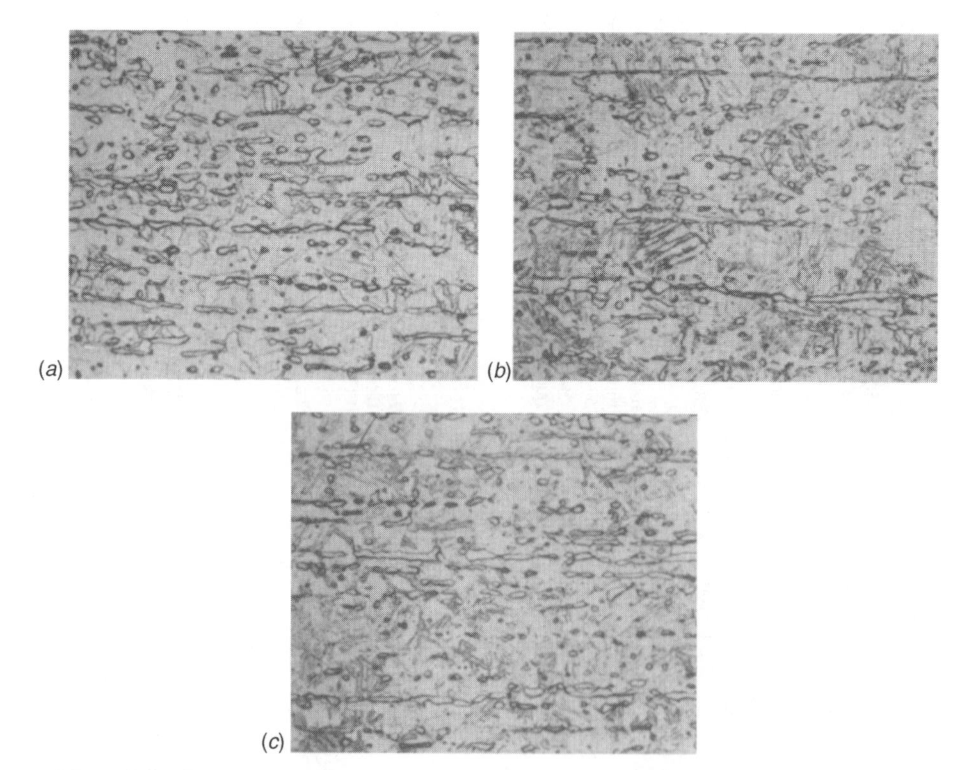

Figure 8.3 Hot-rolled 17-7PH (UNS S17700) semi-austenitic stainless steel: (*a*) annealed at 955°C (1750°F), ferrite in austenite matrix with some martensite; (*b*) annealed, then refrigerated to −73°C (−100°F), ferrite in martensite matrix; (*c*) annealed, refrigerated, then aged at 510°C (950°F), ferrite in precipitation-hardened martensite. (From Linnert [3].)

For the semi-austenitic PH stainless steels, the hardening treatment is more complex than for the martensitic PH stainless steels. There are normally three steps in the heat treatment, as can be seen in Table 8.3. First, the steel is heated to a temperature to *condition* the austenite. Conditioning the austenite consists of time at a temperature where some of the carbon in solution at condition A will precipitate as carbides, which results in a higher M_S temperature [8]. For example, if 17-7PH, initially in condition A, is heated to 760°C (1400°F) for 90 minutes, enough carbides will precipitate to raise the M_S temperature well above room temperature. Then, as the second step, cooling to 15°C (55°F) results in nearly complete transformation to martensite. The third step consists of heating again, this time to a lower temperature, such as 566°C (1050°F), where the main precipitation reaction takes place. Then, once cooled, the alloy is at very high strength, as can be seen in Table 8.3.

An alternative hardening approach for some semi-austenitic PH stainless steels, such as 17-7PH and 15-7Mo, is to condition the austenite at a higher temperature, such as 955°C (1750°F), which results in less carbide precipitation. The resulting M_S temperature is lower, and refrigeration to −73°C (−100°F) is necessary to obtain a nearly complete martensite transformation. Then aging at a lower temperature than that of the third step above, such as 510°C (950°F), is used to induce precipitation and fully harden the alloy. This approach produces higher strength than the first approach (see Table 8.3).

Another approach to hardening the semi-austenitic PH stainless steels is to severely cold-work (cold roll) the steel that was originally in condition A. This produces martensite transformation without austenite conditioning. The steel mill can supply product in this state, known as *condition C*. In condition C, the steel still has about 5% tensile elongation, with a yield strength of about 1310 MPa (190 ksi). So it can be bent or otherwise formed before aging. Only one heat treatment is necessary, a relatively low-temperature aging at typically 480°C (900°F). This approach seems applicable only to sheet metal product forms, but it produces the highest strength of all approaches in 17-7PH and in 15-7Mo steels, with achievable yield strength of typically 1830 MPa (265 ksi) [15,16].

AM350 and AM355 are somewhat different from the other semi-austenitic PH stainless steels. Both require refrigeration to −73°C (−100°F) to promote martensite transformation, as do the other semi-austenitic PH stainless steels. ASTM A693 specifies an austenite conditioning treatment for AM355 but not for AM350. Both alloys are considered to precipitate nitrides during aging [8]. Their tensile strength falls, instead of rising, as a result of aging (see Table 8.3), but their yield strength rises.

The main strengthening precipitates in the other alloys appear to depend on composition. The precipitates in 17-7PH are not reported to be β-NiAl but to be an ordered body-centered cubic phase [8,17]. Those in 15-7Mo are reported to be β-NiAl and Ni_3Al [8]. As with the martensitic PH stainless steels, the semi-austenitic PH stainless steels appear to experience austenite reversion during heat treatment at the higher aging temperatures. Underwood et al. [17] noted nearly constant magnetic response of 17-7PH stainless steel as a function of time at aging temperature after up to 500 hours at 425°C (800°F) or 480°C (900°F), but slightly declining magnetic response with increasing time at 540°C (1000°F) and sharply declining magnetic response with increasing time at 595°C (1100°F). The strength is lower after these higher aging temperatures, as with the martensitic PH stainless steels, so the conclusion of Hochanadel et al. [9], that austenite reversion, not precipitate coarsening and loss of coherency, explains the lower strength, is probably valid for the semi-austenitic stainless steels as well.

8.2.3 Austenitic Precipitation-Hardening Stainless Steels

The physical metallurgy of the austenitic PH stainless steels is conceptually simpler than that of the martensitic or semi-austenitic PH stainless steels. There is no transformation to martensite that must take place, and there is no austenite conditioning treatment. The austenitic PH stainless steels solidify as austenite and the matrix remains austenite at all temperatures, even down to −196°C (−320°F), due to their high nickel content [18]. The austenitic PH stainless steels are normally supplied by the steel mill in the solution-treated condition. Solution treatment is normally done between about 900°C (1650°F) and 980°C (1800°F) for 1 to 2 hours, followed by oil or water quench. In this condition, the steels are quite soft. The ASTM A638 standard that covers Types 660 and 662 does not specify requirements for the strength in the solution-annealed condition, but a typical yield strength for commercial product in this condition is expected to be about 275 MPa (40 ksi) [18].

The austenitic PH stainless steels can be aged in the approximate temperature range 675 to 760°C (1250 to 1400°F) to harden them. The aging time is much longer for these steels, 16 to 20 hours, apparently because diffusion of alloy elements to form precipitates is much slower in the austenite matrix than it is in the martensite matrix of the martensitic and semi-austenitic PH stainless steels. The precipitates that form are commonly called *gamma prime* and consist of Ni_3Ti or $Ni_3(Ti,Al)$ intermetallic compounds [8,19]. Since there is on the order of 10 times as much Ti as Al in these steels, the precipitates should be primarily Ni_3Ti. The Ni atom is over 20% heavier than the Ti atom, and as a result, formation of Ni_3Ti removes Ni from solid solution at a rate of over 4.5 times as much as Ti. With 2% Ti present, over 9% Ni can be removed from solid solution during precipitation. However, with 25% Ni or more, the austenitic PH stainless steels remain stable austenite even after extensive precipitation.

Table 8.3 shows required minimum mechanical properties for some of these steels in the aged condition. It can be seen that the fully hardened strength level of the austenitic PH stainless steels is considerably less than that of the martensitic or semi-austenitic PH stainless steels. The mechanism by which these precipitates harden the steel is in some dispute. Thompson and Brooks [19] cite earlier work indicating that hardening is due to coherency of the precipitates with the matrix, but their work supports the conclusion that it is predominately ordering in the precipitates, not coherency with the matrix, that is responsible for the hardening in A-286. Similarly, Underwood et al. [17] indicate ordering of precipitates is a strengthening mechanism in 17-7PH stainless steel.

8.3 WELDING METALLURGY

The martensitic and semi-austenitic precipitation-hardening stainless steels solidify as primary ferrite in either the FA or F mode, as can be seen in Figure 8.1. This can also be understood with reference to the WRC-1992 diagram, directly for 17-4PH. However, the WRC-1992 diagram does not include effects for aluminum or titanium in the chromium equivalent, so an estimate of the effects of these alloys is needed to properly portray solidification of other alloys, such as 13-8Mo, 17-7PH (Type 631) or Type 635. Hull [20] evaluated these two elements as part of a study of ferrite content in chill cast pins, and found chromium-equivalent coefficients of 2.48 for aluminum and 2.20 for titanium. This means that each 1% Al adds 2.48 to the chromium equivalent, and each 1% Ti adds 2.20 to the chromium equivalent. Using these relations, a modified WRC-1992 diagram is shown in Figure 8.4, with the position of several martensitic and semi-austenitic stainless steels shown. It can be seen that they all solidify as virtually 100% ferrite. The ferrite largely transforms to austenite during cooling, but some residual ferrite in the microstructure at ambient temperatures should be expected. Note that the modified WRC-1992 in Figure 8.4 can be used to predict the solidification conditions, but not the final weld metal microstructure, since much of the austenite will transform to martensite, as described below.

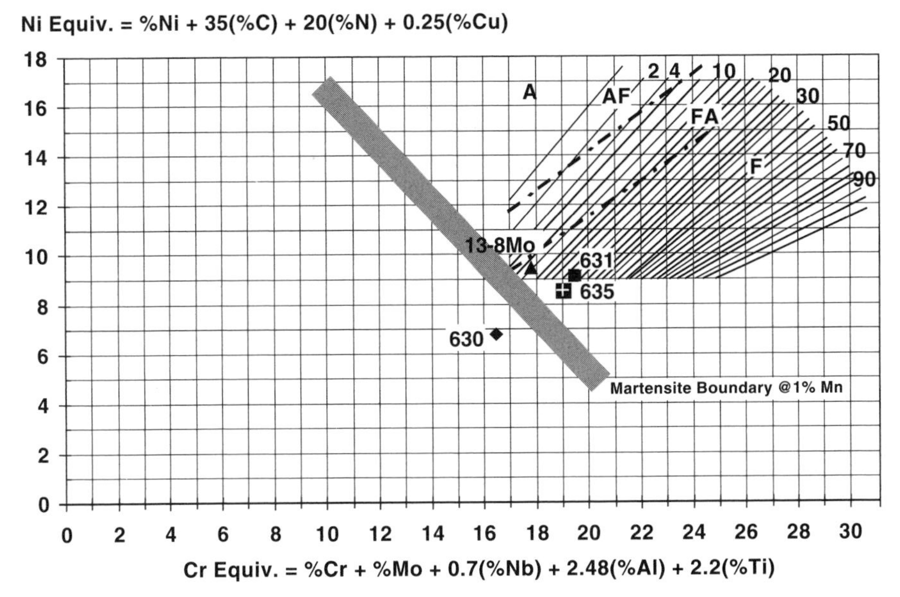

Figure 8.4 WRC-1992 diagram modified with Al and Ti coefficients from Hull [20], with 630, 631, 635, and 13-8Mo PH.

8.3.1 Microstructure Evolution

Microstructure evolution in the martensitic and semi-austenitic grades can occur along several paths. Most of the alloys solidify as ferrite and are either fully ferritic or a mixture of ferrite and austenite at the end of solidification. The majority of the ferrite transforms to austenite at elevated temperature, as noted above. Upon cooling to room temperature, the austenite either fully or partially transforms to martensite. If the transformation is nearly complete, the alloy is classified as a martensitic PH stainless steel. In alloys where the transformation is largely incomplete, and some large volume fraction of retained austenite remains untransformed, the alloys are classified as semi-austenitic PH stainless steel. In both cases, some high-temperature ferrite is normally retained in the structure to ambient temperatures.

In the austenitic alloys, the on-cooling transformation is much simpler. The microstructure is completely austenitic at the end of solidification and this austenite remains stable upon cooling to room temperature. Because of the ferrite-to-austenite and austenite-to-martensite transformations that occur in the martensitic and semi-austenitic alloys, there is usually no evidence of weld solidification substructure (cells and dendrites) in the weld metal microstructure of these alloys. The austenitic PH grades exhibit very distinct solidification structure, as shown in Chapter 6 for A-type solidification.

8.3.2 Postweld Heat Treatment

It is most common to weld the PH stainless steels in the solution-treated condition, before precipitation takes place [8,10,11,15,16,18,21]. In this condition, the martensitic PH stainless steels are somewhat hard but still moderately ductile. The semi-austenitic and austenitic PH stainless steels are quite soft and ductile. Because the weld metal cools rapidly, virtually no precipitation normally occurs in the weld metal, so that in the as-welded condition, the weld metal is not unlike the solution-annealed base metal in terms of its microstructure and properties. However, the weld metal is not as homogeneous.

As discussed in Section 8.2, the strengthening treatments for the PH stainless steels can be quite complex and require careful control to optimize properties. The martensitic alloys, after welding, are usually given a single PWHT in the range 480 to 620°C (900 to 1150°F) [8]. This heat treatment both tempers the martensite and promotes precipitation strengthening. At high PWHT temperatures [over 540°C (1000°F)], some austenite may re-form in the structure [9,14].

Because the semi-austenitic alloys may contain a high volume fraction of stable austenite in the as-welded structure, a conditioning heat treatment is usually performed to precipitate carbides at elevated temperature and render the austenite less stable [8,15,16]. Upon cooling from the conditioning temperature, the austenite then transforms to martensite. After conditioning at the lower temperatures [730 to 760°C (1345 to 1400°F)], effectively all the austenite transforms, whereas at the higher temperatures [930 to 955°C (1705 to 1750°F)], some austenite may remain after cooling to room temperature. In this case, refrigeration is used to transform the austenite to martensite. Cold work may also be used to raise the martensite formation temperature range and promote transformation to martensite above room temperature, although this is seldom applied to weldments.

The austenitic alloys are hardened in the temperature range 700 to 750°C (1260 to 1350°F). Since the austenite is very stable, there is no change in microstructure at these temperatures or on cooling to room temperature. As discussed previously, these alloys are normally strengthened with a gamma-prime Ni_3Ti precipitate.

If the PH stainless steel is in the solution-annealed condition prior to welding, and if the full hardening treatment (including the austenite conditioning treatment for the semi-austenitic PH stainless steels) is applied, weld strength very similar to that of the base metal is normally obtained, although the ductility may be somewhat lower [8,22]. If the base metals are already in the fully hardened condition at the time of welding, there is significant risk of cracking. After aging to peak strength, the martensitic and semi-austenitic PH stainless steels have limited ductility, and strains associated with weld shrinkage may be sufficient to produce cracks around the weld [21]. The ductility of the austenitic PH stainless steels in the fully hardened condition is somewhat better than that for the other PH stainless steels, but they have much more severe weld cracking problems in general. So, in any event, welding of a PH stainless steel in the fully hardened condition is best avoided.

Welding may also be done in the somewhat overaged condition, in which case the steel has higher ductility. However, if this is done, the weldment cannot be aged to the fully hardened condition unless a solution treatment is applied after welding, followed by austenite conditioning for semi-austenitic PH stainless steels, and a new aging treatment for all of the PH stainless steel types.

8.4 MECHANICAL PROPERTIES OF WELDMENTS

The advantage of the PH steels is that they can be strengthened to levels approaching those of high-strength structural steels while maintaining good corrosion resistance. Achieving optimum properties requires careful control of microstructure and heat treatment conditions. For the martensitic and semi-austenitic alloys, excessive levels of retained ferrite can reduce toughness and ductility. It is usually desirable to limit ferrite to below about 10%. The presence of retained austenite will reduce strength in these alloys.

Specifications for PH stainless steel welding filler metals generally do not include mechanical property requirements. The exception is the AWS A5.4 standard that includes requirements for E630-XX (17-4PH) stainless steel weld metal. These are shown in Table 8.4. Note that the minimum requirements for the weld metal in the H1150 condition match those for the 17-4PH base metal, shown in Table 8.3, except that the tensile ductility requirement is reduced for the weld metal.

Peak strength in the martensitic PH alloys is achieved by PWHT in the range 450 to 550°C (840 to 1020°F), as shown in Table 8.3. Note, however, that toughness requirements are not shown for the lowest aging temperatures. Achieving a good balance of strength and toughness, which is required in many applications, requires heat

TABLE 8.4 AWS A5.4 Requirements for Mechanical Properties of E630-XX (17-4PH) Weld Metal[a]

Minimum Tensile Strength		Minimum Elongation
MPa	ksi	(%)
930	135	7

[a]After H1150 heat treatment consisting of solution anneal at 1025 to 1050°C (1875 to 1925°F), cooling to ambient, then aging 4 hours at 610 to 630°C (1135 to 1165°F).

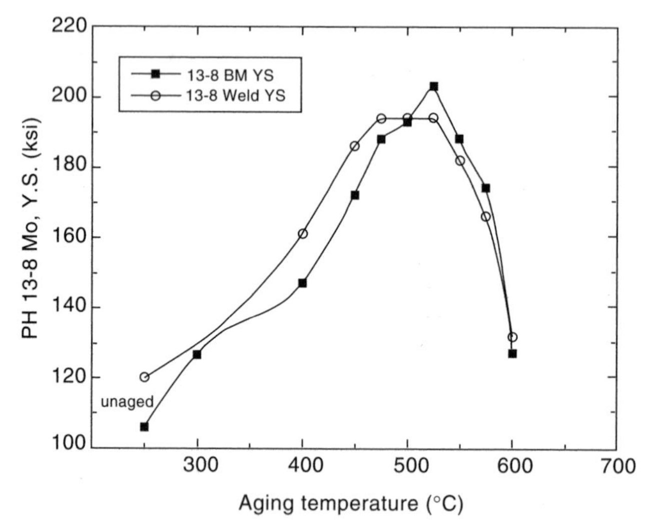

Figure 8.5 Yield strength versus aging temperature for 13-8Mo base metal and weld metal. (From Brooks and Garrison [7]. Courtesy of the American Welding Society.)

treatment in a higher temperature range [e.g., 550 to 600°C (1020 to 1110°F)]. Brooks and Garrison [7] provide plots of weld metal strength and toughness for several martensitic PH stainless steels. Representative plots are shown in Figures 8.5 and 8.6. In general, the peak weld metal strength and the peak base metal strength are quite similar. The weld metal toughness tends to be lower than the base metal toughness except when the weldment is aged to the highest strength, as can be seen in Figure 8.6.

8.5 WELDABILITY

The heat-treatment condition prior to welding should be chosen to minimize defects and optimize properties, and depends on postweld treatments that will be applied to the

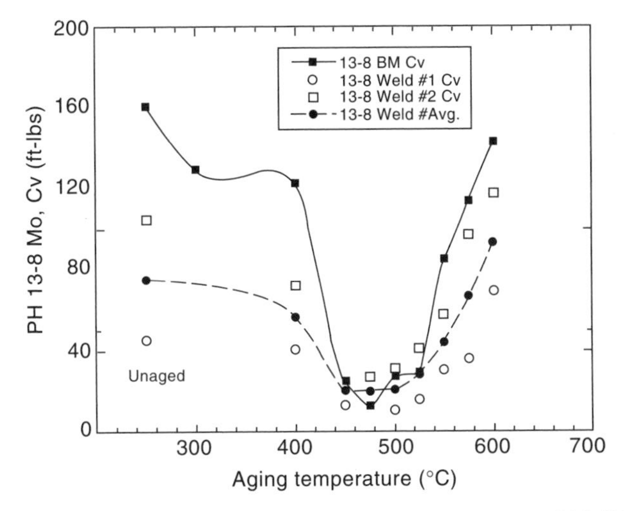

Figure 8.6 Charpy V-notch impact energy versus aging temperature of 13-8Mo base metal and weld metal. (From Brooks and Garrison [7]. Courtesy of the American Welding Society.)

welded assembly. The martensitic types are typically welded in the annealed condition for thin sections and the overaged condition for thick sections, which inherently have high restraint. Welding a material in the overaged condition will reduce the stresses generated due to the materials slightly weakened state. Semi-austenitic types are typically welded in the solution-treated or annealed condition. The austenitic types are the most difficult types to weld because of solidification, liquation, and ductility-dip cracking problems. They are normally welded in a solution-annealed condition, to minimize the intrinsic restraint.

Vagi and Martin [21] examined weldability by making and testing heavily restrained GTA welds in 50-mm (2-in.) plates of 17-4PH (martensitic steel), 17-7PH (semi-austenitic steel), and A-286 (austenitic steel). They found some cracking along ferrite stringers in the 17-4PH base metal near the weld, but not in the weld metal. The appearance of this cracking is similar to that of a lamellar tear, as shown in Figure 8.7. Vagi and Martin [21] found no cracking in the 17-7PH weldment. However, every A-286 weldment exhibited cracking. Cracking in the A-286 was found in the weld metal (solidification cracking) and in the HAZ (liquation cracking).

Brooks [23] performed hot ductility testing and Varestraint testing on multiple heats of A-286 and found both liquation cracking and intergranular cracking. He considered the intergranular cracking to be a form of liquation cracking but could not find evidence of liquation in samples tested at 1150°C (2100°F) or 1175°C (2150°F), and it appears more likely that this was an instance of ductility-dip cracking (DDC). Liquation evidence was found at higher temperatures, and the liquation was concluded to be due mainly to melting of Laves phase. Figure 8.8 shows intergranular cracking that appears to be DDC that occurred during hot ductility testing at 1175°C (2150°F). Figure 8.9 shows the beginning of liquation observed after hot

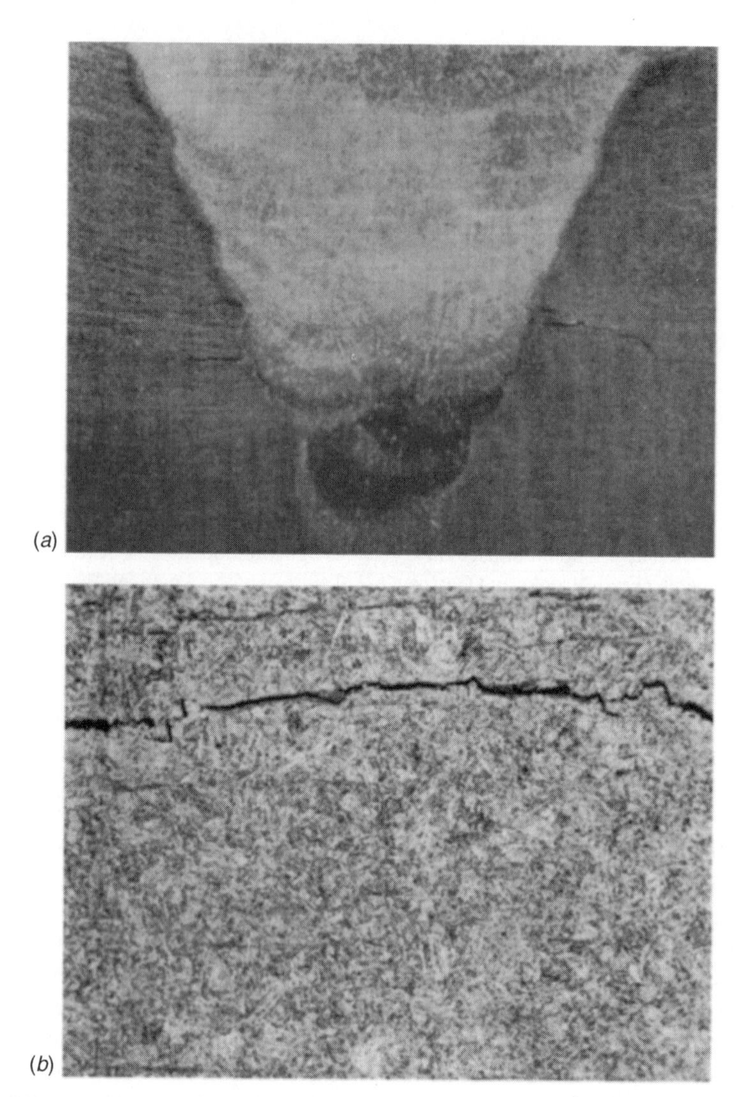

Figure 8.7 Cracking beside the weld in 2-in.-thick 17-4PH. (From Vagi and Martin [21]. Courtesy of the American Welding Society.)

ductility testing at 1205°C (2200°F). Figure 8.10 shows complete grain boundary liquation observed after Varestraint testing of A-286.

A dissimilar filler metal, such as type 309, can be used to prevent weld metal cracking in A-286, but it is not heat treatable, so the joint efficiency will be low. TiC, which is normally present in A-286, has been identified as a particle that is subject to constitutional liquation. This phenomenon is responsible for liquation cracking in the HAZ/PMZ. Minimizing grain size and using low heat inputs will allow for steep

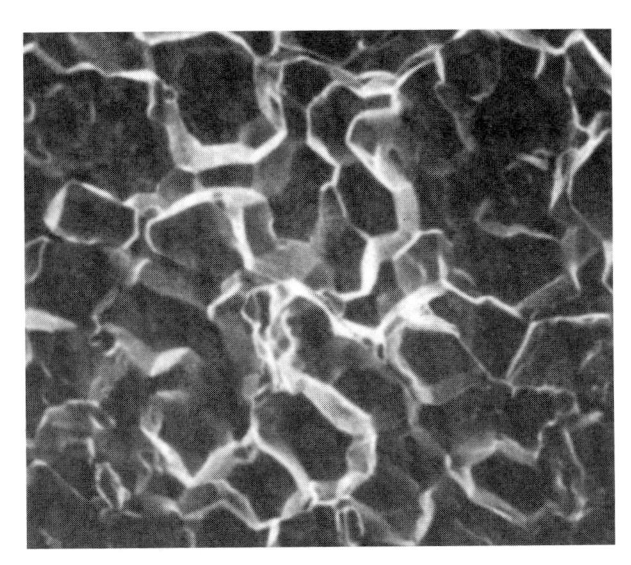

Figure 8.8 Intergranular cracking in A-286 hot ductility sample tested at 1175°C (2150°F). (From Brooks [23]. Courtesy of the American Welding Society.)

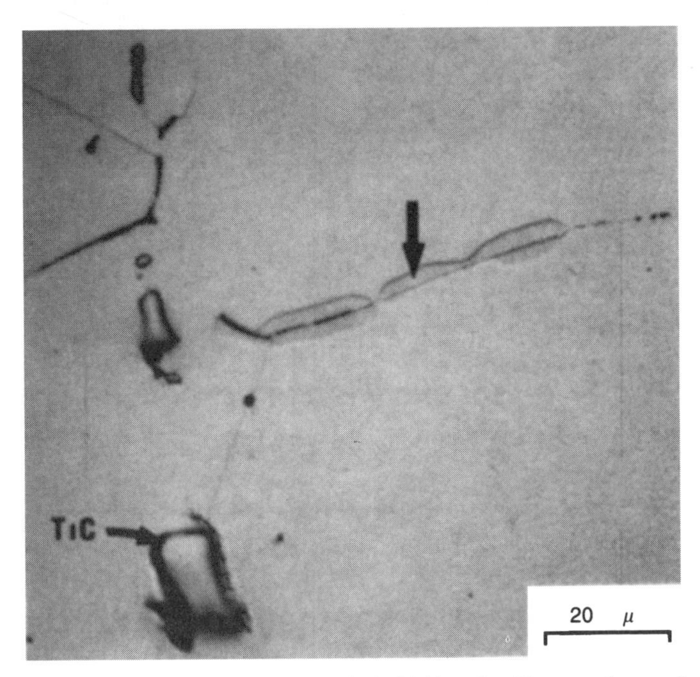

Figure 8.9 Liquation along grain boundary in A-286 hot ductility sample tested at 1205°C (2200°F). (From Brooks [23]. Courtesy of the American Welding Society.)

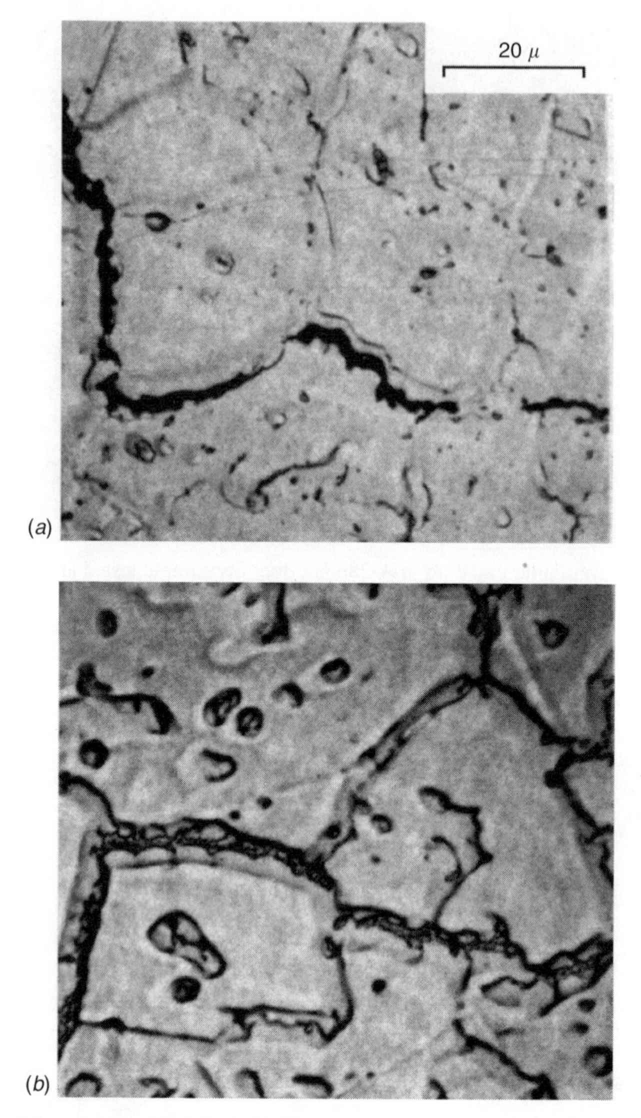

Figure 8.10 Liquation cracking in A-286 Varestraint test sample: (*a*) intergranular cracking in HAZ; (*b*) Laves phase responsible for cracking. (From Brooks [23]. Courtesy of the American Welding Society.)

temperature gradients that would minimize the amount and extent of liquation that occurs.

Weld solidification cracking in austenitic PH alloys, such as A-286, is a persistent problem. Similar to the austenitic stainless steels that solidify as austenite (A mode), the fusion zone of these alloys are fully austenitic with no delta ferrite. A-286 is a fully austenitic material, and adjusting the composition to promote FA-type solidification is not possible. Reducing impurity elements, particularly sulfur, phosphorus, and silicon, which produce low-melting-point compounds, will help prevent this type of cracking. Using stringer techniques and convex bead contours will help reduce stresses and cracking.

As noted previously, Brooks and Krenzer [4] devised a modification of A-286 to overcome problems with solidification cracking, liquation cracking, and ductility-dip cracking. Carbon, manganese, phosphorus, sulfur, silicon, and boron are reduced in this alloy, JBK-75, compared to A-286. Further, the nickel content is increased, which is claimed to promote backfilling of potential solidification and liquation cracks.

Hydrogen-induced cracking is usually not a problem for the PH stainless steels, due to the low carbon contents and resulting low hardness. Stress concentrations can aggravate cracking due to lack of sufficient ductility, so they should be avoided at the design stage. Low-hydrogen practice is recommended, however, especially in thick sections of martensitic and semi-austenitic PH stainless steels because they may be transformed to martensite before any PWHT occurs. PWHT allows hydrogen release. Normally, preheat is not required for welding the PH stainless steels [8].

8.6 CORROSION RESISTANCE

Corrosion resistance of PH stainless steels appears to depend on the condition of heat treatment [10,11,15,16,24]. In general, the corrosion resistance is optimized when the steel is in the fully hardened condition. There are exceptions to this general rule for specific corrosive media. Consequently, it is best to consult the recommendations of the steel supplier. Less information is available concerning corrosion resistance of weldments. The 17-4PH alloy is claimed to show welded corrosion resistance comparable to the unwelded condition when the aging heat treatment is applied after welding [10].

REFERENCES

1. Funk, C. W., and Granger, M. J. 1954. Metallurgical aspects of welding precipitation-hardening stainless steels, *Welding Journal*, 33(10):496s–508s.
2. Smith, R., Wyche, E. H., and Gorr, M. W. 1946. A precipitation-hardening stainless steel of the 18 chromium, 8 nickel type, *Transactions of the AIME*, 167:313–
3. Linnert, G. E. 1957. Welding precipitation hardening stainless steels, *Welding Journal*, 36(1):9–27.
4. Brooks, J. A., and Krenzer, R. W. 1975. Weldable age hardenable austenitic stainless steel, U.S. patent 3,895,939.
5. Dalder, E. N. C. 2004. Private communication.
6. AIRCO Superconductors Specification SMG-80-002 for JBK-75 Stainless Steel Sheet.
7. Brooks, J. A., and Garrison, W. M., Jr. 1999. Weld microstructure development and properties of precipitation-strengthened martensitic stainless steels, *Welding Journal*, 78(8): 280s–291s.

8. Pollard, B. 1993. Selection of wrought precipitation-hardening stainless steels, in *ASM Metals Handbook*, 10th ed., Vol. 6, ASM International, Materials Park, OH, pp. 482–494.
9. Hochanadel, P. W., Robino, C. V., Edwards, G. R., and Cieslak, M. J. 1994. Heat treatment of investment cast PH 13-8Mo stainless steel; I: mechanical properties and microstructure, *Metallurgical and Materials Transactions A*, 25A(4):789–798.
10. AK Steel. 2000. *17-4PH Stainless Steel Product Data Bulletin*, AK Steel Corporation, Middleton, OH.
11. AK Steel. 2000. *15-5PH Stainless Steel Product Data Bulletin*, AK Steel Corporation, Middleton, OH.
12. ASTM. 2002. *Standard Specification for Precipitation-Hardening Stainless and Heat-Resisting Steel Plate, Sheet, and Strip*, ASTM A693-02, American Society for Testing and Materials, West Conshohocken, PA.
13. ASTM. 2000. *Standard Specification for Precipitation Hardening Iron Base Superalloy Bars, Forgings, and Forging Stock for High-Temperature Service*, ASTM A638/A638M-00, American Society for Testing and Materials, West Conshohocken, PA.
14. Anthony, K. C. 1963. Aging reactions in precipitation hardenable stainless steel, *Journal of Metals*, 15(12):922–927.
15. AK Steel. 2000. *17-7PH Stainless Steel Product Data Bulletin*, AK Steel Corporation, Middleton, OH.
16. AK Steel. 2000. *PH 15-7 Mo Stainless Steel Product Data Bulletin*, AK Steel Corporation, Middleton, OH.
17. Underwood, E. E., Austin, A. E., and Manning, G. K. 1962. The mechanism of hardening in 17-7 Ni–Cr precipitation-hardening stainless steels, *Journal of the Iron and Steel Institute*, 200(8):644–651.
18. Allegheny Ludlum. 1998. *Allegheny Ludlum Altemp® A286 Iron-Base Superalloy*, Allegheny Ludlum Corporation, Pittsburgh, PA.
19. Thompson, A. W., and Brooks, J. A. 1982. The mechanism of precipitation strengthening in an iron-base superalloy, *Acta Metallurgica*, 30:2197–2203.
20. Hull, F. C. 1973. Delta ferrite and martensite formation in stainless steels, *Welding Journal*, 52(5):193s–203s.
21. Vagi, J. J., and Martin, D. C. 1956. Welding of high-strength stainless steels for elevated-temperature use, *Welding Journal*, 35(3):137s–144s.
22. Smallen, H. 1961. Welding PH 15-7 Mo precipitation hardening stainless steel, *Welding Journal*, 40(7):324s–329s.
23. Brooks, J. A. 1974. Effect of alloy modifications on HAZ cracking of A-286 stainless steel, *Welding Journal*, 53(11):324s–329s.
24. Allegheny Ludlum. 2003. *Stainless Steel AL 13-8™*, Allegheny Ludlum Corporation, Pittsburgh, PA.

Missing! Harkins, WRC Bull 103, 1965

CHAPTER 9

DISSIMILAR WELDING OF STAINLESS STEELS

9.1 APPLICATIONS OF DISSIMILAR WELDS

Dissimilar metal welding (DMW) is frequently used to join stainless steels to other materials. This approach is most often used where a transition in mechanical properties and/or performance in service are required. For example, *austenitic stainless steel piping* is often used to contain high-temperature steam in power generation plants. Below a certain temperature and pressure, however, carbon steels or low-alloy steels perform adequately, and a transition from stainless to other steels is often used for economic purposes (carbon steel, or low-alloy steel, is much less expensive than stainless steel).

Cladding is another common form of DMW. Large carbon steel or low-alloy steel pressure vessels are often clad with stainless steels to provide corrosion protection at lower cost. Piping may also be internally clad to provide similar protection. The bulk of DMW with stainless steels is in conjunction with carbon steels and low-alloy steels. The power generation industry uses DMW extensively to reduce material costs and enhance performance in elevated-temperature applications.

One stainless steel may be joined to another stainless steel, and varying degrees of dissimilarity are possible. For example, steels with different alloy content but similar microstructure may be joined, or steels of different alloy content and microstructure may be joined. Since stainless steels can have martensitic, ferritic, austenitic, or duplex microstructure, there are numerous microstructural combinations. Reasons for these

Welding Metallurgy and Weldability of Stainless Steels, by John C. Lippold and Damian J. Kotecki
ISBN 0-471-47379-0

combinations may be economic and/or property considerations, or sometimes just convenience. A common convenience issue is to identify a single filler metal suitable for a job involving more than one combination of base metals in order to eliminate the risk of interchanging filler metals or reduce costs by reducing the required inventory of filler metals.

Stainless steels may also be welded to nickel-base alloys in demanding applications where a transition is required to accommodate more corrosive environments or superior strength requirements at elevated temperature. The economics of replacing Ni-base alloys with stainless steels whenever possible is often the controlling factor (Ni-base alloys may be three to 10 times more expensive). Stainless steels are sometimes joined to other nonferrous alloy systems, such as aluminum and copper. These welds are usually made with a solid-state welding process, due to cracking problems or the formation of intermetallic phases during solidification or in the solid state that can embrittle the joint. This chapter deals only with fusion welds, not with solid-state welds.

Because of the potentially large transition in composition and properties that can occur across dissimilar metal fusion welds, there are a number of fabrication and performance issues that should be considered. Perhaps not surprisingly, a number of different failure mechanisms have been associated with these joints.

9.2 CARBON OR LOW-ALLOY STEEL TO AUSTENITIC STAINLESS STEEL

DMW of stainless steels to carbon steel or low-alloy steel presents a number of metallurgical and engineering challenges. Control of the weld metal microstructure, particularly in the initial or root pass, is normally of critical importance since weld deposits can range from fully martensitic to fully austenitic, or may exhibit mixtures of austenite, ferrite, and martensite [1,2]. In addition, a composition transition region will exist between the bulk weld metal and the base metal. This narrow region may have distinctly different microstructure and properties from the adjacent regions [3]. Since many carbon and low-alloy steels are required by code to be heat-treated following welding, further changes in the as-deposited structures in response to thermal treatment should be considered [4].

A range of engineering issues, most related to the metallurgical nature of the weldment, are of concern. Differences in physical and mechanical properties between the weld metal and base metals will almost certainly exist. For example, differences in the coefficient of thermal expansion may result in locally high stresses that can promote service failures, particularly due to thermal cycling from low to high temperatures [4]. The corrosion resistance may also vary locally in both the weld metal and the transition region, due to differences in composition and microstructure.

9.2.1 Determining Weld Metal Constitution

Concerns about DMW weld metal constitution center around the first weld pass or root pass, primarily because dilution effects are greatest in this pass. In many cases of DMW, a highly desirable goal in selecting the filler metal is to obtain stable austenite with a small amount of ferrite in this first weld pass (see Chapter 6). If this microstructure can be obtained, weld solidification cracking is very unlikely and the weld metal will be sufficiently ductile to pass a *bend test* such as that required by the ASME Code. This bend test most often consists of a sample of weld, HAZ, and base metal bent around a mandrel whose radius is twice the thickness of the bend specimen (usually referred to as a *2T bend test*).

The Schaeffler diagram [2,5] and in many cases the WRC-1992 diagram can be used to predict weld metal and transition region microstructure during DMW of stainless steels. The Schaeffler diagram (Figure 3.4) provides the "big picture," allowing the various stainless steel grades and carbon steels to be plotted in the same space. However, it produces erroneous predictions when high-manganese filler metal is used. The extended version of the original WRC-1992 diagram can also be used but will only be predictive of microstructure if the weld metal composition falls within the austenite, austenite + ferrite, austenite + martensite, or austenite + ferrite + martensite regions. The modified version of the WRC-1992 diagram [6], which includes boundaries for martensite formation, broadens the predictive capability of the diagram to address almost all DMW situations involving stainless steels (see Figure 3.18). These diagrams are discussed more extensively in Chapter 3.

The Schaeffler diagram is useful in providing a microstructure approximation for dissimilar metal welds, since stainless steels, carbon steels, and low-alloy steels can be plotted on the diagram, and microstructure predictions can be made for all of these steels. In the example shown in Figure 9.1, a low-alloy steel such as ASTM A508 is being welded to Type 304L (square symbol), using either a Type 309L (round) or Type 310 (triangle) filler metal. The intersection of the dotted lines along the tie-line between the two base metals represents equal contribution (or dilution) by the base metals to the weld metal composition. The composition represented by the intersection might be viewed as a synthetic base metal composition. The ultimate composition of the first pass weld metal will then fall along the dotted tie-line connecting this point to the filler metal composition (if there is equal base metal dilution). If the total base metal dilution into the first pass of filler metal is 30%, which is normal for welding with covered electrodes, the weld metal composition will lie along this second tie-line, 30% of the distance from the 309L point toward the synthetic base metal composition. That composition is approximately the place where this second tie-line crosses the isoferrite line labeled "5% ferrite." Subsequent passes will lie closer to the filler metal composition. Note that for the 309L filler metal, normal dilution of the filler metal by the base metal will result in a two-phase austenite + ferrite microstructure, which is usually highly resistant to weld solidification cracking.

The same approach can be followed in predicting the first-pass weld microstructure if 310 filler metal is used. No portion of the tie-line from the 310 composition in Figure 9.1 to the synthetic base metal composition is within the area of the diagram indicating that ferrite will be present. With Type 310 filler metal, the weld deposit will be almost certainly fully austenitic, unless dilution by the base metals is extremely high, in which case, martensite, not ferrite, will appear. In the case of 310

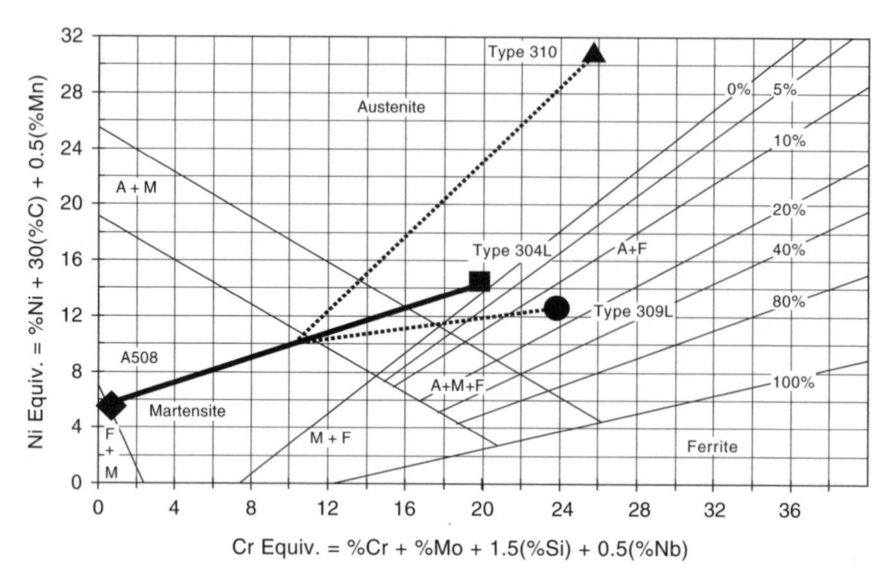

Figure 9.1 Use of the Schaeffler diagram to predict the weld metal constitution of a dissimilar weld between Type 304L and A508 using either 309L or 310 filler metal.

filler metal, then, there will be some sensitivity to solidification cracking in the weld metal.

The tie-line between the filler metal composition and the base metal tie-line can also be used to predict the microstructure in the transition region at the fusion boundary. For example, the tie-line to Type 309L transects the martensite, austenite + martensite, and austenite + ferrite regions. All these microstructures can be expected in a narrow region between the base metal HAZ and the fully mixed weld metal. Similarly, the tie-line between the synthetic base metal and the 310 filler metal transects the martensite, austenite + martensite, and austenite regions. All of these microstructures can be expected in the narrow transition region along the fusion boundary in the 310/304L combination. This is considered more extensively below.

Using a similar example as for the Schaeffler diagram shown in Figure 9.1, Figure 9.2 presents the combination of ASTM A36 and 304L welded with 309L filler metal on the WRC-1992 diagram modified to include the martensite boundary at 1% Mn, a normal level of Mn in 309L weld metal. Note that at less than about 45% dilution of the filler metal by the base metals (tie-line between the points labeled "309L" and "mixture of base metals"), the weld metal composition falls in the austenite + ferrite region with primary ferrite (FA) solidification. Further, over 60% dilution would be required before martensite would appear in the first-pass weld metal. By estimating, or controlling, this dilution, the weld metal FN can be predicted.

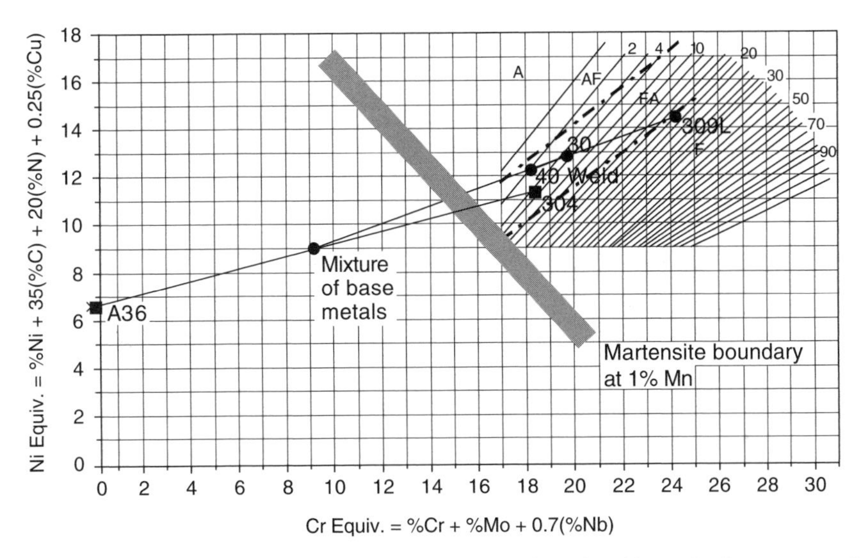

Figure 9.2 Use of the WRC-1992 diagram for prediction of weld metal microstructure for the Type 304L/A36 combination welded with Type 309L filler metal.

9.2.2 Fusion Boundary Transition Region

Predicting the transition region microstructure can be difficult, since it may change dramatically over a very short distance [about 1 mm (0.04 in.)]. Within this region the microstructure may differ significantly from both the bulk weld metal and HAZ and is subject to local compositional gradients and diffusional effects. For example, if the base metal has a higher carbon content than the weld metal (which is usually the case), carbon will diffuse (or "migrate") from the HAZ to the fusion zone during welding or PWHT [4]. This can potentially result in a narrow martensitic region at the fusion boundary that exhibits high hardness [3]. If the weld metal is high in chromium and the base metal has little or no chromium, the tendency for carbon to migrate from the HAZ into the weld metal during PWHT is very great.

Microstructure evolution along the fusion boundary in dissimilar metal welds made with stainless steels can be quite complex. In situations where the base metal is ferritic at temperatures near the melting point (as with most carbon and low-alloy steels) and the weld metal is austenitic, normal epitaxial growth may be suppressed. This can result in the formation of what are called *Type II boundaries*, which run roughly parallel to the fusion boundary [7]. These are in contrast to the *Type I boundaries*, which result from columnar growth from base metal grains into the weld metal and are oriented roughly perpendicular to the fusion boundary.

A schematic of the weld fusion boundary is shown in Figure 9.3 under "normal" conditions (top) and the situation where the base metal and weld metal have

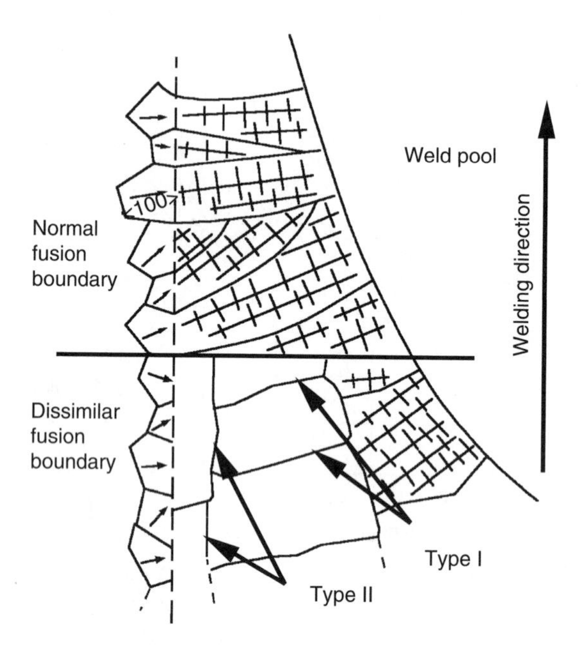

Figure 9.3 Formation of Type II boundaries when austenitic weld metal solidifies in contact with ferritic base metal. (From Nelson et al. [8]. Courtesy of the American Welding Society.)

different crystal structures (BCC versus FCC) at the solidification temperature. Note that distinct boundaries are present in the dissimilar weld, the Type I boundaries running roughly perpendicular to the fusion boundary (along the original solidification direction) and the Type II boundaries running parallel to the fusion boundary. Type II boundaries do not exist under normal weld solidification conditions.

In certain dissimilar combinations, martensite may be present along the fusion boundary due to the transition in composition between a carbon steel or low-alloy steel and stainless steel or Ni-base alloy. As described previously, the Schaeffler diagram or the WRC-1992 diagram can be used to predict the presence of martensite. Rowe et al. [9] and others have demonstrated that hydrogen in the arc can reach this thin zone and produce hydrogen-induced cracking.

The photomicrograph in Figure 9.4 shows the fusion boundary region of an A508 pressure vessel steel that has been clad with Type 309L. Following cladding, the pressure vessel is then subjected to a PWHT at 610°C (1125°F). Note the dramatic change in microstructure across the fusion boundary. Type II boundaries are clearly evident just adjacent to the fusion boundary in the weld metal. These form in the solid state during weld cooling when both the weld metal and HAZ are austenitic, allowing austenite grain growth across the fusion boundary. These boundaries may be sites for carbide precipitation, particularly if significant carbon migration from the base metal has occurred.

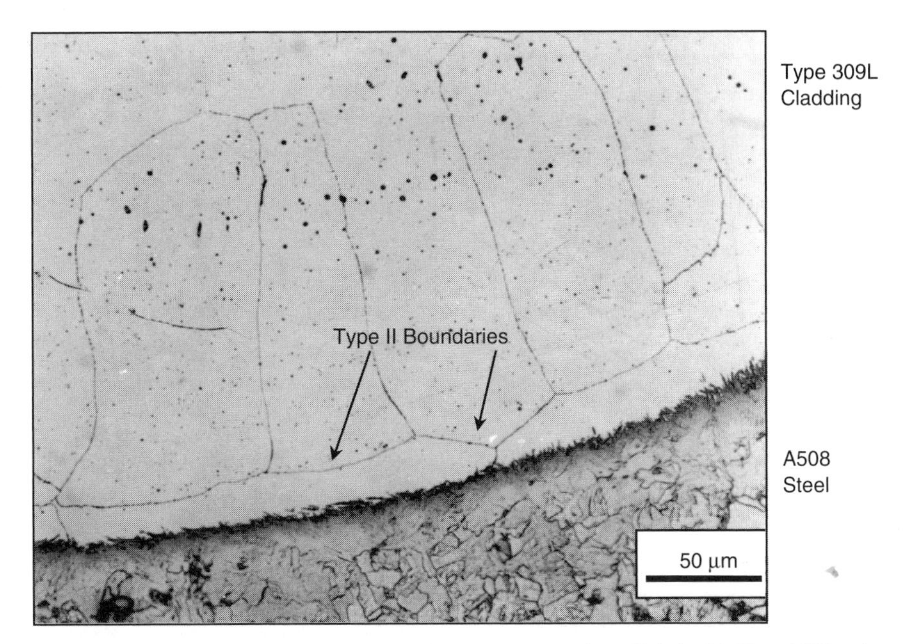

Figure 9.4 Fusion boundary region of A508 pressure vessel steel clad with Type 309L, after PWHT at 610°C (1125°F).

Carbon migration has been observed from carbon steel base metals across the fusion boundary into the transition region [4,11]. This results from the carbon gradient that is present (higher C in the base metal) and the affinity of carbon for the higher-Cr weld metal. This migration can result in local microstructural changes in both the HAZ and transition region. A carbon-depleted zone in the HAZ, immediately adjacent to the fusion boundary, results. Soft ferrite forms in this carbon-depleted zone, which has led to premature failure in creep [12]. This failure mechanism is discussed in Section 9.3.3.

Because of the transition in composition between the A508 steel and the diluted 309L filler metal, a narrow region of martensite can exist along the fusion boundary. This is predicted by using the Schaeffler diagram in Figure 9.1. This results in a sharp increase in hardness that occurs across the fusion boundary region, as shown in Figure 9.5. Note that this hard region still exists after PWHT. Similar fusion boundary microstructures can form in other dissimilar combinations. Figure 9.6 is an example of a dissimilar multipass weld between an A36 steel and 2205 duplex stainless steel using a 2209 filler metal. At the bottom of the weld, a distinct martensitic band is evident along the fusion boundary, while at the top of the weld the transition region is much narrower because dilution is lower. Again, the Schaeffler diagram or the WRC-1992 diagram can be used to predict the weld metal and transition region microstructures. The latter diagram will allow weld metal FN to be estimated.

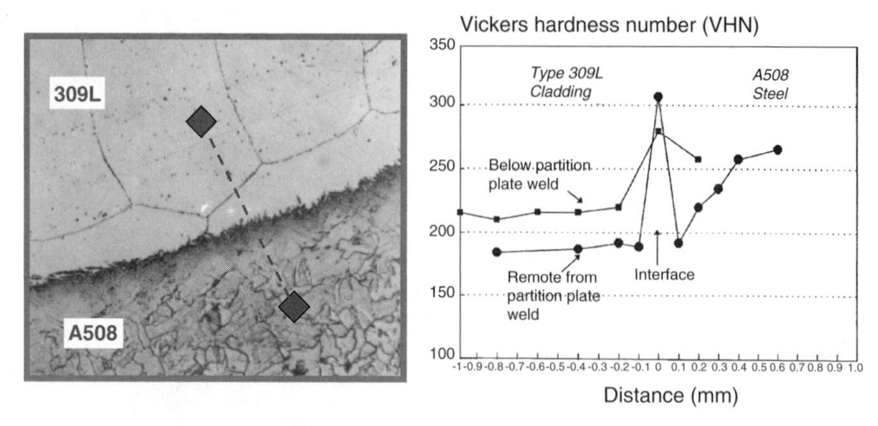

Figure 9.5 Hardness profile across A508/Type 309L fusion boundary following PWHT. (From Lippold [10].)

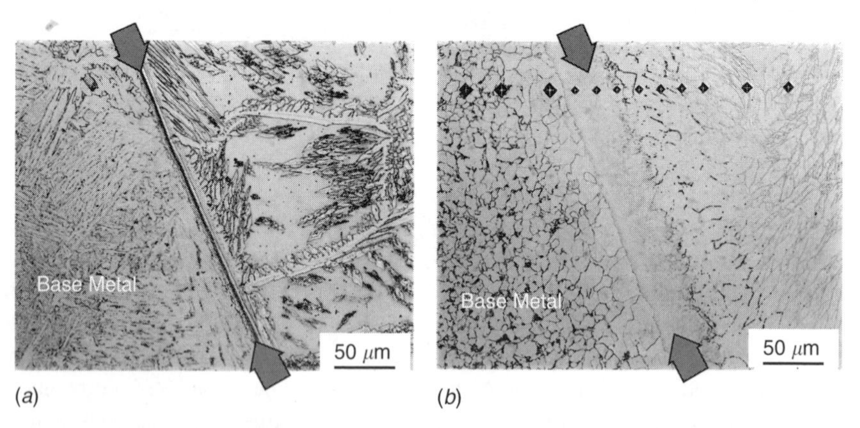

Figure 9.6 Fusion boundary microstructure between A36 structural steel and 2209 filler metal: (*a*) top; (*b*) bottom. Note the traverse of Vickers hardness impressions across the fusion boundary in the bottom of the weld area. The impressions are much smaller, indicating higher hardness, due to martensite, in the transition region. (From Barnhouse and Lippold [13].)

9.2.3 Nature of Type II Boundaries

Type II boundaries are of special interest because a number of instances of in-service cracking, sometimes termed *disbonding*, have been associated with them. As shown in Figures 9.3 and 9.4, a Type II boundary is essentially a grain boundary that runs approximately parallel to the fusion boundary but located a few microns into the fusion zone. It continues across many grains, staying approximately parallel to the fusion boundary, as shown in Figure 9.4. It has been reported that cracking or disbonding can occur along these boundaries, particularly in situations where stainless steel cladding is applied to carbon steels. This mechanism is described in more detail in Section 9.3.2.

Nelson et al. have studied the mechanism of Type II boundary formation in cladding of ferritic steels with austenitic or other FCC alloys [8,9,14,15]. They have concluded that such boundaries form when the substrate alloy (mild steel or low-alloy steel) exists as delta ferrite at the temperature where the FCC cladding alloy is solidifying. This suppresses normal epitaxial nucleation at the fusion boundary and necessitates heterogeneous nucleation of the FCC weld metal. However, shortly after solidification, cooling causes the carbon steel substrate to transform to austenite. Then the fusion boundary, formerly a BCC-to-FCC interface, becomes an FCC-to-FCC interface with considerable orientation mismatch across that interface. The fusion line is then a high-energy boundary which is mobile. The boundary migrates into the FCC cladding a short distance, driven by the temperature gradient, composition gradient, and strain energy produced by differences in lattice parameter between the FCC cladding alloy and the FCC substrate. It is then is locked in place as cooling proceeds.

A schematic of this mechanism from the work of Nelson et al. [14,15] is shown in Figure 9.7. Note that the Type II boundaries form within the temperature range when the carbon steel, or substrate alloy, is in the austenite phase field. Thus, the weld heat input and temperature gradient in the HAZ will have some influence over Type II boundary formation, since this dictates the time over which both the weld metal and HAZ are FCC and boundary migration can occur. Nelson et al. verified this mechanism by using other material combinations, such as Monel™ filler metal (austenitic, 70Ni–30Cu) on pure iron and Monel™ on Type 409 ferritic stainless steel. In the former case, Type II boundaries were present as long as there was not sufficient base metal dilution to cause the weld metal to solidify as ferrite. In the latter case, no Type II boundaries form since the Type 409 is fully ferritic from room temperature to melting. Since austenite is never present in the base metal substrate, the mechanism described in Figure 9.7 is inoperable.

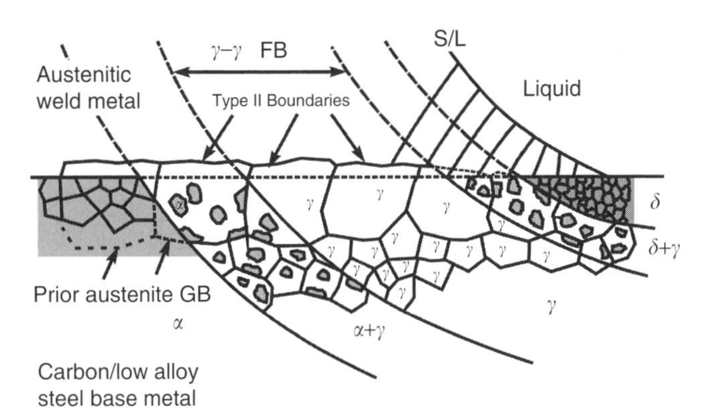

Figure 9.7 Mechanism for Type II boundary formation during dissimilar welding between an austenitic weld metal and a carbon steel. (From Nelson et al. [15]. Courtesy of ASM International.)

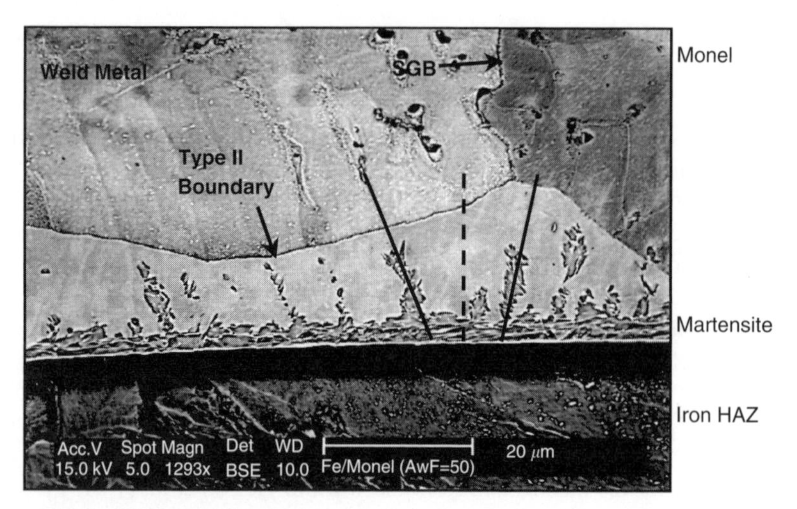

Figure 9.8 Transition region of 70% Ni–30% Cu cladding over essentially pure iron (56% base metal dilution). (From Rowe et al. [9]. Courtesy of the American Welding Society.)

Subsequent to the formation of the Type II boundary, the substrate transforms to ferrite or other decomposition products, and the compositional transition zone, or a part of it, transforms to martensite. The martensite may extend to the Type II boundary or may stop short of it. Figure 9.8 shows an example of martensite stopping short of the Type II boundary. There is a mismatch in coefficient of thermal expansion between the BCC or martensite side of the transition region and the FCC cladding side. Thermal cycling sets up strain in this region, and the Type II boundary is a weak approximately planar interface, which makes it a preferential cracking site (see Section 9.3.2). In addition, the martensite is a potential location for hydrogen-induced cracking if the service environment produces hydrogen charging [16,17].

9.3 WELDABILITY

A number of cracking mechanisms are associated with dissimilar metal welds between carbon steels and stainless steels. These include solidification cracking, clad disbonding along Type II boundaries, creep failure in the HAZ of the carbon steel, and reheat, or PWHT, cracking.

9.3.1 Solidification Cracking

Solidification cracking most commonly occurs when nominally austenitic filler metals, such as Types 308L and 309L, are used. If the root pass weld deposits, or other welds, are heavily diluted by the carbon steel, they will solidify as primary austenite because there was insufficient ferrite potential in the filler metal and/or there was excessive dilution. Figure 9.9 shows an example of solidification cracking in a submerged arc fillet

Figure 9.9 Solidification cracking in a 309L submerged arc fillet weld between A36 steel and 304L stainless steel. (From Kotecki and Rajan [18].)

weld between A36 structural steel and Type 304L stainless steel. Note the differing etching appearance between the portion of the weld metal in contact with the 304L versus that in contact with the A36 steel. The light etching appearance reflects metal that solidified as primary ferrite. The darker etching appearance reflects metal that solidified as primary austenite. A solidification crack lies along the centerline, not of the entire weld, but of the region of primary austenite solidification. The overall weld measured 0.8 FN, due to excessive dilution.

Another example of solidification cracking during dissimilar welding is shown schematically in Figure 9.10. This was a thick-section weld between an A508 pressure vessel steel and Type 347 stainless steel, using Type 308L filler metal in conjunction with the GTAW process. The bulk of the weld solidified in the FA mode and contained ferrite in the range FN 6 to 8. The final weld shown in Figure 9.10 was heavily diluted by the A508 base metal, resulting in solidification in the fully austenitic A mode. This resulted in centerline solidification cracking. Better control of the weld process (torch positioning and heat input) was successful in minimizing this dilution and avoiding cracking by ensuring FA solidification.

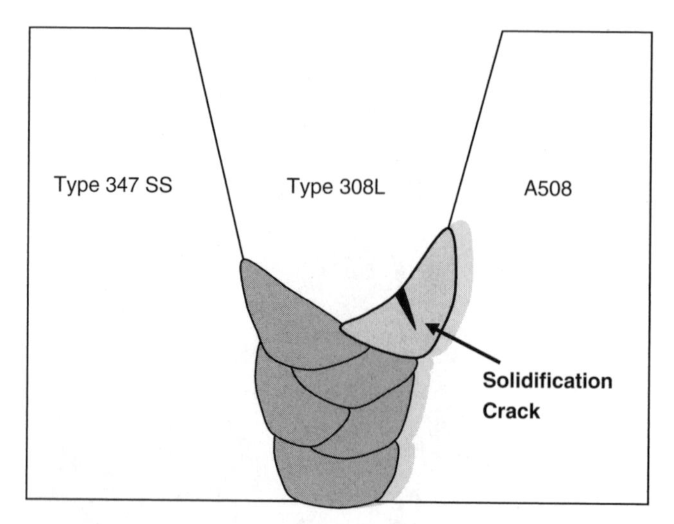

Figure 9.10 Solidification cracking in a dissimilar weld between Type 347 stainless steel and an A508 pressure vessel steel using Type 308L filler metal. Note that excessive dilution by the A508 steel has shifted solidification from primary ferrite to primary austenite.

As described in Section 9.2.1, the Schaeffler and WRC-1992 diagrams can be used to estimate the amount of dilution that can be tolerated without resulting in a shift to fully austenitic weld metals that can be susceptible to cracking. The extended WRC-1992 diagram can be particularly useful if a certain level of ferrite is specified or desired in the dissimilar weld.

9.3.2 Clad Disbonding

Clad disbonding typically occurs along Type II boundaries in the clad deposit. The precise mechanism for this type of failure is not known, but it may involve carbide precipitation, impurity segregation, the orientation of the boundary perpendicular to the direction of principal stress, hydrogen-induced cracking in a thin martensite zone in the compositional transition zone, or a combination of these. The metallographic section in Figure 9.11 shows the fracture surface profile of a clad disbonding failure in Type 309L clad onto A508 steel. The A508 side of the fracture could not be shown since it was still attached to a 20,000-lb pressure vessel. Based on the location and orientation of the fracture, it is clear that failure occurred along Type II boundaries in the clad layer (see Figure 9.4). Although the precise mechanism for this type of failure is not known, it is clear that the nature of this boundary and its presence in a microstructural and compositional transition zone leads to these types of failures. Weld deposits that do not contain Type II boundaries are immune to this type of failure.

Disbonding may occur during the actual cladding operation, during subsequent postweld heat treatment, or in service. It is often difficult to know when the actual failure occurs during fabrication, since inspection is usually not carried out until

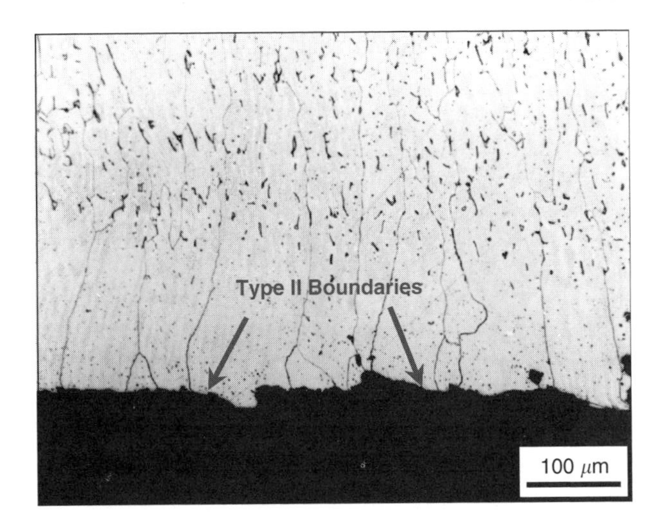

Figure 9.11 Disbonded cladding of 309L from A508 pressure vessel steel.

after the PWHT is completed. Unfortunately, there is no way to avoid Type II boundary formation in many of the common dissimilar combinations used in the power generation industry and elsewhere.

9.3.3 Creep Failure in the HAZ of Carbon or Low-Alloy Steel

Failure in the steel HAZ adjacent to the fusion boundary has been observed in thick-section structural welds. Carbon migration from the HAZ to the weld metal during welding, PWHT, or service exposure results in a soft ferritic HAZ microstructure [4,11,12]. Under imposed residual and thermal stresses (CTE mismatch between HAZ and weld metal), creep cracks can occur along the ferrite grain boundaries. The microstructure along the fusion boundary of a 2.25Cr–1Mo steel clad with Type 309L filler metal is shown in Figure 9.12. This structure had been given a PWHT at 720°C (1330°F) for 10 hours. The hardness indents (in VHN) indicate a high residual hardness in the martensite band at the fusion boundary. In contrast, the 2.25Cr–1Mo HAZ is extremely soft and consists of large ferrite grains. Under elevated temperature exposure, it is this region that is subject to creep failure. This is shown in Figure 9.13 from the work of Klueh and King [12], who studied the failure of a transition joint between 2.25Cr–1Mo steel and Type 321 made with Ni-base Inconel 182 (AWS A5.11 ENiCrFe-3 covered electrode classification) filler metal. These failures typically occur after 10 to 15 years of elevated-temperature service.

The average linear CTE for carbon steel in the temperature range 20 to 600°C (70 to 1110°F) is on the order of 7.5 to 8 μin./in.-°F, while austenitic stainless steels are on the order of 9.5 to 10 μin./in.-°F. This difference in CTE can result in large strains when a dissimilar weld is heated to an elevated temperature. Because the carbon steel has a lower CTE, it will try to constrain the stainless steel weld from expanding. This results

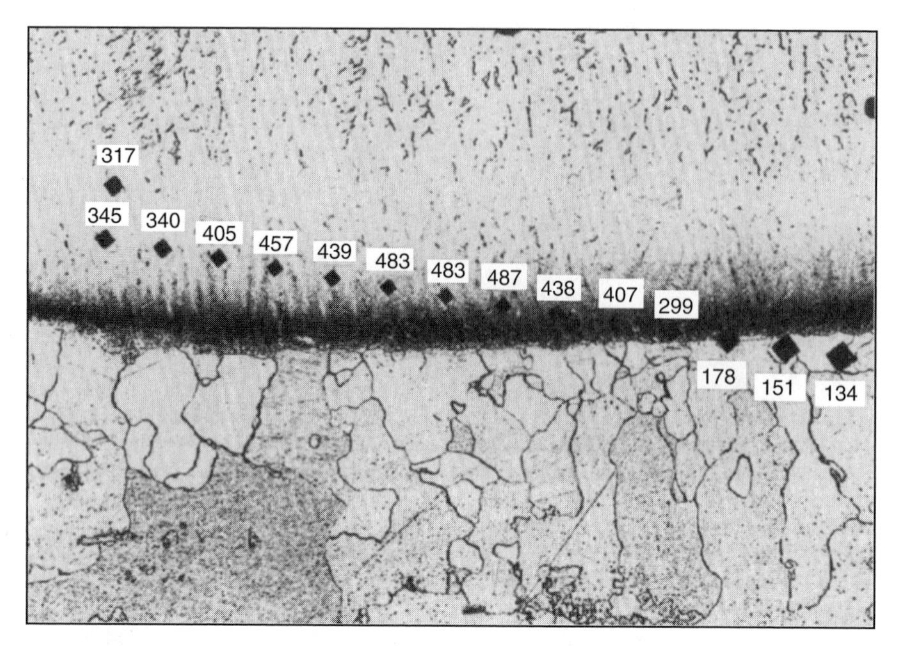

Figure 9.12 Fusion boundary region between 2.25Cr–1Mo base metal and Type 309L filler metal after PWHT at 720°C (1330°F) for 10 hours. Hardness values are Vickers scale (VHN). (From Gittos and Gooch [11]. Courtesy of the American Welding Society.)

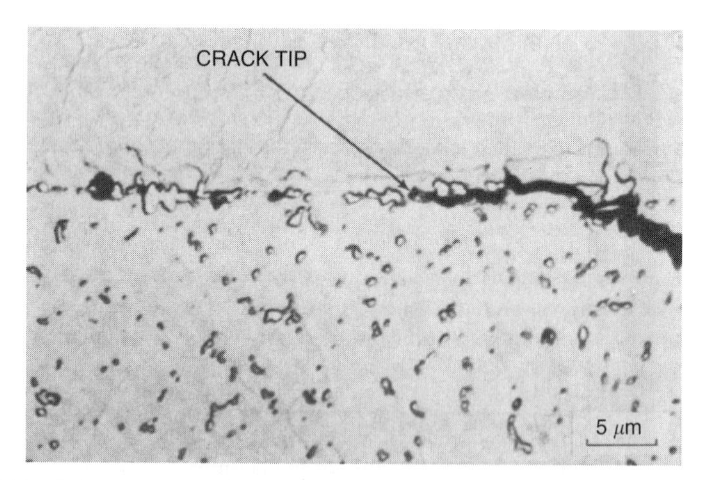

Figure 9.13 Failure along fusion boundary in dissimilar weld between 2.25Cr–1Mo steel and Type 347 stainless steel made with Type 309 filler metal. Failure is in the pressure vessel steel HAZ. (From Klueh and King [12]. Courtesy of the American Welding Society.)

in locally high stresses at the interface. The carbon migration leads to the formation of a very soft ferritic region between the stronger weld metal and base metal, and high local strains will be concentrated in this region. Over time, grain boundary sliding occurs, eventually leading to a creep, or stress rupture, failure in the carbon steel HAZ.

The use of Ni-base filler metals, or inserts, has been effective in reducing the thermally induced CTE mismatch stresses, thereby avoiding cracking. The use of Ni-base filler metals is generally recommended for service temperatures exceeding 425°C (800°F) [19]. This is helpful because Ni-base alloys have a CTE that is midway between the carbon steels and austenitic stainless steels, on the order of 8 to 9 μin./in.-°F in the same temperature range as noted above. Thus, a gradation in CTE occurs across the weld joint rather than an abrupt change at the fusion boundary. Because Ni-base filler metals are compatible with both carbon steels and austenitic stainless steels, the use of filler metals, such as ENiCrFe-2 (INCO A) or ERNiCr-3 (alloy 82), may be useful in avoiding creep failures in service. These weld deposits will be fully austenitic, so precautions against weld solidification cracking should be taken, as discussed in Section 9.4.9.

9.4 OTHER DISSIMILAR COMBINATIONS

Stainless steels may be involved in joints of varying degrees of dissimilarity. Each situation requires some analysis on the part of the engineer or other person responsible for the integrity of the joint. It is especially important to make an appropriate selection of filler metal to produce a sound joint that will give satisfactory service performance. Examples are given in the following sections.

9.4.1 Nominally Austenitic Alloys Whose Melted Zone Is Expected to Include Some Ferrite or to Solidify as Primary Ferrite

This joint has perhaps the least dissimilarity and the least potential difficulty. Examples include joining 304L to 316L or 304H to 347. There is generally wide latitude in selecting a filler metal. The least expensive or most readily available filler metal that matches one or the other base metal is likely to be quite suitable. From the point of view of serviceability of the joint, it matters little whether the filler metal is 308L or 316L for the 304L to 316L joint. Similarly, it matters little whether the filler metal is 308H or 347 for the 304H to 347 joint. Furthermore, an overalloyed filler metal, such as 309L, would also be suitable for the 304L to 316L joint. There should be no weld solidification cracking issues in such joints.

9.4.2 Nominally Austenitic Alloys Whose Melted Zone Is Expected to Contain Some Ferrite, Welded to Fully Austenitic Stainless Steel

Examples include 316L welded to a superaustenitic 6% Mo stainless steel, or 304H welded to 310 stainless steel. In these situations there can be a risk of weld solidification cracking if the weld metal composition is such that it solidifies as primary austenite, but a weld deposit with appreciable ferrite may be unacceptable due to corrosion considerations in the former example or sigma phase formation concerns at high temperature in the latter example. If ferrite in the weld is acceptable in the 316L to 6% Mo stainless steel joint, an intermediate-alloy filler metal such as 309LMo is usually the best choice. But if ferrite is not acceptable, a low-ferrite 316L filler metal or Type 385 filler metal, both of which have good solidification cracking resistance despite lack of ferrite, is likely to be the best choice. In the 304H to 310 joint, a 308H or 309 filler metal may be acceptable if high-temperature service is not anticipated, but this combination of base metals is likely to be used at high temperature. A 310 filler metal is usually a better choice than a 308H filler metal because without ferrite the weld metal has reasonably good solidification cracking resistance if applied correctly. In cases where the weld metal is expected to be fully austenitic, correct application includes (1) small, relatively low-heat-input weld beads; (2) a convex profile; and (3) care to fill craters.

9.4.3 Austenitic Stainless Steel Joined to Duplex Stainless Steel

An example of this situation is joining 316L stainless steel to 2205 duplex stainless steel. In this case, a filler metal selection to obtain ferrite is easy. Either 316L filler metal or 2209 filler metal is acceptable, and the choice is made based on cost and availability. On the other hand, 309L filler metal would also be expected to provide ferrite, but because 309L contains no molybdenum, it might be an inappropriate choice. The two base metals contain Mo, which makes them more resistant to localized corrosion (particularly pitting) in some environments relative to Mo-free 309L.

9.4.4 Austenitic Stainless Steel Joined to Ferritic Stainless Steel

An example of this situation is joining 444 stainless steel to 316L stainless steel. Because ferritic stainless steel filler metals are not readily available, the most common choice, with acceptable results, is likely to be the filler metal that is appropriate for the austenitic base metal, 316L in this case.

9.4.5 Austenitic Stainless Steel Joined to Martensitic Stainless Steel

The filler metal of choice depends on whether or not the weldment will subsequently see PWHT or elevated-temperature service. If no elevated temperature exposure is expected after welding, a filler metal expected to produce weld metal with primary ferrite solidification is usually acceptable. For example, 410 stainless might be joined to 304 stainless with 309 filler metal. However, if elevated temperature exposure is anticipated, a crack-resistant fully austenitic filler metal such as 310 would be a more appropriate selection.

9.4.6 Martensitic Stainless Steel Joined to Ferritic Stainless Steel

An example might be welding 410 to 409 stainless steel. One might be tempted to select either a ferritic stainless steel filler metal or a martensitic stainless steel filler metal. That may be acceptable if mechanical property demands on the joint are not high. Dilution of the filler metal with both base metals is likely to produce a weld metal of mixed ferrite–martensite microstructure. The Balmforth diagram (Figure 3.22) can be used to estimate the weld metal microstructure. Ordinarily, a mixed ferrite–martensite weld metal microstructure has poor ductility, and if a ductile weld is required, an austenitic stainless steel filler metal like 309L, or a nickel-base alloy filler metal like NiCr-3 may be appropriate. If high-temperature thermal cycling is likely, the nickel-base alloy filler metal may be a better choice than the 309L because of a better match in the coefficients of thermal expansion.

9.4.7 Stainless Steel Filler Metal for Difficult-to-Weld Steels

The austenite phase in nominally austenitic stainless steel filler metals, or in ferritic–austenitic filler metals, has the ability to greatly limit diffusion of hydrogen from the weld metal into the HAZ of a ferritic or martensitic base metal. Furthermore, such weld metal is generally ductile and tough, in contrast to high-carbon-steel weld metal. As a result, nominally austenitic stainless steel filler metals and ferritic–austenitic stainless steel filler metals have been used to overcome HAZ cracking problems in welding a number of difficult-to-weld steels. Examples of difficult-to-weld steels are alloy steels with over 0.25% C, including armor steels used in military hardware such as tanks, abrasion-resisting steels used for handling ores in the mining and minerals processing industries, tool and die steels used in the metal-working industry, and the like. Stainless steel filler metals used for these applications include Types 307, 308HMo, 309Mo, 310, 312, and the European 18 8 Mn composition, which is often mistakenly referred to in Europe as 307. The 18 8 Mn composition differs from the 307 composition in that 307 contains Mo whereas 18 8 Mn does not, and 307 contains less Mn than does 18 8 Mn.

The Schaeffler diagram and the WRC-1992 diagram, modified to include martensite boundaries, can be used to predict whether or not the root-pass weld metal will contain ferrite, and whether or not the root-pass weld metal will transform to martensite. Of the filler metals listed, only 312 is likely to produce a root pass containing ferrite when used for joining the difficult-to-weld steels. But the other compositions are quite resistant to solidification cracking even without ferrite, so they find considerable application. The predictions of the Schaeffler diagram and of the WRC-1992 diagram are similar in the case of the example of joining A508 or A36 steel to 304L, discussed in Section 9.2.1. But the two diagrams do not always make similar predictions. An example is the joining of abrasion-resisting steel plate to structural steel using stainless steel filler metal. This choice of filler metal is often made for ore-handling equipment in mining situations to avoid having to use high preheat, because the abrasion-resisting steel is highly sensitive to hydrogen-induced HAZ cracking if ferritic filler metal is used. Austenitic filler metal prevents weld metal hydrogen from reaching the HAZ, thereby greatly reducing preheat requirements. The two most commonly used filler metal for this purpose are 307 and the European 18 8 Mn composition, to make single-pass fillet welds.

Table 9.1 lists compositions typical of A36 structural steel, abrasion-resisting plate (AR), and 18 8 Mn filler metal. It also lists the synthetic base metal consisting of a 50/50

TABLE 9.1 Compositions for the Example of Welding Abrasion Resistant (AR) Plate to A36 Plate with 18 8 Mn Filler Metal with 30% Dilution

Element	A36	AR	50/50 Mixture	18 8 Mn	Fillet Weld
C	0.15	0.3	0.225	0.05	0.103
Mn	0.4	1.4	0.9	6.0	4.47
Si	0.2	0.2	0.2	0.3	0.27
Cr	—	1.4	0.7	19.0	13.5
Ni	—	—	—	9.0	6.3
Mo	—	0.3	0.15	—	0.04
N	—	—	—	0.05	0.035
Schaeffler Cr_{eq}	0.3	2.0	1.15	19.45	13.95
Schaeffler Ni_{eq}	4.7	9.7	7.2	13.5	11.62
WRC-1992 Cr_{eq}	0.0	1.7	0.85	19.0	13.56
WRC-1992 Ni_{eq}	5.25	10.5	7.88	11.75	10.59

Source: Kotecki [6].

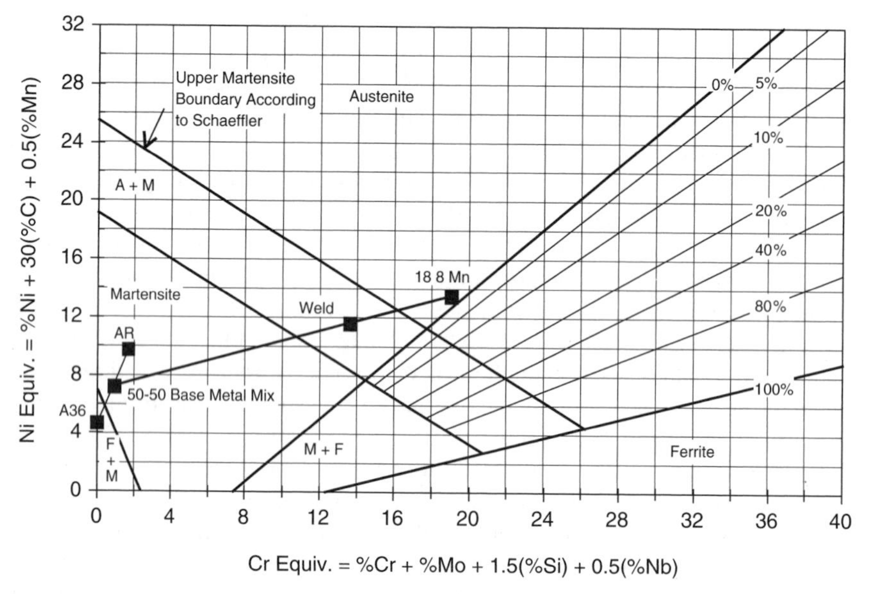

Figure 9.14 Schaeffler diagram, showing the situation of joining A36 structural steel to AR plate with 18 8 Mn filler metal.

mixture of the two actual base metals, and the 30% dilution root-pass composition consisting of 15% A36 steel, 15% AR steel, and 70% 18 8 Mn filler metal. Schaeffler and WRC-1992 chromium and nickel equivalents are calculated in the table. The results are plotted on the Schaeffler diagram in Figure 9.14. The weld metal composition predicted appears in the center of the martensite + austenite field of the diagram. The result is a prediction that the weld metal will be half martensite and therefore brittle.

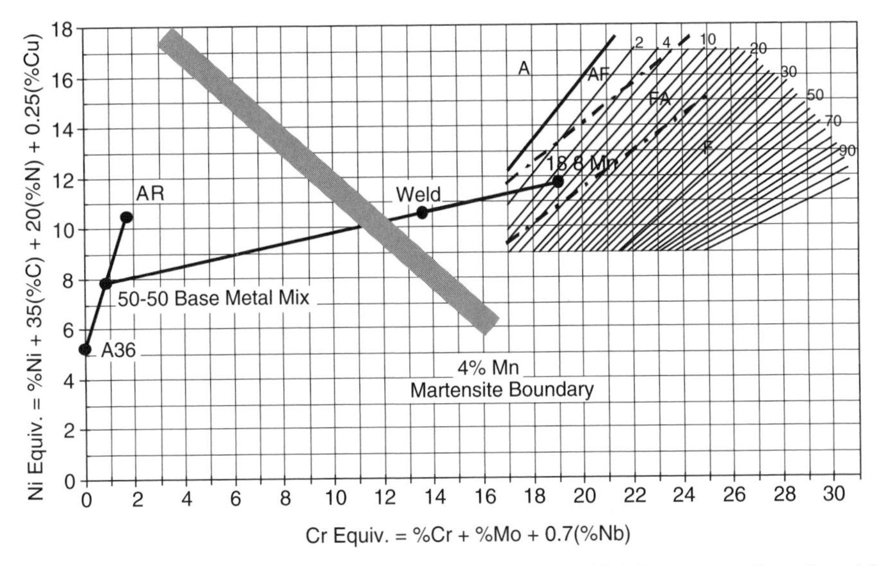

Figure 9.15 WRC-1992 diagram, showing the situation of joining structural steel to AR plate with 18 8 Mn filler metal.

Figure 9.15 plots the compositions on the WRC-1992 diagram modified with extended axes and the martensite boundary for 4% Mn. Note that Table 9.1 indicates that the fillet weld will contain a bit over 4% Mn, so the 4% Mn martensite boundary is appropriate. The weld metal composition predicted lies above and to the right of the martensite boundary, so the microstructural prediction is that there will be no martensite in the weld metal. The weld metal is predicted to be fully austenitic. This composition is highly resistant to solidification cracking without ferrite.

The Schaeffler diagram predicts that the weld will be approximately 50% martensite, whereas the WRC-1992 diagram, modified, predicts no martensite. Which is correct? Figure 9.16 shows the actual fillet weld (top) between $\frac{1}{2}$-in.-thick AR plate and 1-in.-thick A36 structural steel. The weld metal microstructure obtained (bottom) is fully austenitic. So the WRC-1992 diagram prediction is shown to make the more accurate prediction in this case.

9.4.8 Copper-Base Alloys Joined to Stainless Steels

Copper has a tendency to wet austenite grain boundaries in stainless steels, which can lead to severe cracking problems if stainless steel is deposited on copper, or if copper is deposited on stainless steel. As a result, a nonmatching filler metal is usually chosen for such joints. Probably the most common is the over 90% Ni filler metals of AWS classes ENi-1 (covered electrode) or ERNi-1 (bare wire or rod). Small convex weld beads, with attention to filling craters, are usually recommended.

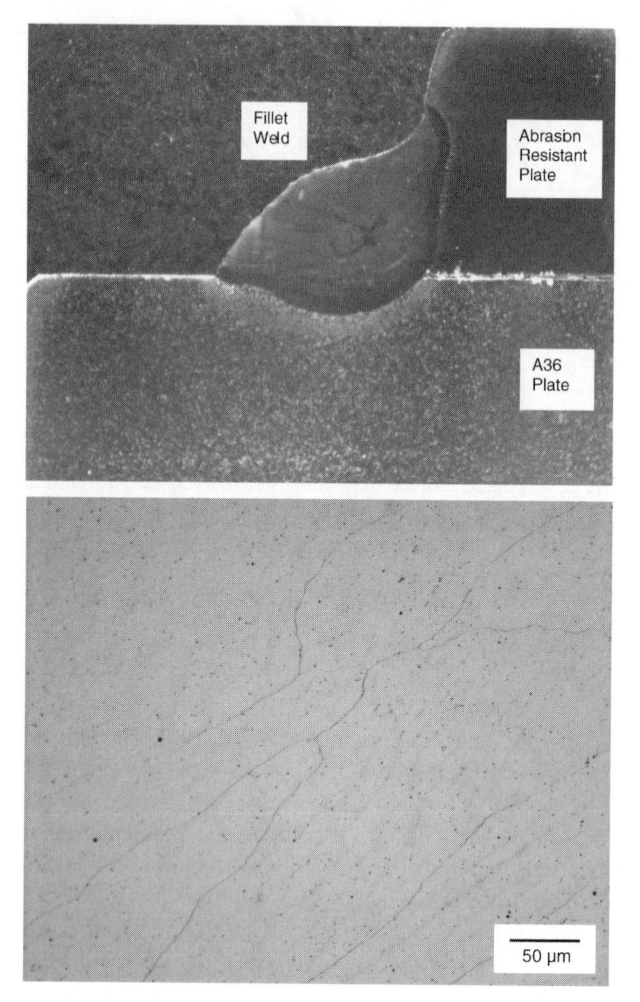

Figure 9.16 Single-pass fillet weld between A36 structural steel and abrasion-resisting steel using 18 8 Mn filler metal (dilute aqua regia etch). (From Kotecki [20].)

9.4.9 Nickel-Base Alloys Joined to Stainless Steels

Because there are so many stainless steels and so many nickel-base alloys, possible combinations from the two alloy systems are very numerous. Problems with solidification cracking often result when stainless steel filler metals are attempted for these joints. As a result, most joining of stainless steel to nickel-base alloys is done with nickel-base alloy filler metals. In most cases, the nickel alloy filler metal appropriate to joining the nickel-base alloy to itself is also appropriate for joining the nickel-base alloy to the stainless steel. The nickel-base alloy filler metals are found in AWS A5.11, *Specification for Nickel and Nickel-Alloy Welding Electrodes for*

TABLE 9.2 Approximate Limit of Diluting Elements (wt%) in Dissimilar Welds Made with Ni-Base Filler Metals

Weld Metal	Diluting Element			
	Iron	Nickel	Chromium	Copper
Nickel	30	—	30	Unlimited
Ni–Cu	25 SMAW, 15 GMAW	Unlimited	8	Unlimited
Ni–Cr–Fe[a]	25	Unlimited	30	15

Source: Avery [19].

[a]Silicon should be limited to less than 0.75%.

Shielded Metal Arc Welding, and in AWS A5.14, *Specification for Nickel and Nickel-Alloy Bare Welding Electrodes and Rods.* So, for example, if the Ni–Cr–Mo–W alloy UNS N 10276 (commonly known as Alloy C-276) were to be joined to 316L stainless steel, a suitable filler metal would be ENiCrMo-4 or ERNiCrMo-4, which is a composition match for the C-276 alloy. Certain nickel-base alloy filler metals have a well-earned reputation for being able to join a wide variety of nickel-base alloys successfully to a wide variety of stainless steels. An example is the NiCrMo-3 filler metal. Controlling dilution of nickel-base filler metals may be critical for preventing solidification cracking. Table 9.2 provides recommendations for the limits of diluting elements from stainless or carbon steel base metals to avoid solidification cracking.

It is not possible to address herein all possible combinations of stainless steels to nickel-base alloys. The suppliers of the nickel-base alloy filler metals can almost always provide a recommendation for a filler metal appropriate to a given combination. In most cases, the techniques of depositing small, somewhat convex weld beads, and filling craters, are helpful in avoiding solidification cracking.

REFERENCES

1. Schaeffler, A. L. 1947. Selection of austenitic electrodes for welding dissimilar metals, *Welding Journal*, 26(10):601s–620s.
2. Thielsch, H. 1952. Stainless-steel weld deposits on mild and alloy steels, *Welding Journal*, 31(1):37s–64s.
3. Pan, C., Wang, R., and Gui, J. 1990. Direct TEM Observation of microstructures of the austenitic/carbon steels welded joint, *Journal of Materials Science*, 25:3281–3285.
4. Lundin, C. D. 1982. Dissimilar metal welds: transition joints literature review, *Welding Journal*, 61(2):58s–63s.
5. Schaeffler, A. L. 1949. Constitution diagram for stainless steel weld metal, *Metal Progress*, 56(11): 680–680B.
6. Kotecki, D. J. 2001. *Weld Dilution and Martensite Appearance in Dissimilar Metal Joining, IIW Document II-1438-01*, American Council of the International Institute of Welding, Miami, FL.

7. Matsuda, F., and Nakagawa, H. 1984. Simulation test of disbonding between 2.25%Cr–1%Mo steel and overlaid austenitic stainless steel by electrolytic hydrogen charging technique. *Transactions of JWRI*, 13(1):159–161.
8. Nelson, T. W., Lippold, J. C., and Mills, M. J. 1999. Nature and evolution of the fusion boundary in ferritic–austenitic dissimilar metal welds, 1: nucleation and growth, *Welding Journal*, 78(10):329s–337s.
9. Rowe, M. D., Nelson, T. W., and Lippold, J. C. 1999. Hydrogen-induced cracking along the fusion boundary of dissimilar metal welds, *Welding Journal*, 78(2):31s–37s.
10. Lippold, J. C. Unpublished research conducted in conjunction with Westinghouse Electric Corporation.
11. Gittos, M. F., and Gooch, T. G. 1992. The interface below stainless steel and nickel-alloy claddings, *Welding Journal*, 71(12):461s–472s.
12. Klueh, R. L., and King, J. F. 1982. Austenitic stainless steel–ferritic steel weld joint failures, *Welding Journal*, 61(9):302s–311s.
13. Barnhouse, E. J., and Lippold, J. C. 1998. Microstructure/property relationships in dissimilar welds between duplex stainless steels and carbon steels, *Welding Journal*, 77(12): 477s–487s.
14. Nelson, T. W., Lippold, J. C., and Mills, M. J. 2000. Nature and evolution of the fusion boundary in ferritic–austenitic dissimilar metal welds, 2: on-cooling transformations, *Welding Journal*, 79(10):267s–277s.
15. Nelson, T. W., Lippold, J. C., and Mills, M. J. 1998. Investigation of boundaries and structures in dissimilar metal welds, *Science and Technology of Welding and Joining*, 3(5):249.
16. Sakai, T., Asami, K., Katsumata, M., Takada, H., and Tanaka, O. 1982. Hydrogen induced disbonding of weld overlay in pressure vessels and its prevention, in *Current Solutions to Hydrogen Problems in Steels, Proceedings of the First International Conference on* Washington, DC, November 1–5, C. G. Interrante and G. M. P. ASM International, Materials Park, OH.
17. Matsuda, F., et al. 1984. Disbonding between 2.25%Cr–1%Mo steel and overlaid austenitic stainless steel by means of electrolytic hydrogen charging technique, *Transactions of JWRI*, 13(2):263–272.
18. Kotecki, D. J., and Rajan, V. B. 1997. Submerged arc fillet welds between mild steel and stainless steel, *Welding Journal*, 76(2):57s–66s.
19. Avery, R. E. 1991. Pay attention to dissimilar metal welds: guidelines for welding dissimilar metals, *Chemical Engineering Progress*, May; also, Nickel Development Institute Series 14-018.
20. Kotecki, D. J. 2003. Unpublished research.

CHAPTER 10

WELDABILITY TESTING

10.1 INTRODUCTION

The definition of the term *weldability* varies widely and is often used to connote a material's ability to be readily fabricated, to perform in service, or both. The AWS definition represents a global approach to the term which includes aspects of both fabrication and service.

Weldability is most often used to define a material's resistance to cracking during fabrication. Thus, a material that has "good weldability" would be resistant to various forms of cracking during welding, requiring essentially no rework or repair. This chapter will review the concepts of material weldability from the standpoint of weld cracking, concentrating primarily on cracking that occurs during welding or due to postweld processing. A number of weldability tests are then described that allow quantification of a material's weldability.

Fabrication-related defects include cracking phenomena that are associated with the metallurgical nature of the weldment, and process- and/or procedure-related

Definition of Weldability (American Welding Society)

The capacity of a material to be welded under fabrication conditions imposed into a specfic, suitably designed structure and to perform satisfactorily in the intended service.

Welding Metallurgy and Weldability of Stainless Steels, by John C. Lippold and Damian J. Kotecki
ISBN 0-471-47379-0

defects. A number of different cracking mechanisms have been identified in stainless steels and have been described in the chapters dealing with the particular alloy systems. These can be broadly grouped by the temperature range in which they occur. *Hot cracking* includes those cracking phenomena associated with the presence of liquid in the system and is localized in the fusion zone and partially-melted zone (PMZ) region of the HAZ. *Warm cracking* occurs at elevated temperature in the solid state (i.e., no liquid is present in the system). These defects may occur in both the fusion zone and HAZ. *Cold cracking* occurs at or near room temperature and is usually synonymous with hydrogen-induced cracking.

10.1.1 Weldability Test Approaches

Because the concept of weldability is so broad, a wide variety of test techniques can be used to assess or quantify weldability. These tests can address issues associated with the process and/or welding procedures, cracking during welding and postweld processing, and service performance and structural integrity. Mechanical testing is commonly required by specification to qualify welding procedures to meet strength, ductility, and toughness requirements. Other tests, including fatigue, fracture toughness, and corrosion tests are often used to ensure the performance of welded structures. Almost all of these test procedures have been standardized and can be found in publications of the American Welding Society (AWS), American Society for Testing and Materials (ASTM), and other professional societies and authorized bodies. In addition, many specialized tests have been developed to address fabrication issues, especially cracking. Few of these latter tests have been standardized, and poor correlation among tests designed to evaluate or quantify the same phenomenon is common.

10.1.2 Weldability Test Techniques

Weldability test techniques can be broadly grouped into four categories: mechanical, nondestructive, service performance, and specialty. With the exception of the inspection (nondestructive) techniques, these tests are destructive in nature, requiring sectioning of welds and special sample preparation. The destructive evaluation techniques are obviously not applicable in a manufacturing or quality control environment, but should be implemented at the material selection and procedure qualification stage.

Specialty tests to evaluate susceptibility to various metallurgically related defects, such as solidification cracking or hydrogen cracking, should be conducted in the alloy development or selection stage. Failure to address issues related to susceptibility to cracking during fabrication can result in considerable cost and development time penalties.

Tests for evaluating the susceptibility to hot cracking can be separated into three categories. *Representative* or *self-restraint tests* use the inherent restraint of the test assembly to cause cracking. They are termed *representative* because they can be used to represent the actual joint configurations and restraint levels of the parts to be welded. The drawback of these tests is that there is usually little quantification of cracking susceptibility—the specimen either cracks or it does not. As a result, no specific data, such as cracking temperature range or stress–strain to cause cracking is obtained, making it difficult to compare different materials. Numerous examples exist of good resistance to cracking in representative tests, or mockups, that did not translate to resistance under actual fabrication conditions.

Simulative tests usually involve the use of tension or bending to simulate high restraint levels in welds. These tests operate under strain control, strain rate control, or stress control. Most are capable of providing some measure of cracking susceptibility that can be used to compare different materials and that may provide metallurgical information.

Hot ductility tests simply measure the strength and ductility of materials at elevated temperatures. Since weld cracking is associated with insufficient ductility relative to the imposed strain, an understanding of the material's hot ductility will often provide insight into the potential for cracking.

10.2 VARESTRAINT TEST

The *Va*riable *Restraint*, or *Varestraint*, *test* was developed in the 1960s by Savage and Lundin at Rensselaer Polytechnic Institute (RPI) [1]. It was devised as a simple, augmented strain type test that would isolate the metallurgical variables that cause hot cracking. Since its introduction, numerous modifications have been made to the original test and a number of variants are used worldwide. The test itself has never been fully standardized and a variety of test techniques are in use today. In the 1990s, Lippold and Lin at Ohio State University developed test techniques that allow the magnitude of the crack susceptible region (CSR) to be determined using Varestraint testing [2].

There are three basic types of Varestraint tests, as shown in Figure 10.1. The original test was a longitudinal type, whereby bending is applied along the length of the weld. This type of test produces cracking in both the fusion zone and adjacent HAZ. Since in some cases it is desirable to separate HAZ liquation cracking from solidification cracking, the spot and transverse tests were developed. The transverse test applies bending strain across the weld and generally restricts cracking to the fusion zone (little or no strain is applied to the HAZ). The spot Varestraint test uses a small spot weld to develop a susceptible HAZ microstructure. The sample is then bent when the weld is still molten in order to isolate cracking in the HAZ.

The gas tungsten arc welding process is most commonly used for these tests, although other fusion welding processes can also be used. Welds are usually autogenous, although filler metal deposits can be tested using special samples. Bending is usually very rapid, but some slow bending versions of the test have been developed. The augmented strain is controlled by the use of interchangeable die blocks, where the strain (ε) in the outer fibers of a sample is given by

$$\varepsilon = \frac{t}{2R + t} \tag{10.1}$$

where t is the sample thickness and R is the radius of the die block.

A number of metrics have been developed to quantify cracking susceptibility using the Varestraint test. Most of these require the measurement of crack lengths

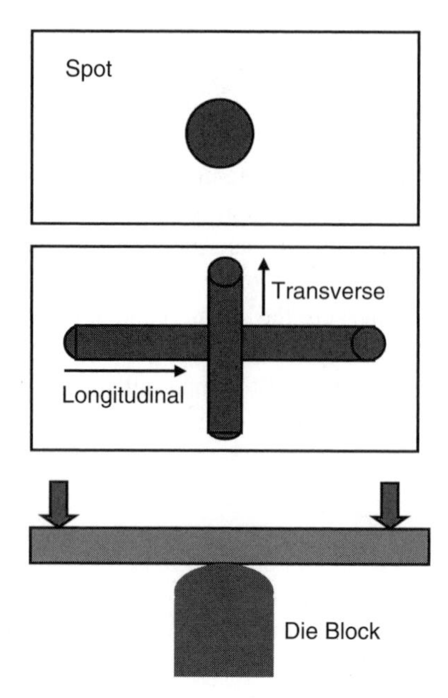

Figure 10.1 Different types of Varestraint tests.

on the surface of the as-tested sample using a low-powered (20 to 50×) microscope. *Total crack length* is the summation of all crack lengths; the *maximum crack length* is simply the length of the longest crack observed on the sample surface. More recently, the concept of *maximum crack distance* (MCD) has been adopted to allow more accurate determination of the cracking temperature range. The strain to initiate cracking, called the *threshold strain*, has also been used to quantify cracking susceptibility. The *saturated strain* represents the level of strain above which the MCD does not change. It is important to test at saturated strain when determining the maximum extent of the crack susceptible region, as described in the following sections.

10.2.1 Technique for Quantifying Weld Solidification Cracking

This section describes a technique developed by Lippold and Lin [2] for quantifying weld solidification cracking susceptibility using the Transvarestraint test. A schematic illustration of the test is shown in Figure 10.2. Above a critical strain level, designated the saturated strain, the MCD does not increase with increasing strain. This indicates that the solidification crack has propagated the full length of the crack-susceptible region. Representative cracking as it appears at the trailing edge of the weld pool is shown in Figure 10.3 for a Type 310 sample tested at 5% strain. By testing over a range of augmented strain, an MCD versus strain plot such as that shown

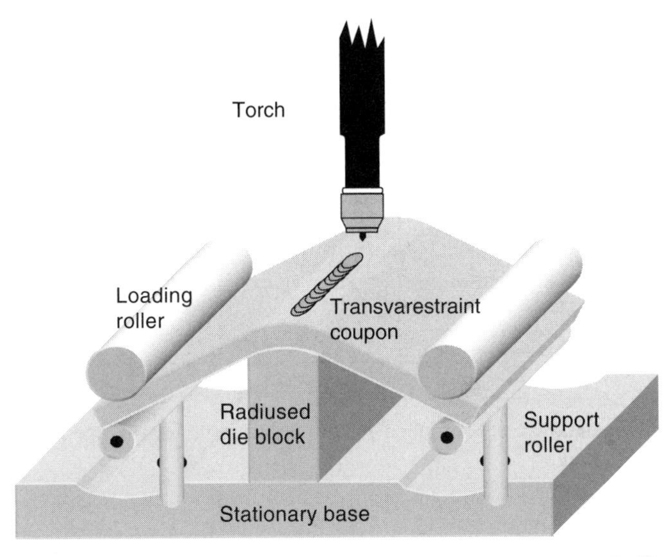

Figure 10.2 Schematic of the Transvarestraint test for evaluating weld solidification cracking susceptibility. (From Finton [3].)

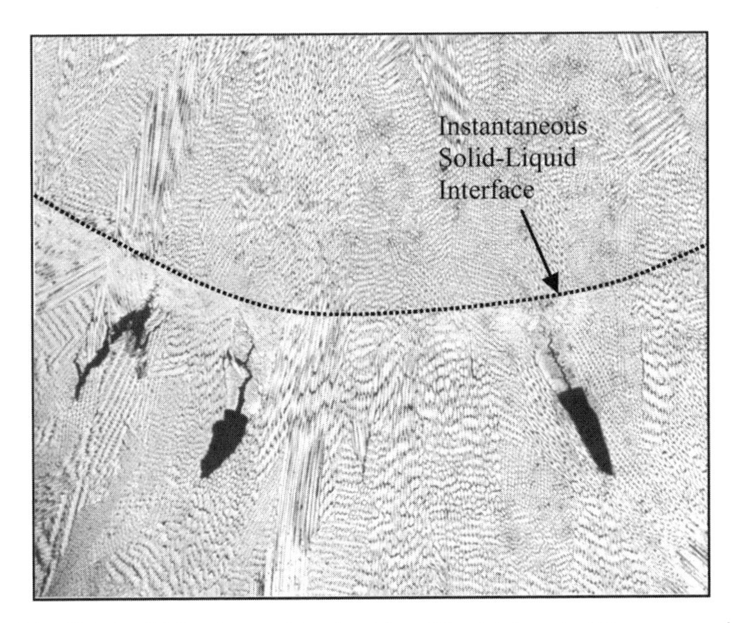

Figure 10.3 Weld solidification cracking in a Type 310 stainless steel Transvarestraint specimen. View is looking down on the surface of the sample. (From Finton [3].)

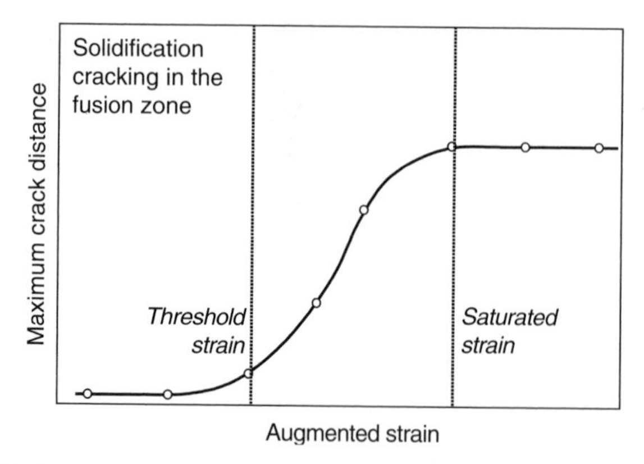

Figure 10.4 Maximum crack distance in the fusion zone versus applied strain during Transvarestraint testing.

in Figure 10.4 can be generated. In this manner, the threshold strain for cracking to occur and the saturated strain above which the MCD does not increase can be identified. Typical strain ranges over which samples are tested are 0 to 7%.

Most stainless steels exhibit saturated strain levels between 5 and 7%. Threshold strain levels are generally in the range 0.5 to 2.0%, depending on the alloy and its solidification behavior. Although the threshold strain may, in fact, be an important criterion for judging susceptibility to weld solidification cracking, the MCD above saturated strain is much easier to determine and provides a measure of the solidification cracking temperature range (SCTR). To determine the SCTR, the cooling rate through the solidification temperature range is determined by plunging a thermocouple into the weld pool. The time over which cracking occurs is approximated by the MCD above the saturated strain divided by the solidification velocity. Using this approach, the SCTR can be calculated using the relationship

$$\text{SCTR} = \text{cooling rate} \times \frac{\text{MCD}}{V} \tag{10.2}$$

where V represents the welding velocity. The concept for determining the SCTR using this approach is shown in Figure 10.5. By using a temperature rather than a crack length as a measure of cracking susceptibility, the influence of welding variables (heat input, travel speed, etc.) can be eliminated. SCTR then represents a metallurgically significant, material-specific measure of weld solidification cracking susceptibility.

SCTR values for a number of austenitic and duplex stainless steels are shown in Table 10.1. Alloys that solidify as primary ferrite (duplex stainless steels 2205 and 2507, and Types 304 and 316L) have low SCTR values, typically less than 50°C.

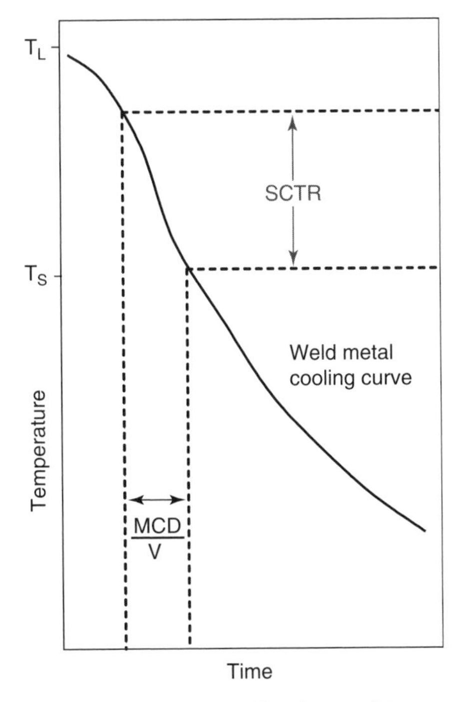

Figure 10.5 Method for determining the solidification cracking temperature range (SCTR) using the cooling rate through the solidification temperature range and maximum crack distance (MCD) at saturated strain.

TABLE 10.1 Solidification Cracking Temperature Range (SCTR) Values Obtained Using the Transvarestraint Test

Alloy	Solidification Mode	Ferrite Number	SCTR (°C)
Duplex, 2205	F	85	26
Austenitic, 304L	FA	6	31
Duplex, 2507	F	75	45
Austenitic, 316L	FA	4	49
Superaustenitic, AL6XN	A	0	115
Austenitic, 310	A	0	139
Austenitic PH, A-286	A	0	418

Alloys that solidify as austenite exhibit SCTR values above 100°C. Alloy A-286, which is notoriously susceptible to solidification and liquation cracking, has a very high SCTR value. The SCTR data allow a straightforward comparison of cracking susceptibility. These values may also allow alloy selection based on restraint conditions. For example, in high-restraint situations, SCTR values below 50°C may be required to prevent cracking, whereas for low-restraint weldments, 150°C may be sufficient.

TABLE 10.2 Variables and Variable Ranges for Transverse Varestraint Testing of Stainless Steels and Ni-Base Alloys

Variable	Range
Arc length range (in.)	0.05–0.15
Maximum voltage changes (V)	$\pm$1–1.5
Minimum specimen length (in.)	3.5
Minimum specimen width (parallel to welding direction) (in.)	3.0
Current range (A)	160–190
Travel speed range (in./min)	4–6
Augmented strain range (%)	3–7
Ram travel speed range (in./sec)	6–10

Recently, Finton [3] used a statistical approach to evaluate the variables associated with transverse Varestraint testing. This study used both austenitic stainless steels (Type 304 and 310) and Ni-base alloys (Alloys 625 and 690) to determine the statistical importance of different variables and to establish variable ranges in which testing should be conducted to give reproducible results. Based on this study, they recommended the variable ranges given in Table 10.2 for use with stainless steel and Ni-base alloys.

10.2.2 Technique for Quantifying HAZ Liquation Cracking

Susceptibility to HAZ liquation cracking can be quantified using both the Varestraint test and the hot ductility test, as described by Lin et al. [4]. The Varestraint test technique for HAZ liquation cracking differs from that described previously for weld solidification cracking, since it uses a stationary spot weld to generate a stable HAZ thermal gradient and microstructure. The technique developed by Lin et al. [4] used both an "on-heating" and an "on-cooling" approach to quantify HAZ liquation cracking. A schematic of the test is shown in Figure 10.6.

The *on-heating test* is conducted by initiating the GTA spot weld, ramping to the desired current level, and then maintaining this current level until the weld pool size stabilizes and the desired temperature gradient in the HAZ is achieved. For the Type 310 and A-286 stainless steels evaluated by Lin et al. [4] a spot weld surface diameter of approximately 12 mm was achieved after a weld time of 35 seconds. The arc is then extinguished and the load applied immediately to force the sample to conform to the die block. By using no delay time between extinction of the arc and application of load, HAZ liquation cracks initiate at the fusion boundary and propagate back into the HAZ along liquated grain boundaries (Figure 10.7).

By plotting the maximum crack length (MCL) versus strain, a saturated strain can be determined which defines the strain above which the maximum crack length does not change. Examples of on-heating MCL versus strain plots for Type 310 and

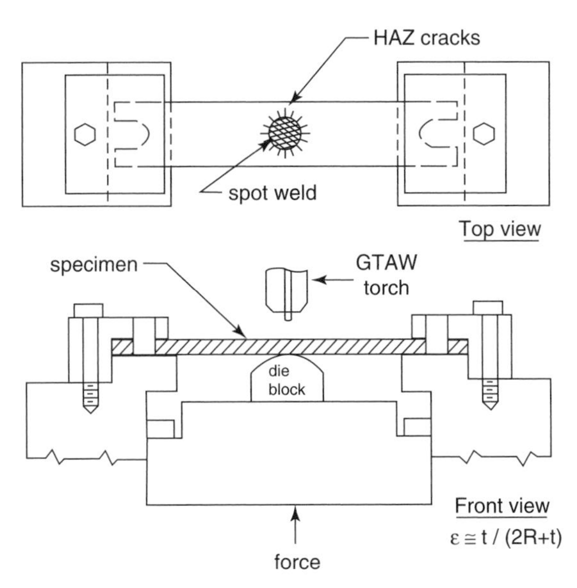

Figure 10.6 Schematic illustration of the spot Varestraint test. (From Lin et al. [4]. Courtesy of the American Welding Society.)

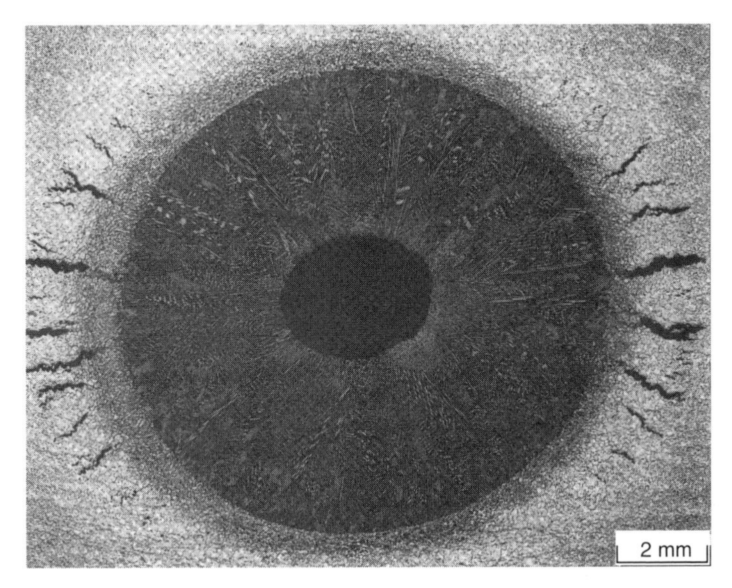

Figure 10.7 Spot Varestraint test sample (plan view) of A-286 stainless steel tested at 5% strain.

A-286 stainless steels are shown in Figure 10.8*a*. Note that a threshold strain for cracking cannot be identified for A-286 and that both materials achieve saturated strain at 3%.

For the *on-cooling test*, the same procedure as described above is used, but after the arc is extinguished there is a delay before the sample is bent. By controlling the delay, or cooling, time, the weld is allowed to solidify and the temperature in the PMZ drops until eventually the liquid films along grain boundaries are solidified completely. By plotting MCL versus cooling time, the time required for liquid films in the PMZ to solidify can be determined. This is shown in Figure 10.8*b*. Note that over 4 seconds is required before cracking disappears in A-286, indicating that the grain boundary liquid films persist to quite low temperatures.

By measuring the temperature gradient in the HAZ using implanted thermocouples, it is possible to determine the thermal crack-susceptible region (CSR) surrounding the weld within which HAZ liquation cracking is possible. As described by Lin et al. [4]

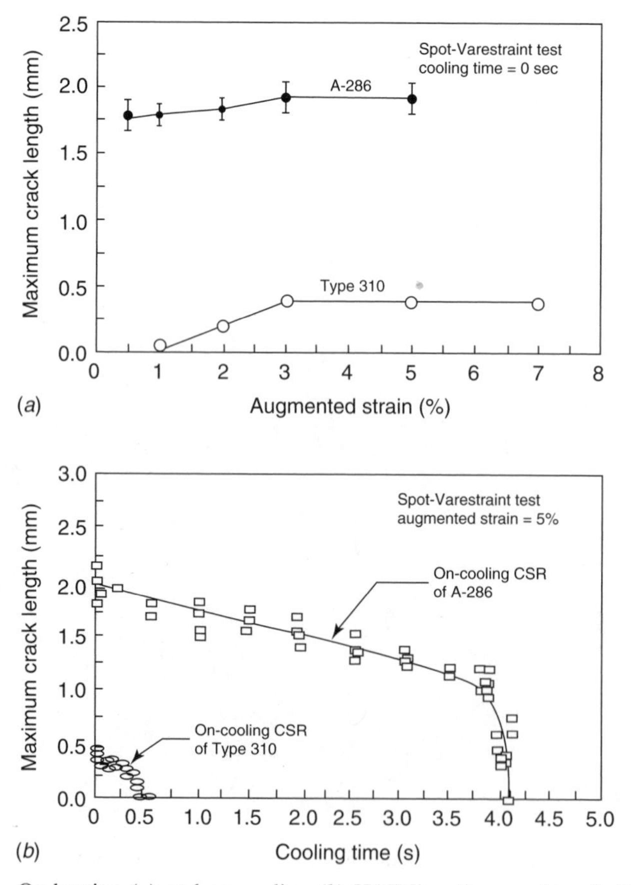

Figure 10.8 On-heating (*a*) and on-cooling (*b*) HAZ liquation cracking in Type 310 and A-286 as measured by the spot Varestraint test. (From Lin et al. [4]. Courtesy of the American Welding Society.)

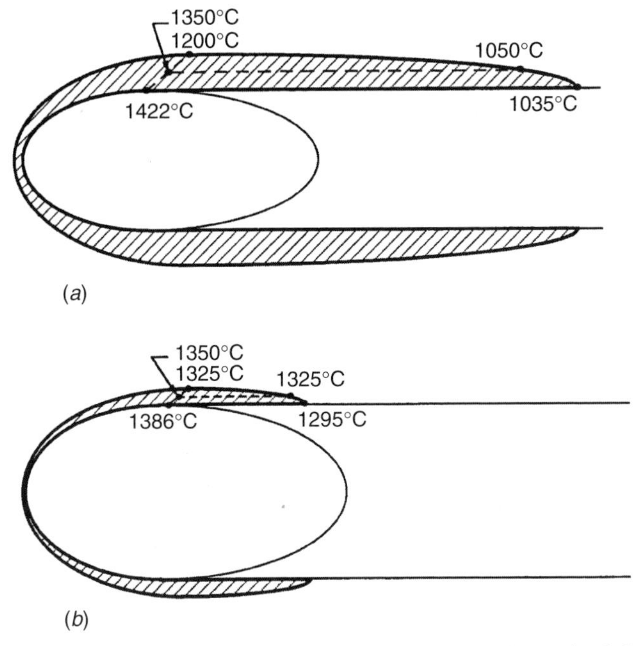

Figure 10.9 The thermal crack-susceptible region (CSR) as determined from the spot Varestraint test for (*a*) A-286 and (*b*) Type 310. (From Lin et al. [4]. Courtesy of the American Welding Society.)

this is done by converting the MCL at saturated strain for both on-heating and on-cooling spot Varestraint tests to a temperature by multiplying the MCL by the temperature gradient in the HAZ. Using this approach, it is possible to describe the region around a moving weld pool that is susceptible to HAZ liquation cracking.

Plots of the thermal CSR for A-286 and Type 310 are shown in Figure 10.9. Note that the width of the CSR at the periphery of the weld is 222°C for A-286 and only 61°C for Type 310. It is also interesting that based on the on-cooling data, the PMZ in A-286 does not fully solidify until the temperature reaches 1035°C, whereas for Type 310 it is 1295°C. This technique provides a quantifiable method to determine the precise temperature ranges within which cracking occurs and allows differences in HAZ liquation cracking susceptibility among materials to be measured readily.

10.3 HOT DUCTILITY TEST

Elevated temperature ductility can provide some indication of a material's weldability, since cracking is usually associated with an exhaustion of available ductility. Most hot ductility testing involves both heating (or "on heating") and cooling (or "on cooling") tests. To properly simulate the heating and cooling rates associated with welding, special equipment has been developed to heat and cool small laboratory samples rapidly. The most widely used test machine for accomplishing this was developed by Savage and Ferguson at RPI in the 1950s [5]. This machine was called the Gleeble and is now produced commercially by DSI, Inc. The Gleeble uses resistance heating (I^2R) to heat small samples under precise temperature control and to cool them by conduction through water-cooled copper grips. Heating rates up to 10,000°C/s are possible. The Gleeble also is capable of mechanically testing samples at any point along the programmed thermal cycle.

Hot ductility testing will allow a ductility "signature" to be developed for a material. This signature will exhibit several distinct features, as shown schematically in Figure 10.10. When ductility is determined on heating, most materials will exhibit an increase in ductility with increasing temperature, followed by a fairly abrupt drop. This drop is associated with the onset of melting. The temperature at which ductility drops to zero is termed the *nil ductility temperature* (NDT). At the NDT, the material still has measurable strength. Additional tests are conducted to determine the point at which the strength is zero, termed the *nil strength temperature* (NST).

To determine the on-cooling ductility curve, samples are heated to the NST (or a temperature between NDT and NST), cooled to a preprogrammed temperature, and tested. The point at which measurable ductility is observed is termed the *ductility recovery temperature* (DRT). At the DRT, liquid formed upon heating above the NDT has solidified to the point that the sample exhibits some ductility when pulled in tension.

Examples of hot ductility test results for Type 310 and A-286 from the work of Lin [6] are shown in Figure 10.11. These are the same materials that were described previously for the spot Varestraint test. Note that the on-heating and on-cooling ductility curves for Type 310 are nearly identical and that the NDT and DRT are essentially equivalent. The NST is only 25°C above the NDT. This indicates that grain boundary liquid films that form between the NDT and NST on-heating solidify on cooling at 1325°C. Because of the narrow temperature range over which liquid films are present in the HAZ/PMZ, this material would have good resistance to HAZ liquation cracking.

A photomicrograph of an A-286 sample heated to the NST temperature is shown in Figure 10.12. Note the complete coverage of the grain boundaries by liquid films. The

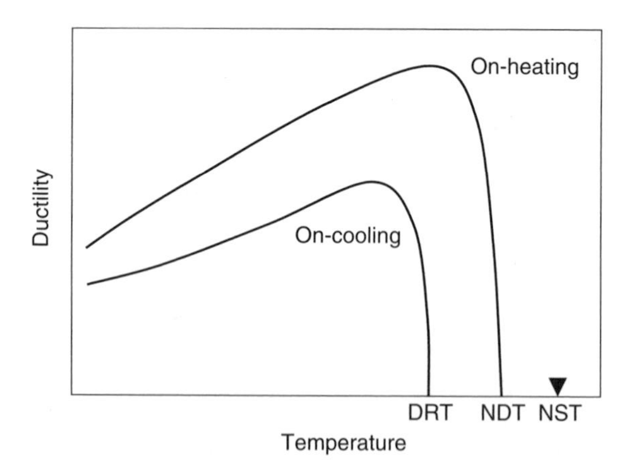

Figure 10.10 Schematic illustration of a hot ductility signature showing on-heating and on-cooling ductility curves and the location of the nil ductility temperature (NDT), nil strength temperature (NST), and ductility recovery temperature (DRT).

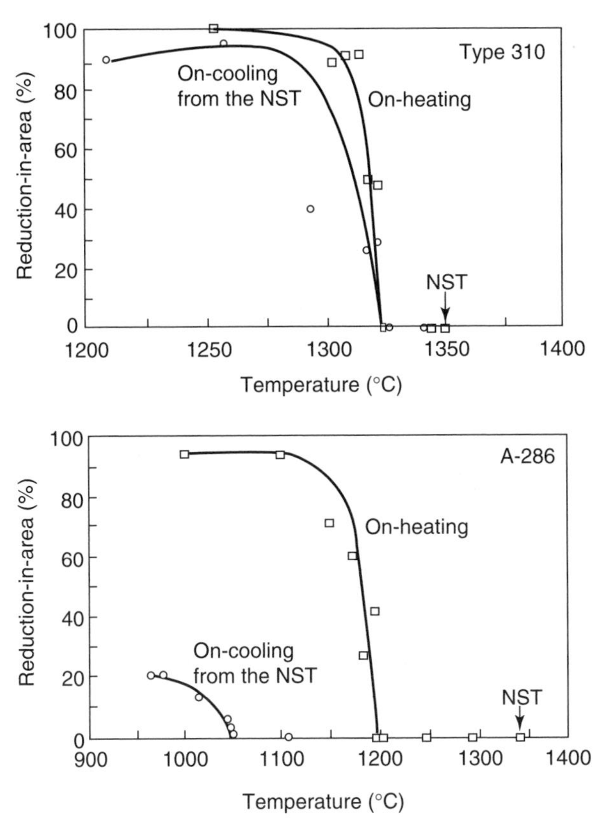

Figure 10.11 Hot ductility test results for Type 310 and A-286 austenitic stainless steels. (From Lin et al. [4]. Courtesy of the American Welding Society.)

mechanism for this was described in Chapter 8. In contrast to the 310 case, the on-heating and on-cooling ductility curves are quite different for A-286 (Figure 10.11, bottom). The NDT for A-286 is approximately 1200°C while the NST is 1350°C. After heating to the NST, on-cooling ductility does not recover until approximately 1050°C. Thus, grain boundary liquid films are present over a 300°C temperature range below the NST. This produces a very wide HAZ/PMZ region and renders the material susceptible to HAZ liquation cracking.

Similar to the spot Varestraint test, the hot ductility results (NDT, NST, and DRT from the hot ductility curves) can be used to define the crack-susceptible region (CSR) in the HAZ. In advance of the weld, liquation will begin when the material is heated above the NDT. The difference between the NDT and the liquidus temperature at the fusion boundary then defines the width of the on-heating liquation temperature range. For convenience, the NST is used to approximate the liquidus temperature.

In the cooling portion of the HAZ, liquid films in regions heated between NDT and NST will solidify. The value of DRT varies as a function of the peak temperature and is the lowest at the point closest to the fusion boundary. The difference between NST and DRT will then represent the maximum temperature range over which liquid films are

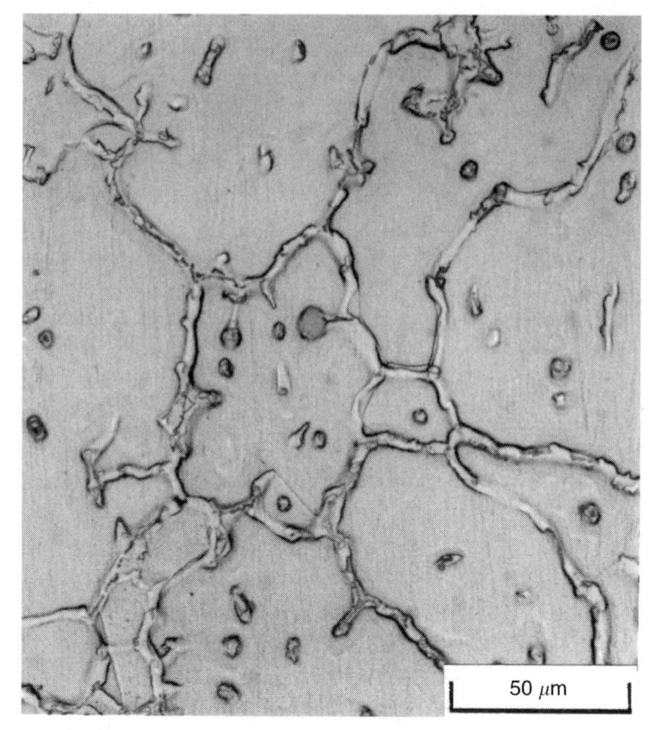

Figure 10.12 A-286 after heating to the nil strength temperature (1350°C). (From Lin [6].)

present in the HAZ. This value can then be used to define the liquation cracking temperature range (LCTR). The LCTR is a measure of liquation cracking susceptibility, similar to SCTR for the weld metal. For Type 310 the LCTR is 25°C, while for A-286 it is 300°C.

Unfortunately, there is no standardized procedure for generating hot ductility curves. The following summarizes the procedure used by Lin [6] to generate the data shown in Figure 10.11.

1. Specimens were 0.25 in. (6.35 mm) in diameter and 4.0 in. (100 mm) long with threaded ends.
2. The specimen free span (Gleeble jaw spacing) was 1.0 in. (25 mm). All tests were run in argon.
3. The NST was determined by heating at a linear rate of 200°F/sec (111°C/s) under a static load of 10 kg until the sample failed.
4. On-heating tests were conducted by heating to the desired peak temperature in 12 seconds and then pulling the samples to failure at a rate of 2 in./sec (50 mm/s). The on-heating time was based on the NST test.
5. On-cooling tests were performed after heating the sample to the NST in 12 seconds and then cooling to the desired temperature at 50°C/s. On-cooling samples were also pulled to failure at 2 in./sec.

10.4 FISSURE BEND TEST

Liquation cracks in the reheated zones of fully austenitic stainless steel weld deposits were known from the 1940s, and even then were referred to as *fissures*. These short cracks, typically less than 2 mm in any direction, have little effect on the tensile or yield strength of weld metal, but in sufficient numbers can reduce the tensile elongation [7]. These cracks are usually very tight and are not seen on the surface of an all-weld metal tensile specimen until after yielding is well under way. Depending on the exact relationship between reheated zones and the surface of the tensile specimen, a given specimen may show many fissures, or only relatively few.

Lundin et al. [8] realized that consistent location and quantification of fissures required consistent search of the reheated zones just below the weld surface, along with plastic deformation to open the fissures and make them visible. The fissure bend test they devised is well suited to locating fissures in deposited weld metal. The test consists of two layers of overlapping weld beads deposited under high-restraint conditions. Figure 10.13 shows a plan view of fissure bend specimens on the restraint fixture. Lundin et al. used fillet welds to anchor the test specimen to the restraint plate to determine if that increased the number of fissures and concluded that it did not [8]. Mechanical restraint is easier to remove, so they recommend it. Figure 10.14 shows the specimen after removal from the welding fixture, with specimen dimensions. They recommended a specific weld deposition sequence and number of beads per layer, completing the first layer before beginning the second layer, as shown in Figure 10.15*a*, to obtain reproducible results. Lundin and Spond [9] subsequently noted that stacking two passes before completing the first layer led to a

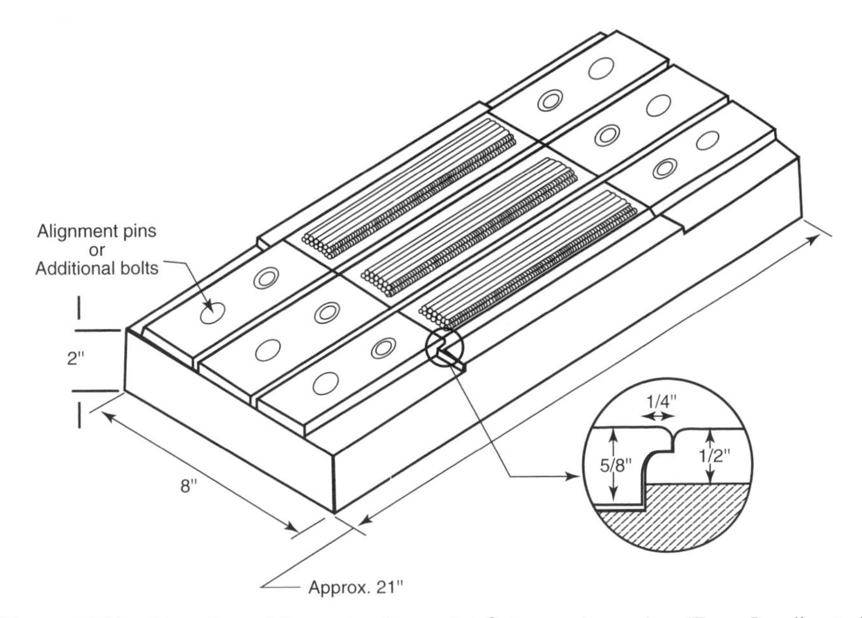

Figure 10.13 Plan view of fissure bend restraint fixture and samples. (From Lundin et al. [8]. Courtesy of the American Welding Society.)

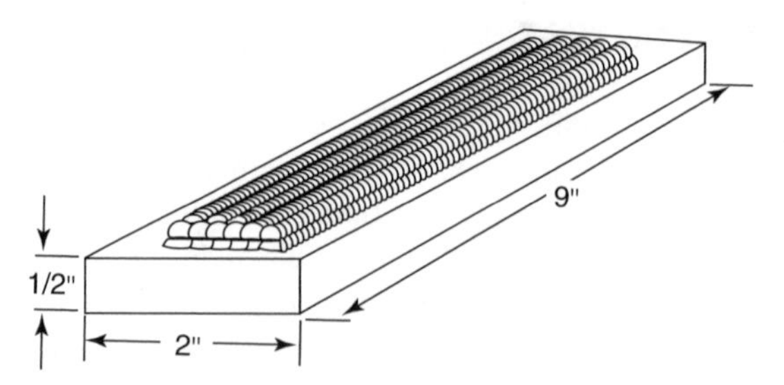

Figure 10.14 Welded fissure bend test specimen after removal from restraint fixture. (From Lundin et al. [8]. Courtesy of the American Welding Society.)

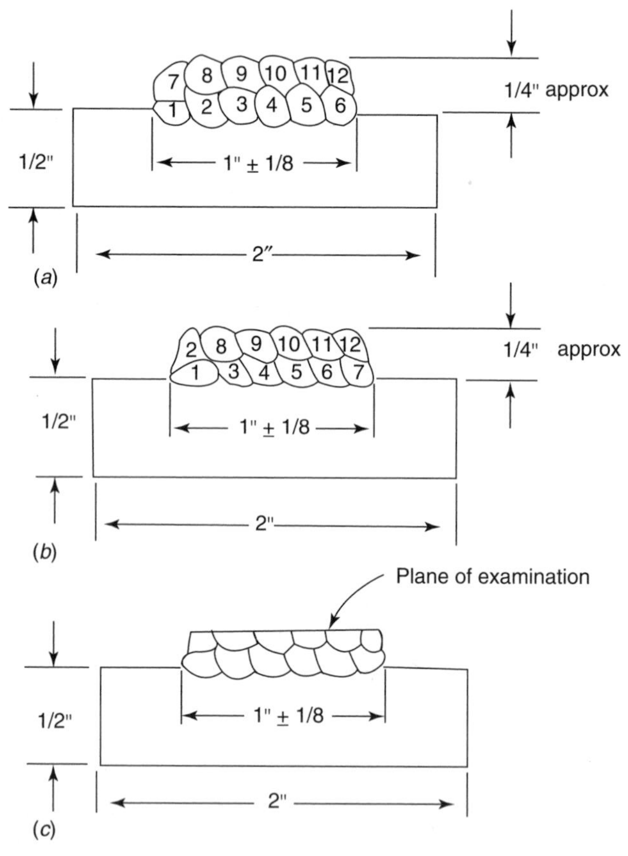

Figure 10.15 Fissure Bend test cross section showing weld pass sequences evaluated and plane of examination. (From Lundin et al. [8,9]. Courtesy of the American Welding Society.)

double-reheat cycle for the top bead in the stack, one reheat when bead 3 is deposited and a second reheat when bead 8 is deposited, as can be seen in Figure 10.15*b*. The double-reheat cycle produced a great increase in the number of fissures found in the reheated portions of bead 2 compared to bead 8 in Figure 10.15*a*.

After welding, the sample surface is ground, removing only enough metal to smooth most of the surface, with the grinding marks in the longitudinal direction. The desired plane of examination is just below the original deposited surface, as shown in Figure 10.15*c*. After grinding smooth, the sample surface is then inspected using dye penetrant to reveal open fissures, which are counted in the central 4 in. (100 mm) of the specimen length. A grid is then laid out on the surface for Ferrite Number determinations, as shown in Figure 10.16. After FN determinations, the specimen is bent over a mandrel until a stop is reached, using the bending fixture of Figure 10.17 to open all

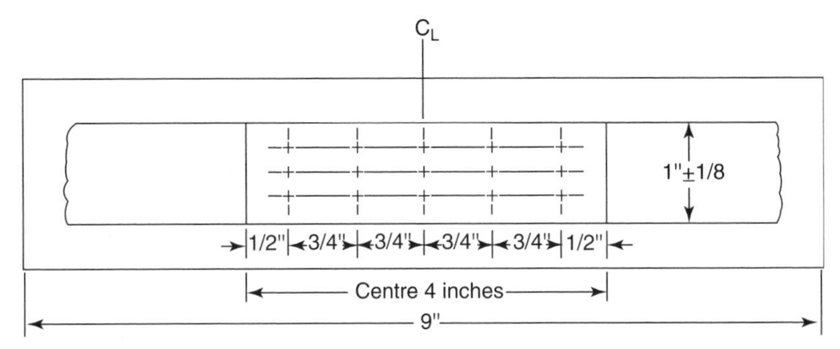

Figure 10.16 Grid layout for location of Ferrite Number determinations. (From Lundin [10]. Courtesy of the American Welding Society.)

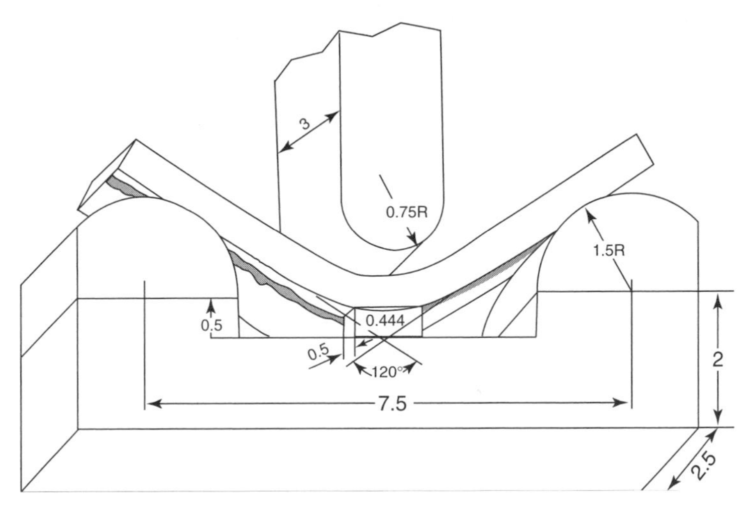

Figure 10.17 Fissure Bend test specimen and Bend test fixture with stop. All dimensions in inches. (From Lundin [10]. Courtesy of the American Welding Society.)

fissures. Again, the fissures found in the central 4 in of weld length are counted, using a 10× stereomicroscope. Normally, there is a significant increase in the fissure count from that observed in the dye penetrant inspection before bending.

Done properly, the Fissure Bend test is capable of being quite discriminating of fissuring tendencies. Lundin [10] used the test to characterize stainless steel weld metals of various low ferrite contents to find the minimum FN needed to avoid fissures with each alloy type. Figure 10.18 shows results obtained for AWS E308L-16-covered electrodes. At 1 FN, nearly 40 fissures were counted. The fissure count decreases continuously with increasing FN until, at about 3 FN and above, no fissures are found. Comparison results are shown in Figure 10.19 for several covered electrode deposits. It can be seen that near 0 FN, all alloys tested produced numerous fissures. As the FN increased, the fissure count decreased until no fissures could be found. The decrease in fissures for increasing FN is monotonic, which would seem to indicate good reproducibility of the test. However, a statistical evaluation of reproducibility using a round robin is lacking.

Lundin and Spond [9] also experimented with two and three reheat cycles by using repeated GTA passes, aligned precisely with the original weld penetration pattern, without filler metal, over the surface of a fissure bend specimen. The fissure count in any given location continued to increase from the second to the third reheat

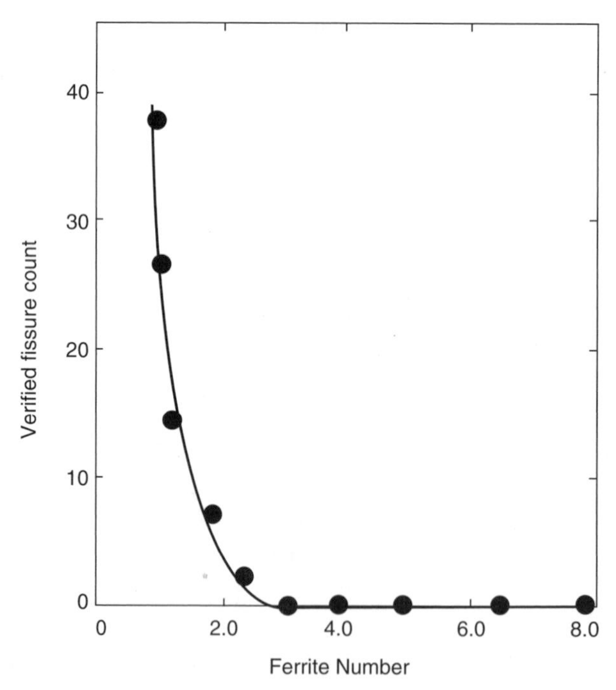

Figure 10.18 Fissure Count versus Ferrite Number for AWS E308L-16 weld metal. (From Lundin [10]. Courtesy of the American Welding Society.)

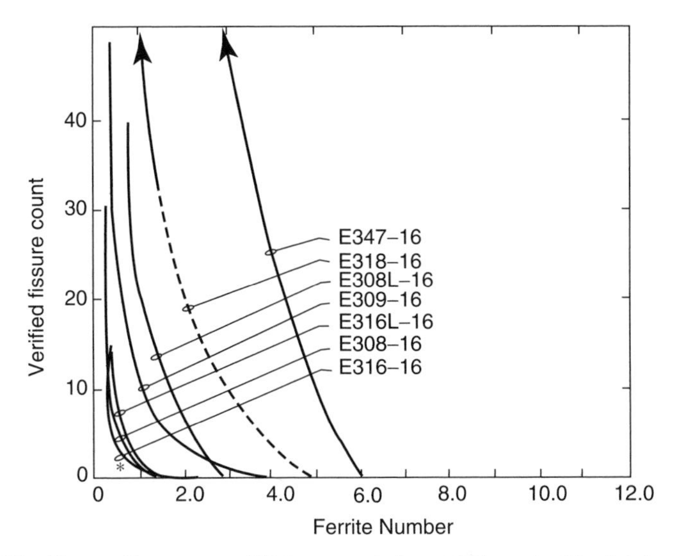

Figure 10.19 Fissure Count versus FN for several almost fully austenitic stainless steel weld filler metals. (From Lundin [10]. Courtesy of the American Welding Society.)

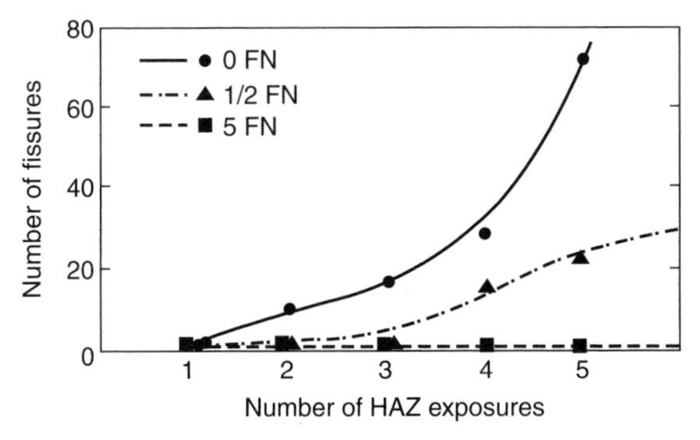

Figure 10.20 Fissure Count versus number of GTA remelt cycles in austenitic stainless steel weld metal. (From Lundin and Chou [11]. Courtesy of the American Welding Society.)

cycle, simulating the situation of defect removal and repair welding. Later, Lundin and Chou [11] reported extending this procedure of GTA remelting to as many as five cycles, and for low-FN weld metals, the number of fissures continued to increase with increasing number of cycles. Figure 10.20 shows results for up to five cycles on weld metals of three initial ferrite levels. This technique can be useful for evaluating the fissuring tendencies of both initial fabrication and multiple repairs. The fissure bend test seems best suited to finding and quantifying liquation cracks, although it is likely that any ductility dip cracks present in the weld metal would also be identified using this test.

10.5 STRAIN-TO-FRACTURE TEST

As discussed in Chapter 6, ductility-dip cracking (DDC) is a solid-state, intergranular cracking mechanism observed most frequently in fully austenitic weld metals and HAZs. Although DDC can occur during Varestraint and Fissure Bend testing, it is difficult to identify and quantify since it occurs just below the temperature range in which liquation cracking occurs. The hot ductility test can also be used to detect susceptibility to DDC, but its sensitivity to the onset of DDC is low.

To better quantify susceptibility to DDC, the strain-to-fracture (STF) test was developed in 2002 by Nissley and Lippold at The Ohio State University [12,13]. The STF test employs a "dogbone" tensile sample with a GTA spot weld applied in the center of the gage section. The spot weld is made under controlled solidification conditions using current downslope control. This results in an essentially radial array of migrated grain boundaries within the spot weld. A schematic of a STF sample is shown in Figure 10.21.

Samples are then tested in a Gleeble™ thermo-mechanical simulator at different temperatures and strains. For stainless steels, temperature and strain ranges are typically from 650 to 1200°C (1200 to 2190°F) and 0 to 20%, respectively. After testing at a specific temperature–strain combination, the sample is examined under a binocular microscope at 50× to determine if cracking has occurred. The number of cracks present on the surface is counted. Using these data, a temperature versus strain envelope is developed that defines the regime within which DDC may occur. Both a threshold strain for cracking (ε_{min}) and ductility-dip temperature range (DTR) can be extracted from these curves. Temperature–strain curves are shown in Figure 10.22 for Type 310, Type 304, and the super-austenitic alloy AL6XN. Based on these curves, Type 310 would be expected to have the highest susceptibility to DDC since the DTR at 15% strain is 400°C and ε_{min} is approximately 5%.

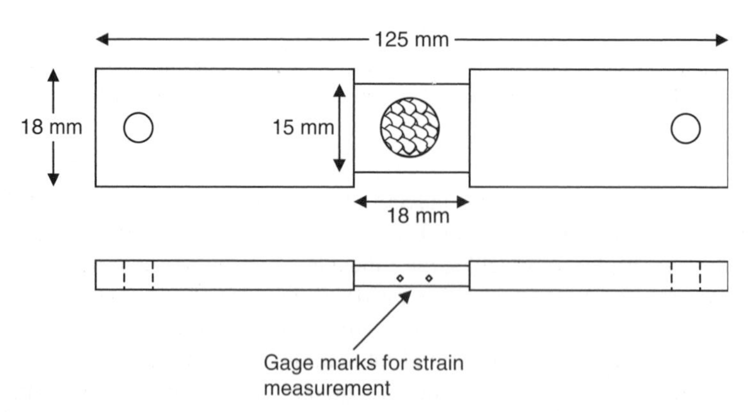

Figure 10.21 Schematic of the strain-to fracture test sample with GTA spot weld in the gage section.

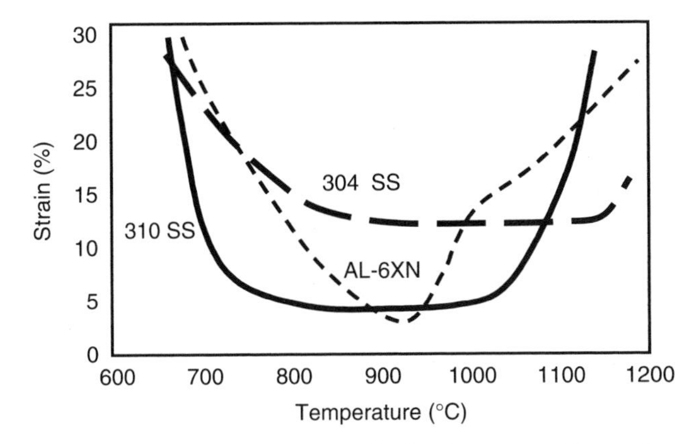

Figure 10.22 Strain-to-fracture test results for three austenitic stainless steels.

This test has been shown to be remarkably sensitive to the onset of grain boundary cracking in the DDC range and has proven to be a valuable tool for studying elevated temperature embrittlement in the weld metal and HAZ. At the time of this writing, work is ongoing at The Ohio State University to further optimize the test. This test is also being used to study the composition and metallurgical variables that affect susceptibility to ductility-dip cracking.

10.6 OTHER WELDABILITY TESTS

There are numerous other weldability tests applicable to stainless steels. Many do not easily lend themselves to producing quantitative results, as the tests described above do. Two that do lend themselves to quantitative results are the Sigmajig test developed by Goodwin [14] and the PVR test described by Folkhard [15]. The Sigmajig test is particularly useful for evaluating the cracking susceptibility of sheet materials. Attempts to correlate measurements of weldability from one test method to another have not been very successful, however. Reasons for this failure to develop satisfactory correlations are not entirely clear. A round robin of hot cracking tests was conducted within the International Institute of Welding, and the results are summarized by Wilken [16], along with attempts to account for the variations observed.

REFERENCES

1. Savage, W. F., and Lundin, C. D. 1965. The varestraint test, *Welding Journal*, 44(10): 433s–442s.
2. Lippold, J. C., and Lin, W. 1996. Weldability of commercial Al–Cu–Li alloys, in *Proceedings of ICAA5, Aluminum Alloys: Their Physical and Mechanical Properties*, J. H. Driver et al., eds., Trans Tech Publications, Enfield, NH, pp.1685–1690.

3. Finton, T. 2003. Standardization of the transvarestraint test, M.S. thesis, Ohio State University, Columbus, OH.
4. Lin, W., Lippold, J. C., and Baeslack, W. A. 1993. An investigation of heat-affected zone liquation cracking, 1: a methodology for quantification, *Welding Journal*, 71(4): 135s–153s.
5. Nippes, E. F., and Savage, W. F. 1955. An investigation of the hot ductility of high-temperature alloys, *Welding Journal*, 34(4):183s–196s.
6. Lin, W. 1991. A methodology for quantifying HAZ liquation cracking susceptibility, Ph.D. dissertation, Ohio State University, Columbus, OH.
7. Campbell, H. C., and Thomas, R. D., Jr. 1946. The effect of alloying elements on the tensile properties of 25–20 weld metal, *Welding Journal*, 25(11):760s–768s.
8. Lundin, C. D., DeLong, W. T., and Spond, D. F. 1976. The fissure bend test, *Welding Journal*, 55(6):145s–151s.
9. Lundin, C. D., and Spond, D. F. 1976. The nature and morphology of fissures in austenitic stainless steel weld metals, *Welding Journal*, 55(11):356s–367s.
10. Lundin, C. D. 1975. Ferrite fissuring relationship in austenitic stainless steel weld metals, *Welding Journal*, 54(8):241s–246s.
11. Lundin, C. D., and Chou, C. P. D. 1985. Fissuring in the "hazard HAZ" region of austenitic stainless steel welds, *Welding Journal*, 64(4):113s–118s.
12. Nissley, N. E., and Lippold, J. C. 2003. Ductility-dip cracking susceptibility of austenitic alloys, in *Trends in Welding Research, Proceedings of the 6th International Conference*, ASM International, Materials Park, OH, pp. 64–69.
13. Nissley, N. E., and Lippold, J. C. 2003. Development of the strain-to-fracture test for evaluating ductility-dip cracking in austenitic alloys, *Welding Journal*, 82(12):355s–364s.
14. Goodwin, G. M. 1987. Development of a new hot-cracking test: the Sigmajig, *Welding Journal*, 66(2):33s–38s.
15. Folkhard, E. 1988. *Welding Metallurgy of Stainless Steels*, Springer-Verlag, New York, pp. 153–157.
16. Wilken, K. 1999. *Investigation to Compare Hot Cracking Tests: Externally Loaded Specimen*, IIW Document IX-1945–99, International Institute of Welding, Paris.

APPENDIX 1

NOMINAL COMPOSITIONS OF STAINLESS STEELS

Welding Metallurgy and Weldability of Stainless Steels, by John C. Lippold and Damian J. Kotecki
ISBN 0-471-47379-0

Nominal Compositions of Stainless Steels

Type	UNS No.	Microstructure	Nominal Composition (%)											
			C	Mn	Si	Cr	Ni	Mo	Nb	Cu	N	Al	Ti	Other
25-12	Casting	Austenitic	0.3	1.2	0.9	25.5	12.0	—	—	—	—	—	—	—
CA6N	Casting	Martensitic	0.04	0.25	0.50	11.50	7.00	—	—	—	—	—	—	—
CA6NM	Casting	Martensitic	0.04	0.50	0.50	12.8	4.0	0.7	—	—	—	—	—	—
CA15	Casting	Martensitic	0.10	0.50	0.75	12.75	—	—	—	—	—	—	—	—
CA15M	Casting	Martensitic	0.10	0.50	0.30	12.75	—	0.60	—	—	—	—	—	—
CA28MWV	Casting	Martensitic	0.24	0.75	0.50	11.75	0.75	1.10	—	—	—	—	—	W: 1.1; V: 0.25
CA40	Casting	Martensitic	0.30	0.50	0.75	12.75	—	—	—	—	—	—	—	—
CA40F	Casting	Martensitic	0.30	0.50	0.75	12.75	—	—	—	—	—	—	—	S: 0.30
CB6	Casting	M + F	0.04	0.50	0.50	16.50	4.50	—	—	—	—	—	—	—
CB7Cu-1	Casting	Martensitic PH	0.04	0.35	0.5	16.6	4.1	—	0.25	2.85	—	—	—	—
CB7Cu-2	Casting	Martensitic PH	0.04	0.35	0.5	14.8	5.0	—	0.25	2.85	—	—	—	—
CB30	Casting	M + F	0.20	0.50	0.75	19.50	—	—	—	—	—	—	—	—
CC50	Casting	M + F	0.30	0.50	0.75	28.00	—	—	—	—	—	—	—	—
CD3MCuN	Casting	Duplex	0.02	0.60	0.55	25.30	6.10	3.35	—	1.65	0.28	—	—	—
CD3MN	Casting	Duplex	0.02	0.75	0.50	22.25	5.50	3.00	—	—	0.20	—	—	—
CD3MWCuN	Casting	Duplex	0.02	0.50	0.50	25.0	7.5	3.5	—	0.75	0.25	—	—	W: 0.75
CD4MCu	Casting	Duplex	0.03	0.50	0.50	25.50	5.20	2.0	—	3.00	—	—	—	—
CD4MCuN	Casting	Duplex	0.03	0.50	0.50	25.50	5.20	2.00	—	3.00	0.18	—	—	—
CD6MN	Casting	Duplex	0.04	0.50	0.50	25.50	5.00	2.10	—	—	0.20	—	—	—
CE3MN	Casting	Duplex	0.02	0.75	0.50	25.00	7.00	4.50	—	—	0.20	—	—	—
CE8MN	Casting	Duplex	0.04	0.50	0.75	24.00	9.50	3.8	—	—	0.20	—	—	—
CE20N	Casting	Austenitic	0.10	0.75	0.75	24.5	9.5	—	—	—	0.14	—	—	—
CE30	Casting	Duplex	0.15	0.75	1.00	28.00	9.50	—	—	—	—	—	—	—
CF3, CF3A	Casting	Austenitic	0.02	0.75	1.00	19.00	10.00	—	—	—	—	—	—	—
CF3M	Casting	Austenitic	0.02	0.75	0.75	19.00	11.00	2.50	—	—	—	—	—	—
CF3MN	Casting	Austenitic	0.02	0.75	0.75	19.50	11.00	2.50	—	—	0.15	—	—	—
CF8, CF8A	Casting	Austenitic	0.04	0.75	1.00	19.50	9.50	—	—	—	—	—	—	—

CF8C	Casting	Austenitic	0.04	0.75	1.00	19.50	10.50	—	0.60	—	—	—	—	—
CF8M	Casting	Austenitic	0.04	0.75	1.00	19.50	10.50	2.50	—	—	—	—	—	—
CF10	Casting	Austenitic	0.07	0.75	1.00	19.5	9.5	—	—	—	—	—	—	—
CF10M	Casting	Austenitic	0.07	0.75	1.00	19.5	10.5	2.5	—	—	—	—	—	—
CF10MC	Casting	Austenitic	0.05	0.75	0.75	16.5	14.4	2.0	0.8	—	—	—	—	—
CF10SMnN	Casting	Austenitic	0.05	8.00	4.00	17.00	8.50	—	—	—	0.13	—	—	—
CF16F	Casting	Austenitic	0.08	0.75	1.00	19.50	10.50	—	—	—	—	—	—	Se: 0.28
CF16Fa	Casting	Austenitic	0.08	0.75	1.00	19.50	10.50	0.60	—	—	—	—	—	S: 0.30
CF20	Casting	Austenitic	0.10	0.75	1.00	19.50	9.50	—	—	—	—	—	—	—
CG3M	Casting	Austenitic	0.02	0.75	0.75	19.5	11.0	3.5	—	—	—	—	—	—
CG6MMN	Casting	Austenitic	0.04	5.00	0.50	22.00	12.50	2.2	0.20	—	0.30	—	—	V: 0.20
CG8M	Casting	Austenitic	0.04	0.75	0.75	19.5	11.0	3.5	—	—	—	—	—	—
CG12	Casting	Austenitic	0.06	0.75	1.00	21.50	11.50	—	—	—	—	—	—	—
CH8	Casting	Austenitic	0.04	0.75	0.75	24.0	13.5	—	—	—	—	—	—	—
CH10	Casting	Austenitic	0.06	0.75	1.00	24.00	13.50	—	—	—	—	—	—	—
CH20	Casting	Austenitic	0.10	0.75	1.00	24.00	13.50	—	—	—	—	—	—	—
CK3MCuN	Casting	Austenitic	0.015	0.60	0.50	20.0	18.5	6.5	—	0.75	0.21	—	—	—
CK20	Casting	Austenitic	0.10	1.00	1.00	25.00	20.50	—	—	—	—	—	—	—
CK35MN	Casting	Austenitic	0.02	1.00	0.50	23.00	21.00	6.4	—	—	0.26	—	—	—
CN3M	Casting	Austenitic	0.02	1.00	0.50	21.00	25.00	5.0	—	—	—	—	—	—
CN3MCu	Casting	Austenitic	0.02	0.75	0.50	20.5	29.0	2.5	—	3.2	—	—	—	—
CN3MN	Casting	Austenitic	0.02	1.00	0.50	21.00	24.50	6.5	—	—	0.22	—	—	—
CN7M	Casting	Austenitic	0.04	0.75	0.75	20.50	29.00	2.5	—	3.50	—	—	—	—
CN7MS	Casting	Austenitic	0.04	0.50	3.00	19.00	23.50	2.75	—	1.75	—	—	—	—
CPCA15	Casting	Martensitic	0.08	0.50	0.75	12.8	—	—	—	—	—	—	—	—
CPE20N	Casting	Austenitic	0.10	0.75	0.75	24.5	9.5	—	—	—	0.14	—	—	—
CPF3, CPF3A	Casting	Austenitic	0.02	0.75	1.0	19.0	10.0	—	—	—	—	—	—	—
CPF3M	Casting	Austenitic	0.02	0.75	0.75	19.0	11.0	2.5	—	—	—	—	—	—
CPF8, CPF8A	Casting	Austenitic	0.04	0.75	1.0	19.5	9.5	—	—	—	—	—	—	—

(continued)

Nominal Compositions of Stainless Steels (*Continued*)

Type	UNS No.	Microstructure	Nominal Composition (%)											
			C	Mn	Si	Cr	Ni	Mo	Nb	Cu	N	Al	Ti	Other
CPF8C	Casting	Austenitic	0.04	0.75	1.0	19.5	10.5	—	0.6	—	—	—	—	—
CPF8M	Casting	Austenitic	0.04	0.75	0.75	19.5	10.5	2.5	—	—	—	—	—	—
CPF10MC	Casting	Austenitic	0.05	0.75	0.75	16.5	14.5	2.0	0.7	—	—	—	—	—
CPH8	Casting	Austenitic	0.04	0.75	0.75	24.0	13.5	—	—	—	—	—	—	—
CPH10	Casting	Austenitic	0.05	0.75	1.0	24.0	13.5	—	—	—	—	—	—	—
CPH20	Casting	Austenitic	0.10	0.75	1.0	24.0	13.5	—	—	—	—	—	—	—
CPK20	Casting	Austenitic	0.10	0.75	0.9	25.0	20.5	—	—	—	—	—	—	—
CT15C	Casting	Austenitic	0.10	0.9	1.0	20.0	32.5	—	1.0	—	—	—	—	—
HC	Casting	Ferritic	0.2	0.5	1.0	28.0	—	—	—	—	—	—	—	—
HD	Casting	Duplex/austenitic	0.2	0.8	1.0	28.0	5.5	—	—	—	—	—	—	—
HE	Casting	Austenitic	0.35	1.0	1.0	28.0	9.5	—	—	—	—	—	—	—
HF	Casting	Austenitic	0.30	1.0	1.0	20.5	10.0	—	—	—	—	—	—	—
HH	Casting	Austenitic	0.35	1.0	1.0	26.0	12.5	—	—	—	—	—	—	—
HI	Casting	Austenitic	0.35	1.0	1.0	28.0	16.0	—	—	—	—	—	—	—
HK30	Casting	Austenitic	0.30	1.0	1.0	26.0	20.0	—	—	—	—	—	—	—
HK, HK40	Casting	Austenitic	0.40	1.0	1.0	26.0	20.0	—	—	—	—	—	—	—
HL	Casting	Austenitic	0.40	1.0	1.0	30.0	20.0	—	—	—	—	—	—	—
HN	Casting	Austenitic	0.35	1.0	1.0	21.0	25.0	—	—	—	—	—	—	—
HP	Casting	Austenitic	0.55	1.0	1.2	26	35	—	—	—	—	—	—	—
HT	Casting	Austenitic	0.55	1.0	1.2	17.0	35.0	—	—	—	—	—	—	—
HT30	Casting	Austenitic	0.30	1.0	1.2	15.0	35.0	—	—	—	—	—	—	—
HU	Casting	Austenitic	0.55	1.0	1.2	19.0	39.0	—	—	—	—	—	—	—
HW	Casting	Austenitic	0.55	1.0	1.2	12.0	60.0	—	—	—	—	—	—	—
HX	Casting	Austenitic	0.55	1.0	1.2	17.0	66.0	—	—	—	—	—	—	—
XM-32	K64152	Martensitic	0.12	0.70	0.2	11.75	2.50	1.75	—	—	0.03	—	—	V: 0.32
320	N08020	Austenitic	0.04	1.00	0.50	20.00	35.00	2.50	0.60	3.50	—	—	—	—
AL-6XN	N08367	Austenitic	0.02	1.00	0.50	21.00	24.50	6.50	—	—	0.22	—	—	—

800	N08800	Austenitic	0.05	0.75	0.50	21.0	32.5	—	—	—	—	0.38	0.38	—
800H	N08810	Austenitic	0.07	0.75	0.50	21.0	32.5	—	—	—	—	0.38	0.38	—
	N08811	Austenitic	0.08	0.75	0.50	21.0	32.5	—	—	—	—	0.38	0.38	—
904L	N08904	Austenitic	0.01	1.00	0.50	21.00	25.50	4.50	—	1.50	—	—	—	—
	N08926	Austenitic	0.010	1.00	0.25	20.0	25.0	6.50	—	1.0	0.20	—	—	—
13-8Mo PH	S13800	Martensitic PH	0.03	0.10	0.05	12.75	8.00	2.25	—	—	—	1.12	—	—
XM-13	S13800	Martensitic PH	0.03	0.10	0.05	12.75	8.00	2.25	—	—	—	1.12	—	—
15-5 PH	S15500	Martensitic PH	0.04	0.50	0.50	14.75	4.50	—	0.30	3.50	—	—	—	—
XM-12	S15500	Martensitic PH	0.04	0.50	0.50	14.75	4.50	—	0.30	3.50	—	—	—	—
632 (15-7Mo)	S15700	Semiaustenitic PH	0.05	0.50	0.50	15.00	7.00	2.50	—	—	—	1.00	—	—
17-4PH	S17400	Martensitic PH	0.04	0.50	0.50	16.25	4.00	—	0.30	4.00	—	—	—	—
630	S17400	Martensitic PH	0.04	0.50	0.50	16.25	4.00	—	0.30	4.00	—	—	—	—
635	S17600	Martensitic PH	0.04	0.50	0.50	16.75	6.75	—	—	—	—	0.20	0.80	—
17-7PH	S17700	Semiaustenitic PH	0.05	0.50	0.50	17.00	7.00	—	—	—	—	1.00	—	—
631	S17700	Semiaustenitic PH	0.05	0.50	0.50	17.00	7.00	—	—	—	—	1.00	—	—
XM-34	S18200	Ferritic	0.04	1.25	0.50	18.5	—	2.00	—	—	—	—	—	S: 0.20
	S18235	Ferritic	0.012	0.25	0.50	18.0	—	2.25	—	—	—	—	0.65	—
201	S20100	Austenitic	0.08	6.50	0.50	17.0	4.50	—	—	—	0.20	—	—	—
201L	S20103	Austenitic	0.02	6.5	0.40	17.0	4.5	—	—	—	0.20	—	—	—
201LN	S20153	Austenitic	0.02	7.0	0.40	16.8	4.5	—	—	—	0.20	—	—	—
Gall-Tough	S20161	Austenitic	0.08	5.0	3.5	16.5	5.0	—	—	—	0.14	—	—	—
	S20162	Austenitic	0.08	6.0	3.5	18.8	8.0	1.50	—	—	0.15	—	—	—
202	S20200	Austenitic	0.08	8.8	0.50	18.00	5.00	—	—	—	0.20	—	—	—
XM-1	S20300	Austenitic	0.04	5.8	0.50	17.0	5.8	—	—	2.00	—	—	—	S: 0.20
204	S20400	Austenitic	0.02	8.0	0.50	16.0	2.25	—	—	—	0.22	—	—	—
Nitronic 30	S20400	Austenitic	0.02	8.00	0.50	16.00	2.25	—	—	—	0.23	—	—	—
204Cu	S20430	Austenitic	0.08	7.8	0.50	16.5	2.50	—	—	3.0	0.15	—	—	—
205	S20500	Austenitic	0.18	14.8	0.50	17.2	1.4	—	—	—	0.36	—	—	—
Nitronic 50	S20910	Austenitic	0.04	5.00	0.40	22.00	12.50	2.25	0.20	—	0.30	—	—	V: 0.20

(*continued*)

Nominal Compositions of Stainless Steels *(Continued)*

Type	UNS No.	Microstructure	Nominal Composition (%)											
			C	Mn	Si	Cr	Ni	Mo	Nb	Cu	N	Al	Ti	Other
XM-19	S20910	Austenitic	0.04	5.00	0.40	22.00	12.50	2.25	0.20	—	0.30	—	—	V: 0.20
XM-31	S21400	Austenitic	0.06	15.0	0.65	17.8	0.50	—	—	—	0.40	—	—	—
XM-14	S21460	Austenitic	0.06	15.0	0.40	18.0	5.5	—	—	—	0.42	—	—	—
XM-17	S21600	Austenitic	0.04	8.2	0.40	19.8	6.0	2.50	—	—	0.38	—	—	—
XM-18	S21603	Austenitic	0.02	8.2	0.40	19.8	6.0	2.50	—	—	0.38	—	—	—
Nitronic 60	S21800	Austenitic	0.05	8.00	4.00	17.00	8.50	—	—	—	0.13	—	—	—
Nitronic 40	S21900	Austenitic	0.04	9.00	0.50	20.2	6.50	—	—	—	0.28	—	—	—
XM-10	S21900	Austenitic	0.04	9.00	0.50	20.2	6.50	—	—	—	0.28	—	—	—
XM-11	S21904	Austenitic	0.02	9.0	0.50	20.2	6.5	—	—	—	0.28	—	—	—
Nitronic 33	S24000	Austenitic	0.04	13.00	0.40	18.00	3.00	—	—	—	0.30	—	—	—
XM-29	S24000	Austenitic	0.04	13.00	0.40	18.00	3.00	—	—	—	0.30	—	—	—
XM-28	S24100	Austenitic	0.08	12.5	0.50	17.8	1.50	—	—	—	0.32	—	—	—
Nitronic 32	S28200	Austenitic	0.08	18.00	0.50	18.00	—	1.00	—	1.00	0.50	—	—	—
301	S30100	Austenitic	0.08	1.00	0.50	17.00	7.00	—	—	—	0.05	—	—	—
301L	S30103	Austenitic	0.02	1.00	0.50	17.00	7.00	—	—	—	0.10	—	—	—
301LN	S30153	Austenitic	0.02	1.00	0.50	17.00	7.00	—	—	—	0.15	—	—	—
302	S30200	Austenitic	0.08	1.00	0.40	18.00	9.00	—	—	—	—	—	—	—
302B	S30215	Austenitic	0.08	1.00	2.50	18.0	9.0	—	—	—	0.05	—	—	—
303	S30300	Austenitic FM	0.08	1.00	0.50	18.00	9.00	—	—	—	—	—	—	S: 0.20
XM-5	S30310	Austenitic	0.08	3.5	0.50	18.0	8.5	—	—	—	—	—	—	S: 0.30
303Se	S30323	Austenitic FM	0.08	1.00	0.50	18.00	9.00	—	—	—	—	—	—	Se: 0.2; S: 0.13
XM-2	S30345	Austenitic	0.08	1.00	0.50	18.0	9.0	0.50	—	—	—	0.80	—	S: 0.14
304	S30400	Austenitic	0.04	1.00	0.40	19.00	9.25	—	—	—	—	—	—	—
304L	S30403	Austenitic	0.02	1.00	0.40	19.00	10.00	—	—	—	—	—	—	—
304H	S30409	Austenitic	0.07	1.00	0.40	19.00	9.25	—	—	—	—	—	—	—
253MA	S30415	Austenitic	0.05	0.40	1.50	18.5	9.5	—	—	—	0.15	—	—	Ce: 0.06
302Cu	S30430	Austenitic	0.02	1.00	0.50	18.0	9.0	—	—	3.5	—	—	—	—

304N	S30451	Austenitic	0.04	1.00	0.40	19.0	9.2	—	—	—	0.13	—	—	—
XM-21	S30452	Austenitic	0.04	1.00	0.40	19.0	9.2	—	—	—	0.22	—	—	—
304LN	S30453	Austenitic	0.02	1.00	0.40	19.00	10.00	—	—	—	0.13	—	—	—
	S30454	Austenitic	0.02	1.00	0.50	19.0	9.5	—	—	—	0.23	—	—	—
304B	S30460	Austenitic	0.04	1.0	0.4	19.0	13.5	—	—	—	—	—	—	B: 0.25
304B1	S30461	Austenitic	0.04	1.0	0.4	19.0	13.5	—	—	—	—	—	—	B: 0.40
304B2	S30462	Austenitic	0.04	1.0	0.4	19.0	13.5	—	—	—	—	—	—	B: 0.0.62
304B3	S30463	Austenitic	0.04	1.0	0.4	19.0	13.5	—	—	—	—	—	—	B: 0.88
304B4	S30464	Austenitic	0.04	1.0	0.4	19.0	13.5	—	—	—	—	—	—	B: 1.1
304B5	S30465	Austenitic	0.04	1.0	0.4	19.0	13.5	—	—	—	—	—	—	B: 1.37
304B6	S30466	Austenitic	0.04	1.0	0.4	19.0	13.5	—	—	—	—	—	—	B: 1.62
304B7	S30467	Austenitic	0.04	1.0	0.4	19.0	13.5	—	—	—	—	—	—	B: 2.0
305	S30500	Austenitic	0.06	1.00	0.40	18.00	11.8	—	—	—	—	—	—	—
306	S30600	Austenitic	0.01	1.00	4.00	17.8	14.8	—	—	—	—	—	—	—
AL 611	S30601	Austenitic	0.01	0.65	5.3	17.5	17.5	—	—	—	—	—	—	—
85H	S30615	Austenitic	0.20	1.00	3.6	18.2	14.8	—	—	—	—	1.18	—	—
308	S30800	Austenitic	0.04	1.00	0.50	20.00	11.00	—	—	—	—	—	—	—
253MA	S30815	Austenitic	0.08	0.40	1.70	21.0	11.0	—	—	—	0.17	—	—	Ce: 0.06
309	S30900	Austenitic	0.10	1.00	0.50	23.00	13.50	—	—	—	—	—	—	—
309S	S30908	Austenitic	0.04	1.00	0.40	23.00	13.50	—	—	—	—	—	—	—
309H	S30909	Austenitic	0.07	1.00	0.40	23.00	13.50	—	—	—	—	—	—	—
309Cb	S30940	Austenitic	0.04	1.00	0.40	23.00	14.00	—	0.60	—	—	—	—	—
309HCb	S30941	Austenitic	0.07	1.00	0.40	23.00	14.00	—	0.80	—	—	—	—	—
310	S31000	Austenitic	0.15	1.00	0.75	25.00	20.50	—	—	—	—	—	—	—
310S	S31008	Austenitic	0.04	1.00	0.75	25.00	20.50	—	—	—	—	—	—	—
310H	S31009	Austenitic	0.07	1.00	0.40	25.00	20.50	—	—	—	—	—	—	—
310Cb	S31040	Austenitic	0.04	1.00	0.75	25.00	20.50	—	0.60	—	—	—	—	—
310HCb	S31041	Austenitic	0.07	1.00	0.40	25.00	20.50	—	0.80	—	—	—	—	—
310MoLN	S31050	Austenitic	0.01	1.00	0.25	25.00	22.00	2.10	—	—	0.12	—	—	—

(*continued*)

Nominal Compositions of Stainless Steels *(Continued)*

Type	UNS No.	Microstructure	Nominal Composition (%)											
			C	Mn	Si	Cr	Ni	Mo	Nb	Cu	N	Al	Ti	Other
XM-26	S31100	Duplex	0.03	0.50	0.50	26.0	6.5	—	—	—	—	—	—	—
44LN	S31200	Duplex	0.02	1.00	0.50	25.0	6.0	1.60	—	—	0.17	—	—	—
254SMo	S31254	Austenitic	0.01	0.50	0.40	20.00	18.00	6.25	—	0.75	0.20	—	—	—
27-7MO	S31277	Austenitic	0.01	1.50	0.3	21.8	27.0	7.2	—	1.0	0.35	—	—	—
DP-3	S31260	Duplex	0.02	0.50	0.40	25.0	6.5	3.0	—	0.50	0.20	—	—	W: 0.30
UR B66	S31266	Austenitic	0.02	3.0	0.50	24.0	22.5	5.7	—	1.75	0.48	—	—	W: 2.0
314	S31400	Austenitic	0.15	1.00	2.25	24.50	20.50	—	—	—	—	—	—	—
316	S31600	Austenitic	0.04	1.00	0.40	17.00	12.00	2.20	—	—	—	—	—	—
316L	S31603	Austenitic	0.02	1.00	0.40	17.00	12.00	2.20	—	—	—	—	—	—
316H	S31609	Austenitic	0.07	1.00	0.40	17.00	12.00	2.20	—	—	—	—	—	—
316Ti	S31635	Austenitic	0.04	1.00	0.40	17.00	12.00	2.20	—	—	—	—	0.50	
316Cb	S31640	Austenitic	0.04	1.00	0.40	17.00	12.00	2.20	0.60	—	—	—	—	—
316N	S31651	Austenitic	0.04	1.00	0.40	17.00	12.00	2.20	—	—	0.13	—	—	—
316LN	S31653	Austenitic	0.02	1.00	0.40	17.00	12.00	2.20	—	—	0.13	—	—	—
316L (high N)	S31654	Austenitic	0.02	1.00	0.50	17.0	11.5	2.50	—	—	0.23	—	—	—
317	S31700	Austenitic	0.04	1.00	0.40	19.00	13.00	3.30	—	—	—	—	—	—
317L	S31703	Austenitic	0.02	1.00	0.40	19.00	13.00	3.30	—	—	—	—	—	—
317LM	S31725	Austenitic	0.02	1.00	0.40	19.00	15.50	4.50	—	—	—	—	—	—
317LMN	S31726	Austenitic	0.02	1.00	0.40	18.50	15.50	4.50	—	—	0.15	—	—	—
317LN	S31753	Austenitic	0.02	1.00	0.40	19.00	13.00	3.30	—	—	0.16	—	—	—
2205 (former)	S31803	Duplex	0.02	1.00	0.50	22.0	5.5	3.0	—	—	0.14	—	—	—
	S32001	Duplex	0.02	5.0	0.50	20.5	2.00	—	—	—	0.11	—	—	—
2203	S32003	Duplex	0.02	1.0	0.5	21.0	3.5	1.75	—	—	0.17	—	—	—
	S32050	Austenitic	0.020	0.75	0.50	23.0	21.5	6.4	—	—	0.26	—	—	—
321	S32100	Austenitic	0.04	1.00	0.40	18.00	10.50	—	—	—	—	—	0.40	—
321H	S32109	Austenitic	0.07	1.00	0.40	18.00	10.50	—	—	—	—	—	0.50	—
2205	S32205	Duplex	0.02	1.00	0.50	22.5	5.5	3.2	—	—	0.17	—	—	—

2304	S32304	Duplex	0.02	1.25	0.50	23.00	4.2	0.30	—	0.32	0.12	—	—	—
	S32520	Duplex	0.02	0.75	0.40	25.0	6.8	3.5	—	1.25	0.28	—	—	—
255	S32550	Duplex	0.02	0.75	0.50	25.5	5.5	3.4	—	2.00	0.18	—	—	—
	S32615	Austenitic	0.04	1.00	5.4	18.0	20.5	0.90	—	2.00	—	—	—	—
654SMo	S32654	Austenitic	0.010	3.0	0.25	24.5	22.0	7.5	—	0.45	0.50	—	—	—
2507	S32750	Duplex	0.02	0.60	0.40	25.00	7.00	4.00	—	—	0.28	—	—	—
Zeron 100	S32760	Duplex	0.02	0.50	0.50	25.0	7.0	3.5	—	0.75	0.25	—	—	W: 0.75
	S32803	Ferritic	0.010	0.25	0.30	28.5	3.5	2.15	0.32	—	—	—	—	—
329	S32900	Duplex	0.04	0.50	0.40	25.5	3.5	1.50	—	—	—	—	—	—
	S32906	Duplex	0.02	1.0	0.25	29.0	6.6	2.05	—	—	0.35	—	—	—
7 Mo Plus	S32950	Duplex	0.02	1.00	0.30	27.5	4.4	1.75	—	—	0.25	—	—	—
330	S33000	Austenitic	0.05	1.00	1.20	18.50	35.50	—	—	—	—	—	—	—
	S33228	Austenitic	0.06	0.50	0.20	27.0	32.0	—	0.8	—	—	—	—	Ce: 0.08
334	S33400	Austenitic	0.04	0.50	0.50	19.00	20.00	—	—	—	—	0.38	0.38	—
	S34565	Austenitic	0.020	6.0	0.50	24.0	17.0	4.5	—	—	0.50	—	—	—
347	S34700	Austenitic	0.04	1.00	0.40	18.00	11.00	—	0.60	—	—	—	—	—
347H	S34709	Austenitic	0.07	1.00	0.40	18.00	11.00	—	0.80	—	—	—	—	—
348	S34800	Austenitic	0.04	1.00	0.40	18.00	11.00	—	0.60	—	—	—	—	Ta < 0.10
348H	S34809	Austenitic	0.07	1.00	0.40	18.00	11.00	—	0.80	—	—	—	—	Ta < 0.10
633 (AM350)	S35000	Martensitic PH	0.09	1.00	0.25	16.50	4.50	2.90	—	—	0.10	—	—	—
803	S35045	Austenitic	0.08	0.75	0.50	27.0	34.5	—	—	—	—	0.38	0.38	—
864	S35135	Austenitic	0.04	0.50	0.80	22.5	34.0	4.4	—	—	—	—	0.70	—
353MA	S35315	Austenitic	0.06	1.00	1.60	25.0	35.0	—	—	—	0.15	—	—	Ce: 0.06
634 (AM355)	S35500	Semiaustenitic PH	0.12	1.00	0.25	16.50	4.50	2.90	—	—	0.10	—	—	—
XM-9	S36200	Martensitic PH	0.03	0.25	0.15	14.25	6.6	—	—	—	—	—	0.75	—
	S38031	Austenitic	0.010	1.0	0.2	27.0	31.0	6.5	—	1.2	0.20	—	—	—
XM-15	S38100	Austenitic	0.04	1.00	2.00	18.0	18.0	—	0.30	2.00	—	—	—	—
384	S38400	Austenitic	0.02	1.00	0.50	16.0	18.0	—	—	—	—	—	—	—
	S38660	Austenitic	0.04	2.0	0.75	13.5	15.5	2.0	—	—	—	—	0.25	—

(*continued*)

Nominal Compositions of Stainless Steels *(Continued)*

Type	UNS No.	Microstructure	Nominal Composition (%)											
			C	Mn	Si	Cr	Ni	Mo	Nb	Cu	N	Al	Ti	Other
	S38815	Austenitic	0.02	1.00	6.0	14.0	15.0	1.18	—	1.18	—	—	—	—
	S38926	Austenitic	0.010	1.00	0.2	20.0	25.0	6.5	—	1.0	0.20	—	—	—
	S39277	Duplex	0.01	0.40	0.40	25.0	7.2	3.5	—	1.60	0.28	—	—	W: 1.00
403	S40300	Martensitic	0.08	0.50	0.25	12.25	—	—	—	—	—	—	—	—
405	S40500	Ferritic	0.04	0.50	0.50	13.00	—	—	—	—	—	0.20	—	—
409 (former)	S40900	Ferritic	0.05	0.50	0.50	11.2	—	—	—	—	—	—	0.40	—
409	S40910	Ferritic	0.02	0.50	0.50	11.2	—	—	0.10	—	—	—	0.30	—
409	S40920	Ferritic	0.02	0.50	0.50	11.2	—	—	0.05	—	—	—	0.35	—
409	S40930	Ferritic	0.02	0.50	0.50	11.2	—	—	0.25	—	—	—	0.25	—
	S40940	Ferritic	0.03	0.50	0.50	11.1	—	—	0.40	—	—	—	—	—
	S40945	Ferritic	0.02	0.50	0.50	11.2	—	—	0.29	—	—	—	0.12	—
	S40975	Ferritic	0.02	0.50	0.50	11.2	0.75	—	—	—	—	—	0.50	—
	S40976	Ferritic	0.02	0.50	0.50	11.1	0.88	—	0.40	—	—	—	—	—
	S40977	Ferritic/martensitic	0.02	0.50	0.50	11.5	0.60	—	—	—	—	—	—	—
410	S41000	Martensitic	0.11	0.50	0.50	12.50	—	—	—	—	—	—	—	—
	S41003	Ferritic/martensitic	0.02	0.75	0.50	11.5	—	—	—	—	—	—	—	—
410S	S41008	Martensitic	0.04	0.50	0.50	12.5	—	—	—	—	—	—	—	—
XM-30	S41040	Martensitic	0.10	0.50	0.50	12.0	—	—	0.18	—	—	—	—	—
	S41041	Martensitic	0.16	0.50	0.25	12.25	—	—	0.30	—	—	—	—	—
	S41045	Ferritic	0.02	0.50	0.50	12.5	—	—	0.40	—	—	—	—	—
	S41050	Martensitic	0.03	0.50	0.50	11.5	0.85	—	—	—	—	—	—	—
414	S41400	Martensitic	0.08	0.50	0.50	12.50	2.00	—	—	—	—	—	—	—
	S41425	Martensitic	0.03	0.75	0.25	13.5	5.5	1.75	—	—	0.09	—	—	—
410NiMo	S41500	Martensitic	0.03	0.75	0.30	12.75	4.50	0.75	—	—	—	—	—	—
416	S41600	Martensitic	0.08	0.60	0.50	13.00	—	—	—	—	—	—	—	S: 0.20
	S41603	Ferritic/martensitic	0.04	0.60	0.50	13.0	—	—	—	—	—	—	—	S: 0.20
XM-6	S41610	Martensitic	0.08	2.00	0.50	13.0	—	—	—	—	—	—	—	S: 0.20

416Se	S41623	Martensitic	0.08	0.60	0.50	13.00	—	—	—	—	—	—	—	Se: 0.2
	S41800	Martensitic	0.18	0.25	0.25	13.00	2.00	—	—	—	—	—	—	W: 3.00
420	S42000	Martensitic	0.20	0.50	0.50	13.00	—	—	—	—	—	—	—	—
	S42010	Martensitic	0.22	0.50	0.50	14.2	0.60	0.62	—	—	—	—	—	—
420F	S42020	Martensitic	0.35	0.60	0.50	13.00	—	—	—	—	—	—	—	S: 0.20
420Fse	S42023	Martensitic	0.30	0.60	0.50	13.00	—	—	—	—	—	—	—	Se: 0.2
	S42035	Ferritic/Martensitic	0.04	0.50	0.50	14.5	1.8	0.7	—	—	—	—	0.40	—
422	S42200	Martensitic	0.22	0.75	0.25	11.8	0.75	1.1	—	—	—	—	—	V: 0.25; W: 1.1
	S42300	Martensitic	0.30	1.15	0.25	11.50	—	2.75	—	—	—	—	—	V: 0.25
429	S42900	Ferritic	0.06	0.50	0.50	15.00	—	—	—	—	—	—	—	—
430	S43000	Ferritic	0.06	0.50	0.50	17.00	—	—	—	—	—	—	—	—
430F	S43020	Ferritic	0.06	0.62	0.50	17.0	—	—	—	—	—	—	—	S: 0.20
430F Se	S43023	Ferritic	0.06	0.62	0.50	17.0	—	—	—	—	—	—	—	Se: 0.20
439	S43035	Ferritic	0.03	0.50	0.50	18.00	—	—	—	—	—	—	0.60	—
431	S43100	Martensitic	0.10	0.50	0.50	16.00	2.00	—	—	—	—	—	—	—
434	S43400	Ferritic	0.06	0.50	0.50	17.00	—	1.00	—	—	—	—	—	—
436	S43600	Ferritic	0.06	0.50	0.50	17.00	—	1.00	0.50	—	—	—	—	—
	S43932	Ferritic	0.02	0.50	0.50	18.0	—	—	0.50	—	—	—	—	—
	S43940	Ferritic	0.02	0.50	0.50	18.0	—	—	0.50	—	—	—	0.35	—
440A	S44002	Martensitic	0.70	0.50	0.50	17.00	—	—	—	—	—	—	—	—
440B	S44003	Martensitic	0.85	0.50	0.50	17.00	—	—	—	—	—	—	—	—
440C	S44004	Martensitic	1.10	0.50	0.50	17.00	—	—	—	—	—	—	—	—
440F	S44020	Martensitic	1.10	0.50	0.50	17.00	—	—	—	—	—	—	—	S: 0.20
440Fse	S44023	Martensitic	1.08	0.62	0.50	17.0	—	—	—	—	—	—	—	Se: 0.20
442	S44200	Ferritic	0.10	0.50	0.50	20.5	—	—	—	—	—	—	—	—
444	S44400	Ferritic	0.01	0.50	0.50	18.50	—	2.10	0.20	—	—	—	0.30	—
	S44500	Ferritic	0.01	0.50	0.50	20.0	—	—	0.40	—	—	—	1.10	—
446	S44600	Ferritic	0.10	0.75	0.50	25.00	—	—	—	—	—	—	—	—

(*continued*)

Nominal Compositions of Stainless Steels *(Continued)*

Type	UNS No.	Microstructure	Nominal Composition (%)											
			C	Mn	Si	Cr	Ni	Mo	Nb	Cu	N	Al	Ti	Other
—	S44625	Ferritic	0.005	0.20	0.02	26.00	—	1.00	—	—	—	—	—	—
XM-33	S44626	Ferritic	0.03	0.40	0.40	26.0	—	1.18	—	—	—	—	0.60	—
26-1	S44627	Ferritic	0.002	0.05	0.20	26.0	—	1.0	0.10	—	—	—	—	—
XM-27	S44627	Ferritiic	0.002	0.05	0.20	26.0	—	1.0	0.10	—	—	—	—	—
	S44635	Ferritic	0.01	0.50	0.40	25.2	4.0	4.0	0.15	—	—	—	0.45	—
	S44660	Ferritic	0.02	0.50	0.50	26.5	2.2	3.5	0.20	—	—	—	0.40	—
29-4	S44700	Ferritic	0.005	0.20	0.10	29.00	—	4.0	—	—	—	—	—	C + N < 0.025
	S44735	Ferritic	0.015	0.50	0.50	29.0	—	3.9	0.20	—	—	—	0.40	—
29-4-2	S44800	Ferritic	0.005	0.20	0.10	29.00	2.25	4.0	—	—	—	—	—	C + N < 0.025
XM-25	S45000	Martensitic PH	0.03	0.50	0.50	15.0	6.0	0.75	0.40	1.50	—	—	—	—
XM-16	S45500	Martensitic PH	0.03	0.25	0.25	11.75	8.5	—	0.30	2.0	—	—	1.10	—
	S45503	Martensitic PH	0.005	0.25	0.10	11.75	8.5	—	0.30	2.0	—	—	1.2	—
	S46500	Semiaustenitic PH	0.01	0.12	0.12	11.75	11.00	1.00	—	—	—	—	1.65	—
	S46800	Ferritic	0.015	0.50	0.50	19.0	—	—	0.35	—	—	—	0.18	—
662	S66220	Austenitic PH	0.04	0.75	0.50	13.5	26.0	3.0	—	—	—	0.20	1.80	B: 0.005
660 (A286)	S66286	Austenitic PH	0.04	1.00	0.50	14.75	25.50	1.25	—	—	—	0.20	2.10	V: 0.30; B: 0.005
JBK-75	—	Austenitic PH	0.02	0.000	0.000	15.0	30.0	1.25	—	—	0.000	0.25	2.15	V: 0.25; B: 0.0015; O: 0.000

Source: Extracted from the *ASTM Standards*, 2003 Edition, Volumes 1.02 and 1.03.

APPENDIX 2

ETCHING TECHNIQUES FOR STAINLESS STEEL WELDS

The microstructure of stainless steel welds can be revealed using a variety of etching techniques. Weld metal, the partially melted zone around the weld, and sometimes the HAZ, are not homogeneous, so they tend to etch differently than base metals. A stainless steel weldment may involve two different stainless steels, or even joints of a stainless steel to another ferrous material or to a nickel-base alloy. Stainless steel filler metal may be used to join carbon steels or low-alloy steels, or for cladding of such steels. Stainless steels may also be joined with a nickel-base alloy. Various etchants and etching techniques have been found to provide results of interest to the investigator.

Etching techniques can be divided into chemical methods, electrolytic methods, and staining methods. In general, chemical methods are simpler to apply and require less equipment, so they tend to be favored by the nonspecialist. Electrolytic methods tend to be favored by those who specialize in examination of corrosion-resistant alloys. Various staining methods help to bring out phases of interest at higher contrast (or with color) than possible with the other techniques. Tables A2.1, A2.2, and A2.3 include etching techniques that the authors have found useful in examining the microstructure of stainless steel welds. This is not meant to be an exhaustive list, merely what the authors commonly use for the materials listed. More extensive listings of etchants and etching methods can be found in the ASM *Metals Handbook* [1] and in the *CRC Handbook of Metal Etchants* [2].

Welding Metallurgy and Weldability of Stainless Steels. by John C. Lippold and Damian J. Kotecki
ISBN 0-471-47379-0

TABLE A2.1 Chemical Etches for Stainless Steels

Etchant	Materials	Composition/Use	Notes
Kalling's No. 1	Martensitic	1.5 g $CuCl_2$, 33 mL HCl, 33 mL ethanol, 33 mL H_2O. Immerse or swab at room temperature.	Darkens martensite, colors ferrite, and does not attack austenite.
Villela's	Martensitic	1 g picric acid, 5 mL HCl, 100 mL ethanol, Immerse or swab at room temperature.	Etches martensite, leaves carbides, sigma phase, and ferrite outlined.
Kalling's No. 2	Austenitic, duplex	5 g $CuCl_2$, 100 mL HCl, 100 mL ethanol. Immerse or swab at room temperature.	Darkens ferrite in nominally austenitic and duplex stainless steel weld metal, leaves austenite light. A very versatile etch, but should be used fresh because older etch tends to produce pitting.
Mixed acids	Ferritic, austenitic, duplex	Equal parts HCl, HNO_3, and acetic acids. Use fresh. Swab at room temperature.	A general attack etch will reveal ferrite and austenite, segregation patterns, precipitates, and grain boundaries. This etchant must be used within a few minutes of mixing and then discarded when it turns orange in color.
Glyceregia	Martensitic, ferritic, austenitic, duplex	3 parts glycerol, 2–5 parts HCl, 1 part HNO_3. Use fresh. Immerse or swab at room temperature.	A general-purpose etch similar to mixed acids, but not so aggressive. It outlines ferrite and austenite, attacks martensite and sigma phase. It should also be used fresh, and discarded when it becomes orange in color.
Dilute aqua regia	Dissimilar welds	15 mL HCl, 5 mL HNO_3, 100 mL H_2O, at room temperature.	Useful for etching stainless steel to carbon steel or low-alloy steel joints, for stainless filler metal used in high-carbon steels, and for cladding of stainless steel on carbon steel. It tends to reveal structure also in carbon steel and low-alloy steel heat-affected zones, although exposure time must be limited to avoid overetching the HAZ. It darkens austenite and outlines ferrite, sigma phase, and carbides.
Murakami's	Duplex, superduplex	10 g $K_3Fe(CN)_6$, 10 g KOH or 7 g NaOH, 100 mL H_2O, heated to 80 to 100°C. Immerse in fresh solution. Use under an exhaust hood as cyanide fumes are released.	Should be used fresh, as evaporation changes its strength. It darkens ferrite, outlines sigma phase and colors it light blue, and leaves austenite unattacked.

TABLE A2.2 Electrolytic Etches for Stainless Steels

Etchant	Materials	Composition/Use	Notes
Oxalic acid, 10%	Ferritic, austenitic	10 g oxalic acid, 90 mL H_2O. at room temperature. Etch at 3–6 V for 5–60 s.	Very effective at revealing grain boundaries, particularly if there is carbide precipitation. This technique is often used to detect sensitization.
Ramirez's [3]	Duplex, superduplex	40 vol% HNO_3 in water. *Step 1:* Apply 1–1.2 V for 2 min. *Step 2:* Apply 0.75 V for 7 min.	This technique reveals the various types of austenite that form, without overetching the chromium nitrides.

TABLE A2.3 Staining Techniques for Stainless Steels

Etchant	Materials	Composition/Use	Notes
Modified Murakami's [4]	Duplex	Electrolytic 10% oxalic acid at 6 V for 10–20 s. Immersion in boiling Murakamis's (10 g permanganate, 10 g KOH, 100 mL H_2O) for up to 60 min.	The first two steps reveal the austenite and ferrite phases. The boiling Murakami's produces an interference film on the ferrite that reveals segregation patterns within the ferrite. Austenite remains white.
Ferrofluid [5]	Austenitic, duplex	Apply colloidal suspension of "ferrofluid" (Fe_3O_4) in paraffin. Rinse from sample and dry.	The Fe_3O_4 particles attach to the ferrite phase, due to its remnant magnetism, and color the ferrite. The austenite remains white. This technique provides very good contrast between ferrite and austenite.

REFERENCES

1. ASM. 1985. *Metals Handbook*, 9th ed., Vol. 9, ASM International, Materials Park, OH, pp. 279–296.
2. Walker, P., and Tarn, W. H., eds. 1991. *CRC Handbook of Metal Etchants*, CRC Press, Boca Raton, FL, pp. 1188–1199.
3. Ramirez, A. J., Brandi, S. D., and Lippold, J. C. 2001. Study of secondary austenite precipitation by scanning electron microscopy, *Acta Microscopica*, Vol. 1, Suppl. A, p. 147.
4. Varol, I., Baeslack, W. A., and Lippold, J. C. 1989. Characterization of weld solidification cracking in a duplex stainless steel, *Metallography*, 23:1–19.
5. Ginn, B. J. 1985. A technique for determining austenite to ferrite ratios in welded duplex stainless steels, *Welding Institute Research Bulletin*, 26:365–367.

AUTHOR INDEX

Alexander, D. J., 168, 226
Allen, S. M., 227
Alonso-Falleros, N., 262
Andersson, J.-O., 225
Andrews, K., 43, 54, 62, 76, 85, 86
Anthony, K. C., 271, 286
Apps, R. L., 114, 138
Arata, Y., 155, 179, 188, 225, 226
Asami, K., 308
Austin, A. E., 286
Avery, R. E., 28, 307, 308

Babu, S. S., 55
Baerlacken, E., 10, 18
Baeslack, W. A., 227, 262, 263, 330
Baker, R. G., 228
Balmforth, M. C., 25, 28, 50–52, 55, 71, 72, 81, 86, 111, 112, 139, 303
Bandel, G., 138
Barmin, L. N., 44, 45, 54
Barnhouse, E. J., 294, 308
Baskes, M. I., 182, 227
Beck, F. H., 35, 53
Bennett, L. H., 53
Bernstein, I. M., 18, 52, 137, 204, 225
Binder, W. O., 27, 52, 103, 127, 139
Bloom, F. K., 52
Bond, A. P., 138–140
Boniszewski, T., 228
Boyer, H. E., 86
Brandi, S., 18, 262, 263
Brandis, H., 98, 99, 130, 138
Braski, D. N., 225
Brooks, J. A., 155, 158, 176, 180–182, 225–227, 229, 269, 276, 280, 281, 283–286
Brown, C. M., 252

Campbell, H. C., 27–29, 52, 53, 330
Carpenter, B., 37, 38, 53
Castro, R. J., 7, 11, 18, 28, 41, 54, 55, 60, 65, 68, 86, 92, 131, 138, 139
Charles, J., 237, 262
Chou, C. P. D., 193, 228, 327, 330
Christoffel, R. J., 228
Cieslak, M. J., 286
Cihal, V., 151, 225
Clark, W. A. T., 225
Collins, M. G., 228
Cooper, L. R., 55
Copson, H. R., 21, 52, 206, 207, 228
Crowe, D. J., 86

Welding Metallurgy and Weldability of Stainless Steels, by John C. Lippold and Damian J. Kotecki
ISBN 0-471-47379-0

Dahl, W., 52
Dalder, E. N. C., 225, 285
David, S. A., 55, 155, 167, 225–227, 235, 262, 263
Davis, J. A., 129, 140
De Cadenet, J. J., 18, 28, 54, 65, 86, 131, 138
DeLong, W. T., 28, 33, 34, 38–42, 47, 53, 214, 228, 330
Demo, J. J., 92, 96, 98, 120, 127, 128, 137, 139
DeRosa, S., 139
Dhooge, A., 213, 229
Dickinson, D. W., 124, 139
Dixon, R. D., 53
Duke, J. M., 55
Dundas, H. J., 138
DuPont, J. N., 216, 218, 229
Duren, C., 52

Eagar, T. W., 118, 139, 227
Eaton, N. F., 228
Ebert, H., 211, 229
Edwards, G. R., 54, 86
Eichelman, G. H., 43, 54
Elmer, J. W., 184, 187, 227
Espy, R. H., 28, 39, 53, 214, 216, 229
Evans, T. N., 86

Fadeeva, G. V., 53
Fairhurst, D., 86
Farrar, J. C. M., 80, 81, 82, 86
Fekken, U., 263
Ferree, J. A., 28, 41, 54
Field, A. L., 27–30, 52
Finke, P., 139
Finton, T., 313, 316, 330
Fischer, W., 138
Fisher, D. J., 182, 227
Fisher, R. M., 138
Fleischmann, M., 52
Floreen, S., 52, 96, 138
Flowers, J. W., 53
Folkhard, E., 18, 138, 329, 330
Fontana, M. G., 7, 53
Franks, R., 52
Franson, I. A., 140
Fredriksson, H., 225
Fritz, J. D., 140
Fujimoto, T., 225
Funk, C. W., 264, 270, 285
Furman, D. E., 152, 225

Garner, A., 208, 209, 229
Garrison, W. M., Jr., 269, 280, 281, 285
Giesen, W., 3, 7
Gittos, M. F., 300, 308
Goldschmidt, H, 3, 7
Gooch, T. G., 28, 49, 50, 55, 62, 76, 82–86, 201, 220, 227, 228, 300, 308
Goods, S. H., 200, 228
Goodwin, G. M., 172, 225, 226, 329, 330
Gorr, M. W., 285
Gradwell, K. J., 262
Granger, M. J., 264, 270, 285
Green, N. D., 7
Griffith, A. J., 30, 52
Grigor'ev, S. L., 59
Grobner, P. J., 138, 256, 258, 263
Grubb, J. F., 138, 139
Guaytima, G., 228
Gui, J., 307
Guillet, L., 3, 7
Gullberg, R., 100–102, 138

Hadrill, D. M., 228
Hammar, O., 28, 53, 177, 181, 183, 185, 226
Hansen, M., 9, 18
Harkness, S. D., 55
Hauser, D., 170, 171, 226
Hayden, H. W., 52, 96, 138
Haynes, A. G., 55, 62, 76, 85, 86
Headley, T. J., 226
Hebble, T. M., 227
Heiple, C. R., 179, 226
Hemsworth, B., 228
Hihman, C. R., 55
Hoch, G., 52
Hochanadel, P. W., 269, 271, 276, 286
Honeycombe, J., 227
Hooper, R. A., 131, 140
Howard, R. D., 86
Hull, F. C., 28, 38, 41–43, 53, 54, 277, 286
Hunter, G. B., 118, 139

Irvine, J. J., 52
Irvine, K. J., 62, 76, 85, 86
Iskander, Y. S., 55

Jacobs, M. H., 139
Jansson, B., 225
Jones, D. A., 7, 228
Jones, D. G., 139
Jonsson, P., 263
Juhas, M. C., 225

Kah, D. H., 124, 139
Kakhovskii, N. I., 28, 32, 53
Kaltenhauser, R. H., 28, 47, 48, 55, 111, 139
Kanne, W. R., Jr., 228

Karfs, C. W., 200, 228
Karlsson, L., 255, 263
Katayama, S., 154–156, 158, 160, 225, 226
Katsumata, M., 308
Ke, L., 263
Kerr, H. W., 139, 155, 225
Kiesheyer, H., 98, 99, 130, 138
King, J. F., 299, 300, 308
Klimpel, A., 226
Klueh, R. L., 299, 300, 308
Korolev, N. V., 54
Koseki, T., 39, 53, 234, 236, 241–244, 262
Kotecki, D. J., 28, 39–42, 46, 53–55, 214, 226, 229, 263, 297, 304, 306–308
Kou, S., 184, 227
Krenzer, R. W., 285
Krivobok, V. N., 103, 139
Krysiak, K. F., 120, 139
Kubaschewski, O., 52
Kubbs, J. J., 86
Kujanpää, V., 54, 177, 226
Kung, C. Y., 62, 76, 85, 86
Kurz, W., 182, 227

Lake, F. B., 41, 53, 54
Larson, B., 260, 261, 263
Le, Y., 184, 227
Lee, C. H., 226
Lefevre, J., 28, 47, 48, 55
Leone, G. L., 155, 225
Li, L., 179, 226
Lienert, T. J., 181, 184, 185, 227
Lin, W., 198, 228, 311, 312, 316–322, 329, 330
Linnert, G. E., 52, 264, 265, 269, 270, 274, 285
Lipodaev, V. N., 53
Lippold, J. C., 14, 18, 28, 47, 48, 50, 55, 67, 68, 72, 74, 86, 112, 117, 118, 138, 139, 149, 155, 171, 182–187, 191, 192, 194, 225, 227–229, 240, 244, 250, 255, 257, 262, 263, 294, 308, 311, 312, 328–330
Little, B., 210, 229
Logakina, I. S., 54
Long, C. J., 34, 53
Lorenz, K., 138
Luke, S., 228
Lundin, C. D., 180, 193, 226–228, 307, 311, 323–327, 329, 330
Lundqvist, B., 260, 261, 263

Maguire, M. C., 229
Manakova, N. A., 54
Manning, G. K., 286
Mansfeld, F., 229
Marcincowski, M. J., 138
Marshall, A. W., 80–82, 86
Martin, D. C., 281, 282, 286
Martin, J. W., 86
Matlock, D. K., 53, 54
Matsuda, F., 225, 226, 308
Matsunawa, A., 225
Maurer, E., 4, 7, 25, 26, 29, 52
McCowan, C. N., 39, 52–54
McGannon, H. E., 61, 86
McHenry, H. I., 226
Meelker, H., 263
Mel'Kumor, N., 53
Menon, R., 226
Messler, R. W., Jr., 179, 226
Michael, J. R., 226, 229
Mills, M. J., 308
Miura, A., 138
Moisio, T., 54, 225, 226
Moline, M., 227
Murata, Y., 54
Musch, H., 52
Mushala, M. C., 228

Nakagawa, H., 226, 308
Nakao, Y., 227
Nelson, T. W., 292, 295, 308
Newell, H. D., 26, 27, 29, 30, 52
Nichol, T. J., 129, 140
Nicholson, M. E., 138
Nilsson, J.-O., 245, 263
Nippes, E. P., 228, 330
Nishimoto, K., 227
Nishio, T., 117, 119, 123, 126, 138
Nissley, N. E., 194, 196, 228, 328, 330
North, T. H., 263
Novozhilov, N. M., 28, 39, 40, 53

Oblow, E. M., 55
Ogawa, T., 39, 53, 179, 180, 226, 234, 236, 241–244, 262
Ohmae, T., 138
Okagawa, R. K., 53
Olson, D. L., 25, 52–54, 86
Ostrom, G. A., 53

Pacary, G., 181, 183, 184, 227
Pan, C., 307
Panton-Kent, R., 28, 49, 55
Patriarca, P., 28, 49, 55
Paxton, H. W., 138
Payson, P., 62, 76, 84–86
Peckner, D., 18, 52, 137, 204, 225
Perricone, M. J., 216, 218, 229
Perteneder, E., 263

Pickering, F. B., 86
Plumtree, A., 100–102, 138
Pollard, B., 120, 122, 139, 269, 286
Portevin, A., 3, 7
Potak, Y. M., 28, 35–37, 41, 44, 53
Povich, M. J., 228

Rabensteiner, G., 263
Rajan, V. B., 297, 308
Ramirez, A. J., 16–18, 150, 228, 229, 245, 246, 248, 262, 263
Rao, P., 228
Rayment, J. J., 62, 76, 85, 86
Read, D. T., 172, 226
Reed, R. P., 53
Reid, H. F., 226
Reif, O., 52
Richter, J., 139
Riedrich, G., 26, 52
Robino, C. V., 214, 226, 229, 286
Robinson, F. P. A., 118, 139
Roper, J. R., 179, 226
Roscoe, C. V., 262
Rowe, M. D., 292, 296, 308

Sagalevich, E. A., 28, 35, 36, 37, 41, 44, 53
Sakai, T., 308
Savage, C. H., 62, 76, 85, 86
Savage, W. F., 14, 18, 149, 155, 225, 228, 311, 320, 329, 330
Sawhill, J. M., Jr., 123, 139
Schaeffler, A. L., 28–35, 38–44, 47, 51–53, 111, 214, 216, 227, 289, 290, 292, 293, 298, 303–305, 307
Scherer, R., 26, 52
Schneider, H, 28, 32, 53
Schoefer, E. A., 28, 34, 35, 53
Schwartzendruber, L. J., 53
Scully, J. R., 7
Sedriks, A. J., 7
Seferian, D., 31, 53
Self, J. A., 43–45, 54, 62, 76, 85, 86
Semchysen, M., 100, 138
Sendzimir, M. G., 6, 7, 132, 133
Serna, C. P., 262
Shademan, S., 117, 118, 139
Sherhod, C., 139
Shi, S., 229
Shinozaki, K., 263
Shortsleeve, F. J., 138
Siewert, T. A., 28, 40–42, 52–54
Smallen, H., 286
Smith, R., 264, 285
Spendelow, H. R., 103, 139
Spond, D. F., 227, 228, 323, 326, 330
Steinmeyer, P. A., 226
Steven, W., 62, 76, 85, 86
Strauss, B., 4, 7, 25, 26, 29, 52
Sundman, B., 225
Suutala, N., 28, 42, 54, 177, 179, 180, 182, 183, 189, 225, 226
Svensson, U., 28, 38, 181
Szirmae, A., 138
Szumachowski, E. R., 39, 53, 226

Takada, H., 308
Takalo, T., 54, 225, 226
Takemoto, T., 40, 54
Talbot, A. M., 152, 225
Tanaka, O., 308
Tanaka, T., 54
Thielemann, R. H., 27–29, 46, 47, 49, 52
Thielsch, H., 92, 96, 137, 307
Thomas, C. R., 118, 139
Thomas, R., 114, 116, 138
Thomas, R. D., Jr., 52, 189, 226, 227, 330
Thomas, R. G., 171, 226
Thompson, A. W., 158, 225, 276, 286
Topilin, V. V., 53
Tösch, J., 263
Tofaute, W., 138
Tricot, R., 11, 18, 55, 60, 68, 86, 92, 138, 139
Tsunetomi, E., 179, 180, 226
Tumuluru, M., 225

Underwood, E. E., 275, 276, 286
Unterweiser, P. M., 86

Vagi, J. J., 281, 282, 286
Van der Mee, V., 263
Van der Schelde, R., 263
Van Echo, J. A., 170, 171, 226
Van Nassau, L., 263
Van Wortel, J. C., 229
Varol, I., 227, 247, 262, 263
Verwey, M., 263
Villafuerte, J. C., 139
Vitek, J. M., 51, 55, 167, 226, 227, 235, 262

Wagner, P., 229
Wang, R., 307
Washko, S. D., 139
Wegrzyn, J., 226
Wentrup, H., 52
Wilken, K., 329, 330
Williams, J. C., 225
Williams, R. O., 138
Wilson, A., 263

Wood, J. R., 28, 47, 55, 96, 138
Woolin, P., 55, 86, 218, 229
Worden, C. O., 229
Wright, J. C., 30, 52
Wright, R. N., 28, 47, 55, 96, 120, 138, 139
Wyche, E. H., 285

Yoshida, Y., 285

Zappfe, C. A., 138
Zhang, H., 211, 229
Zhang, W., 227
Zieerhofer, J., 263

SUBJECT INDEX

13-8Mo, 266, 268, 269, 271, 272, 277, 280, 281, 286
15-5PH, 266, 268–272, 286
15-7Mo, 266, 268, 272, 274, 275, 286
16-8-2, 175, 222
17-10 P, 264
17-4PH, 23, 265–272, 277, 279–282, 285, 286. *See also* 630
17-7PH, 266, 268, 273–277, 281, 286. *See also* 631
18 8 Mn, 303–306

2.25Cr-1Mo, 299, 300
209, 175
219, 145, 168, 169
2205, 15, 17, 18, 192, 206, 231, 232, 235–238, 241–246, 248–251, 256–261, 293, 302, 314, 315
2209, 81, 221–223, 231, 293, 294, 302
2304, 231, 232, 235, 238, 248, 261
2507, 232, 234, 235, 238, 248–251, 256, 257, 261, 314, 315
25-4-4, 90, 105, 106, 130
254SMo, 209, 214, 215
255, 192, 232, 235, 238, 244, 251, 258
29-4, 90, 98, 105, 106, 130
29-4-2, 90, 98, 105, 106, 130

304, 4, 21, 123, 124, 142–144, 148, 151, 152, 154, 168, 169, 190–192, 195, 196, 203, 204, 206, 209, 223, 265, 291, 302, 314, 316, 328, 329
304H, 144, 196, 211–213, 301, 302
304L, 4, 6, 142, 144, 152, 164, 165, 172, 183, 184, 186, 190–192, 200, 204, 214, 223, 224, 227, 289–291, 297, 301, 303, 315
304LN, 141, 214, 215
307, 303
308, 38, 133, 144–146, 152, 168–171, 179, 193, 204, 209, 210, 226, 327
308H, 145, 146, 169, 213, 301, 302
308HMo, 303
308L, 125, 145, 146, 168–170, 172, 193, 207, 294, 296, 297, 301, 326, 327
309, 33, 144–146, 152, 169, 193, 254, 282, 293, 300, 302
309L, 80, 145, 146, 169, 221–223, 234, 289–294, 296–303
309LMo or 309MoL, 161, 221–223, 302
309Mo, 303

Welding Metallurgy and Weldability of Stainless Steels, by John C. Lippold and Damian J. Kotecki
ISBN 0-471-47379-0

310, 144–146, 152, 169, 195, 196, 213, 226, 289, 290, 301–303, 312, 313, 315, 316, 318–322, 328, 329
310H, 212, 213
312, 161, 168–170, 221, 303
316, 6, 33, 142–146, 152, 154, 169, 171, 193, 204, 206, 207, 327
316H, 145, 146, 169, 196, 211–213
316L, 33, 142, 144–146, 152, 169, 171, 172, 183, 204, 207, 209, 210, 214, 220–223, 225, 226, 301, 302, 307, 314, 315, 327
316LN, 172, 214, 215
316N, 143
317, 144–146, 152, 169
317L, 145, 146, 169, 209
317LM, 175, 222
317LMN, 215
317LN, 215
321, 142–144, 152, 183, 189, 201, 205, 299
321H, 212, 213
329, 230–232, 238
3311, 264
347, 142–146, 152, 169, 189, 193, 196–198, 201, 205, 206, 228, 297, 298, 300, 301, 327
347H, 212
385, 302

405, 89, 97, 105, 106, 108, 111, 131
409, 12, 47, 89–91, 93, 96, 97, 105, 107, 108, 111–116, 128, 138, 140, 295, 302
409Nb or 409Cb, 91, 106
410, 4, 12, 15, 57, 58, 60, 61, 63, 64, 66, 71, 72, 77–80, 84, 85, 302
410NiMo, 57, 58, 74, 76–79, 85
414, 57, 85
416, 57
420, 57, 58, 64, 71–73, 79, 84, 85
422, 57, 85
423L, 73, 79
430, 4, 12, 89, 91, 93, 94, 105, 106, 108, 109, 111–113, 116, 123, 124, 129–132, 135–137
430Nb, 91, 106, 117, 119, 123, 126, 131, 132
431, 57, 64, 85
434, 89, 105, 108, 116
436, 89, 105, 116–118, 124–126, 132–135
439, 89, 90, 105–107, 109, 111, 112, 116, 137
440, 12, 59
440A, 57, 85
440B, 57
440C, 57, 64
442, 88, 89, 105, 108
444, 90, 105, 106, 109, 116, 118, 302
446, 88, 89, 105, 108, 119, 139
446LMo, 91, 106
468, 89, 105, 106, 109, 137
475°C (885°F) embrittlement, 10, 88, 96–98, 168, 256, 257, 271

625, 316. *See also* NiCrMo-3
630, 266–268, 277, 279, 280. *See also* 17-4PH
631, 266, 277. *See also* 17-7PH
635, 264, 266, 271, 272, 277
660, 265–267, 276. *See also* A-286
662, 266, 267, 273, 276. *See also* Discaloy
690, 316

A-286 or A286, 265–268, 273, 276, 281–286, 315–322. *See also* 660
A36 (ASTM), 290, 291, 293, 294, 297, 303–306
A4.2 (AWS), 34, 182, 226
A5.4 (AWS), 77, 78, 91, 143, 145, 169, 175, 222, 233, 259, 268, 279, 280
A5.9 (AWS), 91, 106, 143, 145, 169, 233, 259, 268
A5.22 (AWS), 91, 143, 146, 169, 233, 259
A508 (ASTM), 289, 290, 292–294, 297–299, 303
A_{C1}, 82–84
AL6XN or AL-6XN, 195, 196, 214, 215, 218, 315, 328, 329
Alloy 82, 301. *See also* NiCr-3
Alpha-prime or alpha prime, 10, 97, 168, 189, 256, 271
AM350, 266, 268, 273–275
AM355, 266, 268, 273–275
AOD or argon-oxygen decarburization, 6, 142, 177, 179
AR or abrasion resisting, 303–306
Augmented strain, 192, 311, 312, 314, 316, 318
Austenite reversion, 271, 275, 276
Austenite/austenite-plus-ferrite-boundary, 26, 27, 30, 43
Austenite-martensite boundary, 43, 44

Bend ductility, 117, 118
BTR or brittle temperature range, 188, 194

C-276, 307
CA-6NM, 57, 58, 74, 76
Carbide precipitation, 45, 67, 110, 142, 143, 166, 167, 201–205, 213, 275, 292, 298
Carbides, 2, 6, 12, 20, 22–24, 37, 59, 63, 67, 70, 72, 73, 75, 78, 88, 93, 95, 96, 99–101, 109, 111–113, 127–129, 137, 143, 147–149, 151, 166, 167, 196, 197, 201, 204, 205, 212, 247, 248, 271, 275, 278
Carbonitrides, 20, 99, 101, 110–113, 118, 201
Carbon-depleted zone, 293
CB-30, 88
CC-50, 88

CCC or copper contamination cracking, 199, 224
CCT or critical crevice temperature, 208
CD4MCu, 230, 231, 232, 238
CD4MCuN, 231, 232
Charpy, 100, 116, 121, 131, 152, 260, 272, 281. *See also* CVN
Chi or χ, 20, 97, 98, 130, 149, 151, 237, 256
Chloride(s), 5, 6, 76, 87, 92, 126, 129, 132, 137, 143, 206, 208, 223, 261
Chromium equivalent or Cr_{eq}, 27, 29–36, 38, 40–42, 47, 49, 82, 153, 175, 177, 178, 181, 182, 185, 221, 234, 235, 254, 255, 277
Clad or cladding, 57, 287, 292–296, 298, 299, 308
Cold cracking, 57, 310
Composition transition region, 288
Condition A, 274, 275
Condition C, 275
CPT or critical pitting temperature, 208, 209
Cr_2N, 17, 20, 23, 110, 118, 127, 150, 151, 166, 237, 240, 245, 246, 253
$Cr_{23}C_6$, 60, 93
$(Cr,Fe)_{23}C_6$, 11
Creep, 141, 171, 194, 211, 213, 226, 293, 296, 299, 301
Crevice or crevice corrosion, 2, 5, 22, 23, 87, 126, 201, 208
Cryogenic, 23, 44, 53, 141, 151, 171–173, 189, 223, 226, 231
CSR or crack susceptible region, 311, 312, 318, 319, 321
CTE or coefficient of thermal expansion, 59, 166, 176, 288, 296, 299, 301,
Custom 450, 265, 269, 271, 272
Custom 455, 269, 271, 272
CVN, 77, 101, 117, 120, 126, 131, 260. *See also* Charpy

DBTT or ductile-brittle (fracture) transition temperature, 100, 101, 115–117, 120, 124, 126, 131, 132, 135
DDC or ductility dip, 173, 187, 188, 194–196, 224, 228, 281, 285, 327–330
Difficult-to-weld steels, 303
Disbonding, 294, 296, 298, 308
Discaloy, 266. *See also* 662
Double temper, 77
DRT or ductility recovery temperature, 320, 321
Ductility dip or DDC, 173, 187, 188, 194–196, 224, 228, 281, 285, 327–330

E-Brite, 106, 111, 123, 124
Effective quench temperature, 235, 241, 244, 262
EN 1600, 210
ENi-1, 305
ENiCrFe-2, 301
ENiCrMo-3, 219
ENiCrMo-4, 219, 307
ENiCrMo-10, 219
ERNi-1, 305
ERNiCr-3, 301
ERNiCrMo-3, 132, 219
ERNiCrMo-4, 219, 307
ERNiCrMo-10, 219
Erosion corrosion, 201
Eta, 149

FeritScope or Feritscope, 182, 253
Ferrite grain growth, 17, 106, 108, 113–115, 130, 244, 248, 250
Ferrite grain size, 108, 116, 248
Ferrite Number or FN, 34, 35, 38–42, 50, 52, 54, 55, 161, 168–172, 174, 175, 177, 178, 182, 189, 190, 193–195, 198, 214, 221–223, 239, 251, 253, 255, 256, 290, 293, 297, 325–327
Ferrite Potential, 190, 214, 224, 296
Ferrite factor, 47, 49, 111
Ferritic/martensitic, 116
Firth, 3, 4
Fissure bend, 90, 193, 323–328, 330
Free machining, 54, 180, 226

Gall-Tough, 214, 215
Galvanic, 201
Gamma loop, 9, 10, 27, 60, 80, 95
Gamma-prime, 149, 267, 271, 276, 279
Gleeble, 124, 198, 320, 322
Grain refinement, 113, 114, 122, 124

H1150, 271, 279, 280
H900, 271
Hazard HAZ, 193, 228, 330
HC, 288
HIC or hydrogen induced cracking, 61, 74, 77, 79, 80, 95, 123, 126, 254, 263, 285, 292, 298, 308, 310
High angle boundary, 163, 164
HK40, 211, 212
Hot cracking, 211, 226, 310, 311, 329, 330
HP45Nb, 211
HTE or high-temperature embrittlement, 96–98, 100–104, 118, 120, 122, 124, 128, 129, 132, 139
Hydrogen, 59, 61, 74–81, 95, 123, 126, 254, 255, 263, 285, 292, 296, 298, 303, 308, 310

IGA or intergranular attack, 2, 5, 127, 142, 143, 173, 201, 202, 205, 223
IGSCC or intergranular stress corrosion cracking, 5, 96, 135–137, 142, 173, 201, 205, 208

INCO A, 301
Inspector Gauge, 253
IGC or intergranular corrosion, 22, 27, 81, 99, 126–130, 138, 140, 142, 143, 149, 151, 201, 203, 204
Intragranular, 93, 100, 102, 109, 113, 118, 127, 212, 246
ISO 3581, 210
ISO 8249, 34, 182
ITE or intermediate temperature embrittlement, 96, 97, 103, 167, 173, 255

JBK-75, 267, 285

K-factor, 47–49, 111
Knifeline attack, 205

Larson-Miller parameter, 74, 83, 171
Laves phase, 20, 44–46, 149, 151, 271, 281, 284
LCTR or liquation cracking temperature range, 322
Liquation, 78, 123, 165, 166, 173, 179, 189–194, 224, 227, 281–285, 311, 315, 316, 318–323, 327, 328, 330
Liquidus, 12–14, 218, 321
LTS or low temperature sensitization, 205, 206

M_7C_3, 67
$M_{23}C_6$, 16, 17, 20, 22, 23, 67, 109, 111, 113, 127, 137, 143, 147, 149–151, 166–168, 201, 204, 205, 237
$M_{23}(C,N)_6$, 20, 109, 111, 113
MagneGage, 182, 253
Marangoni, 179
Martensite boundary, 43, 54, 55, 109, 111, 130, 277, 290, 291, 304, 305
Martensite finish temperature or M_F, 62, 64, 65, 75, 85
Martensite start temperature or M_S, 24, 43–45, 61, 62, 64, 65, 75, 84–86
Massive transformation, 157, 184
MC, 22, 109, 128, 137, 143, 151, 166, 196, 201, 205
MCD or maximum crack distance, 312, 314, 315,
MCL or maximum crack length, 192, 312, 316, 318, 319
MGB or migrated grain boundary, 154, 156, 158, 160, 162–164, 187, 190, 192, 328
MIC or microbiologically induced corrosion, 201, 208–210
Microfissures, 190
Migrate, 163, 164, 291
Migration, 163, 164, 188, 292, 293, 295, 299, 301
MnS, 21
Monel, 295, 296
Mössbauer, 53, 271

NbC, 151, 189, 196, 198, 205
NDT or nil ductility temperature, 320, 321
$Ni_3(Al, Ti)$, 149, 151, 276
Ni_3Al, 23, 151, 275, 276
Ni_3Ti, 23, 151, 271, 276, 279
Nickel equivalent, 27–35, 37–42, 44, 46, 47, 49–51, 82, 153, 175, 177, 178, 181, 182, 185, 221, 234, 235, 254
NiCr-3, 303. *See also* Alloy 82
NiCrMo-3, 132, 219, 307
Nitrides, 20, 24, 88, 99–101, 109–113, 118, 123, 127, 128, 166, 201, 240, 241, 245, 247, 248, 250, 275
Nitronic, 141, 168–170, 214, 215, 217
NST or nil strength temperature, 320–322

On-cooling, 60, 278, 308, 316, 318–322
On-heating, 60, 76, 82, 118, 258, 316, 318–322

Pitting, 2, 5, 22–24, 87, 95, 126, 142, 143, 201, 208, 209, 214, 216, 218, 220, 229, 231, 234, 240, 245, 261, 262, 302
PMZ or partially melted zone, 189, 190, 282, 310, 318–321
PRE_N or pitting resistance equivalent, 215–217, 219, 261
Primary austenite, 12, 24, 152–155, 162, 173, 175, 177, 179, 180, 182–184, 189, 190, 214, 221, 245, 246, 296–298, 301
Primary ferrite, 12, 24, 64, 65, 104, 108, 123, 152, 155, 156, 174, 175, 177, 180, 182–184, 189, 213, 214, 216, 221, 274, 277, 290, 297, 298, 301, 302, 314
Pseudobinary, 10–12, 14, 60, 68, 92, 110, 147, 149, 153, 161, 234, 235, 248, 267, 269
Pteraphthalic acid, 210
PVR test, 329
PWHT or postweld heat treatment, 68, 71–73, 76, 78–85, 120–122, 126, 128–132, 135, 204, 265, 278, 279, 285, 291–294, 296, 299, 300, 302

Recrystallization, 114, 165, 168
Reheat cracking, 78, 79, 196–199, 211–213, 229

Saturated strain, 312, 314–316, 318, 319
Scaling, 22, 95
SCC or stress corrosion cracking, 5, 6, 21, 87, 92, 96, 126, 132, 135, 137, 142, 143, 166, 201, 206–208, 214, 216, 228, 231, 261

Schaeffler, 28–35, 38–44, 47, 50–53, 111, 214, 216, 227, 289, 290, 292, 293, 298, 303–305, 307
SCTR or solidification cracking temperature range, 314, 315, 322
Secondary austenite, 18, 244–247, 263
Sendzimir, 6, 7, 132, 133
Sensitization, sensitize, sensitizing, sensitized, 6, 52, 111, 127–129, 137, 140, 143, 166, 167, 201, 203–206, 208, 223
Severn Gage, 182
SGB or solidification grain boundary, 154, 162–164, 176, 187, 198, 296
Sigma or σ, 9, 10, 17, 20, 22, 59, 73, 96–99, 102, 130, 147, 149, 151, 152, 167, 168, 173, 189, 211, 216, 223, 226, 237, 256, 258, 259, 302
Silicide, 22
Solidification crack(ing), 21, 54, 78, 81, 123, 124, 163, 167, 171, 173–181, 183, 186–189, 211, 213, 214, 216, 220–222, 225–227, 254, 255, 264, 267, 281, 284, 285, 289, 290, 296–298, 301–303, 305–307, 310–316
Solution heat treatment or SHT, 74, 269
Spot Varestraint, 190, 192, 311, 317–321
SSGB or solidification subgrain boundary, 162–164
Stabilized, 98, 109, 111, 116, 117, 120, 123, 128, 132, 137, 139, 142, 143, 166, 201, 204–206
Stainless W, 264
STF or strain-to-fracture test, 328
Strain control, 311
Strain rate control, 311
Stress control, 311
Stress rupture, 171, 189, 301
Stress-assisted, 201
Super austenitic or superaustenitic, 143, 195, 215, 216, 218–220, 229, 301, 315, 328
Super duplex or superduplex, 81, 234, 237, 248, 250, 261–263
Supermartensitic, 49, 55, 57, 80–84, 86
Surface tension, 179

TCL or total crack length, 124, 180, 192, 312
Temper, 64, 71–73, 77, 78, 82, 83, 130
Tempering, 56, 59, 63, 71–74, 76–78, 80, 82–84, 86, 131
Ternary eutectic, 12
Thermal expansion, 5, 59, 135, 166, 230, 231, 288, 296, 303
ThermoCalc, 8, 15–17, 150, 216, 267, 269
Threshold strain, 312, 314, 318, 328
Threshold time, 99
TiC, 80, 151, 189, 205, 282
T_m or melting temperature, 98, 100, 118, 166, 194
Transvarestraint, 312–315, 330
Troostosorbite, 25, 26
TWI or The Welding Institute, 50, 82, 218, 229
Type I, 291, 292
Type II, 291–296, 299

Undercooling, 182
Upper shelf, 116, 120, 124, 135, 168
Urea, 210, 223

V2B, 264, 265
Varestraint, 123, 124, 175, 179, 180, 190, 192, 263, 281, 282, 284, 311–321, 328–330
VOD or vacuum-oxygen decarburization, 6

Warm cracking, 310
WRC-1988, 40, 41, 54
WRC-1992, 40–42, 46, 49, 50, 54, 55, 173, 175, 177, 178, 182–184, 189, 190, 214, 220–224, 234, 253, 254, 277, 289–293, 298, 303–305

ZCC or zinc contamination cracking, 200
Zeron 100, 81

β-NiAl, 271, 275
$\Delta T_{12\text{-}8}$, 248